Electric Energy Conversion and Transmission

SYED A. NASAR

Department of Electrical Engineering
University of Kentucky

ELECTRIC ENERGY CONVERSION AND TRANSMISSION

Macmillan Publishing Company

New York

Collier Macmillan Publishers

London

Printed in the United States of America

Macmillan Publishing Company
866 Third Avenue, New York, New York 10022

Collier Macmillan Canada, Inc.

Library of Congress Cataloging in Publication Data

Nasar, S. A.
Electrical energy conversion and transmission.

Includes bibliographical references and index.
1. Electric power systems. 2. Electric machinery.
I. Title.
TK1001.N37 1985 621.31 84-949
ISBN 0-02-385960-1
Printing: 2 3 4 5 6 7 8 Year: 6 7 8 0 1 2

ISBN 0-02-385960-1

Dedicated to my Daughter
Naheed

PREFACE

Several engineering schools now have a combined course in electric machines and electric power systems. Often an introduction to direct energy conversion is included in such courses.

This book is intended to present an overview of electric energy components and systems. Thus, the topics discussed include: electric energy fundamentals; various methods of electric energy conversion, with emphasis on electromechanical energy conversion; components of electric energy systems, such as rotating electric machines, transformers and transmission lines; and a review of electric power systems covering fault analyses, load flow, and stability studies. Obviously, a broad range of topics is covered, perhaps, at the expense of depth of coverage. However, emphasis is on the fundamentals and rigor. References cited at the end of each chapter may be consulted for supplementing the material presented and for an in-depth study of the pertinent topic.

The book contains sufficient material for a two-semester course sequence combining electric machines, power systems, and energy-related topics. However, the subject matter may be judiciously chosen for a suitable one-semester course in electric energy systems. For prerequisites, traditional mathematics and electric circuits courses through the sophomore level are adequate. Some facility with the digital computer is also assumed. Thus, the book should fit into the usual electrical engineering undergraduate curriculum anywhere between the junior and senior levels.

To briefly review the contents of the book, Chapter 1 is meant to be a general introduction and deals with energy fundamentals, energy-related data, units, and certain basic laws.

Various methods of electric energy conversion—direct as well as electromechanical—are presented in Chapter 2. The topics are not covered in great detail. Rather, the scope of the various methods is discussed, with an emphasis on the governing laws of electric energy conversion processes.

Chapter 3 is a short chapter. It discusses qualitatively the components of electric energy systems. Schematic (or circuit) representations of these components are also included in this chapter.

Transformers—single phase and polyphase—and magnetic circuits are presented more or less in a traditional fashion in Chapter 4. Chapters 5 through 7 respectively treat dc machines, induction machines, and synchronous machines. Constructional details and physical features of transformers and rotating machines are not included. It is hoped that illustrative photographs of various types of machines, transformers, and transmission lines given in Chapter 3 will aid the reader in visualizing the physical appearance of the pertinent items. The emphasis is on analysis and basic operating principles. Operating characteristics of transformers and machines are discussed very briefly. Chapter 8 is devoted to small ac motors, and generalized machine theory is outlined in Chapter 9.

Transmission lines are discussed in Chapter 10. The treatment includes traditional ac transmission lines and an introduction to high-voltage dc transmission.

The electric power system is reviewed in Chapter 11. The presentation includes fault calculations, load flow, and stability studies. Finally, Chapter 12 is exclusively devoted to power electronics and solid-state control of dc and ac motors. With the availability of high-power semiconductors, solid-state motor controllers are in common use. Thus, it is felt that a brief review of power electronics is essential to complete the study of electric energy systems.

Topics such as economic load dispatch and power system protection are considered beyond the scope of this book.

Illustrative solved examples are given throughout the text. The appendixes include various tables of constants and certain review material.

Acknowledgment is made to Mr. A. Doyle Baker, of Kentucky Utilities, for his help in preparation of this book. I appreciate the helpful comments provided by the Macmillan reviewers. They are Richard W. Brown (University of Hartford), Ron Chu (Drexel University), Ward Getty (University of Michigan), Charles A. Gross (Auburn University), Medhat Ibrahim (California State University, Fresno), A. H. Qureshi (Wayne State University), Stephen A. Sebo (Ohio State University), Neal A. Smith (Ohio State University), Stewart Stanton (Agricultural and Technical State University, Greensboro, N.C.), and Salah M. Yousif (California State University, Sacramento).

S.A.N.

CONTENTS

CHAPTER 1

Fundamentals of Electric Energy Systems

1-1

INTRODUCTION

Harnessing and utilizing energy has always been a key factor in improving the quality of life. The use of energy has aided our ability to develop socially, and live with physical comfort. Thus energy demand has been rapidly growing. It was estimated that in just 30 years, between 1971 and 2001, the United States will consume more energy than it has in its entire history [1]. A more recent projection of energy requirements for the United States is shown in Fig. 1-1. An increase of 3.6 percent in annual energy requirements for the period 1981–1990, over the actual 1980 total energy requirement of 2,293,964 gigawatthours,* has been forecast [2].

With the above-mentioned projections in mind, we now recall the following headlines from newspapers:

Con Ed Power Cut 25% by Mishaps; Long Crisis Ahead. Generator Repair May Take up to Month—Consumers Help Save Electricity [3]

*We will define the various units of energy in Section 1-2.

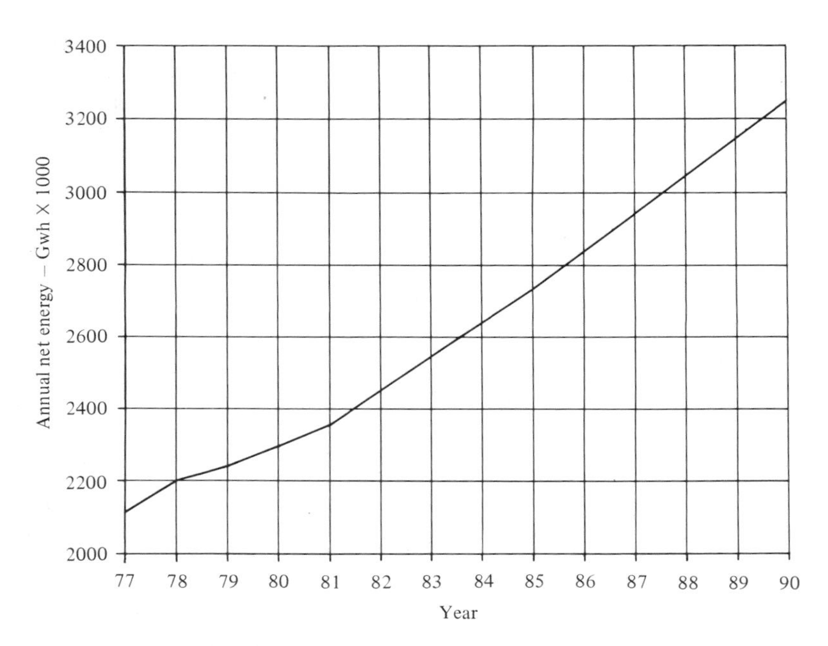

FIGURE 1-1. Net annual energy requirements for the contiguous United States. [From Ref. 2]

Nation's Energy Crisis: It Won't Go Away Soon [4]
Nation's Energy Crisis: Nuclear Future Looms [5]
Nation's Energy Crisis: Is Unbridled Growth Indispensable to the Good Life? [6]
Conservationists Oppose Rise in Supply of Power [7]

From these headlines a picture of the ''energy problem'' in the United States emerges; that is, we have had (and still have) an energy crisis. Various means have been considered to handle this crisis. The two obvious ways to ease the critical situation are by conservation of energy, while utilizing energy-efficient equipment, and by producing more energy. There is a definite limit to the former, whereas the latter is hindered by environmental considerations and is limited by available resources. Nuclear energy conversion is not among the popular choices for providing us with electric energy. In fact, the use of energy in improving the quality of life beyond a certain limit is being questioned. The United States consumes almost 35 percent of the world's energy, but has less than 6 percent of the world population. Figure 1-2 shows the commercial use of energy versus gross national product, for 1970, in the United States compared to that of several other countries. There has been a per capita increase of about 1 percent per year in energy consumption in the United States. If the under-developed countries could by the year 2000 reach the 1970 standard of living

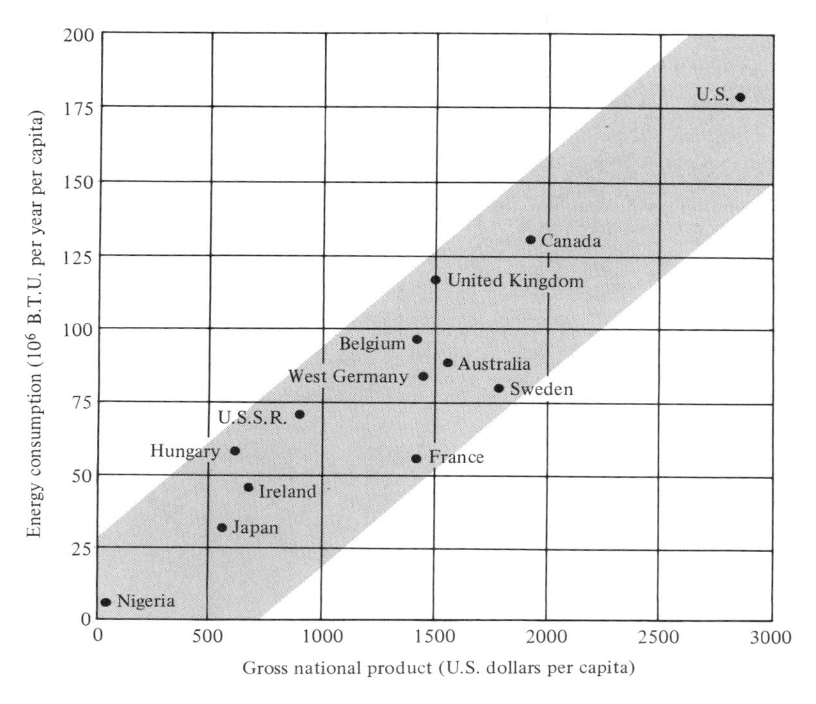

FIGURE 1-2. Commercial energy consumption and gross national product for different countries. [From Ref. 1]

in the United States, the worldwide energy demand would increase to 10 times the present level. Compared with the energy available from fossil-fuel reserves, the problems relating to the energy demands of the future are awesome.

From the preceding remarks, a rather conflicting picture of the energy problem emerges. On the one hand, the demand for energy is ever-increasing. On the other hand, environmental considerations, limited natural resources, and skepticism concerning nuclear energy tend to impede a rapid increase in the conversion of energy to meet the demand. As with most engineering problems, we face the problem of trade-offs. In this book we look at some of these trade-offs, and we will study various methods of energy conversion and their scopes and limitations. After discussing the various methods of electric energy conversion, we will focus on electromechanical energy conversion, which is the basis of the prevailing method of production of electric power.

As the title of the book suggests, our goals are to develop an understanding of the principles of energy conversion—in particular, electromechanical energy conversion—and to study certain aspects of electric power systems, such as its components, transmission of electric power, and operations of electric power systems. In order to assess the energy perspective quantitatively, we first introduce certain units and basic definitions.

1-2 WORK, ENERGY, AND POWER

Let a force, F, be applied to a mass so as to move the mass through a linear distance l in the direction of F. Then the *work,* U, of the force is defined as the product Fl; that is,

$$U = Fl \tag{1-1}$$

If the displacement is not in the direction of F, the work is the product of the displacement and the component of the force along the path; that is,

$$U = Fl \cos \alpha \tag{1-2}$$

where α is the angle of F with l. Work is measured in joules (J). From (1-1), 1 joule is the work done by a force of 1 newton in moving a body through a distance of 1 meter in the direction of the force.

The *energy* of a body is its capacity to do work. Energy has the same units as work, although several other units are also used for different forms of energy. The various forms of energy of greatest interest to us are electrical, mechanical, thermal, chemical, and nuclear. The fundamental unit of electrical energy is the watt-second (W-s), where

$$1 \text{ W-s} = 1 \text{ J} \tag{1-3}$$

More commonly, however, electrical energy is measured in kilowatthours (kWh). From (1-3) we have

$$1 \text{ kWh} = 3.6 \times 10^6 \text{ J} \tag{1-4}$$

The two most important forms of mechanical energy are kinetic energy and potential energy. A body possesses kinetic energy (KE) by virtue of its motion such that the kinetic energy of a mass M (kg), moving with a velocity u (m/s), is given by

$$\text{KE} = \tfrac{1}{2} Mu^2 \quad \text{J} \tag{1-5}$$

The potential energy of a body exists by virtue of its position. Gravitational potential energy (PE), for instance, results from the position at a height h (m) above ground level of a mass M (kg) in the gravitational field, and is given by

$$\text{PE} = Mgh \quad \text{J} \tag{1-6}$$

where g is the acceleration due to gravity in m/s^2.

Thermal energy is usually measured in calories. By definition, a calorie is the amount of heat required to raise the temperature 1 gram of water at 15°C through 1 degree Celsius. A more common unit is the kilocalorie (kcal). Experimentally, it has been found that

$$1 \text{ cal} = 4.186 \text{ J} \tag{1-7}$$

Yet another unit of thermal energy is the British thermal unit, abbreviated Btu, and related to the joule and the calorie as follows:

$$1 \text{ Btu} = 1.055 \times 10^3 \text{ J} = 0.252 \times 10^3 \text{ cal} \tag{1-8}$$

Energy released in both nuclear and chemical reactions is often measured in Btu or kcal. Because the joule and the calorie are relatively small units of energy, thermal energy and electrical energy are generally expressed in terms of Btu and kWh (or even megawatthours), respectively. A still larger unit of energy is the quad (Q), which is an abbreviation of "quadrillion Btu." The mutual relationships among these various units are

$$1 \text{ quad} = 10^{15} \text{ Btu*} = 1.055 \times 10^{18} \text{ J} \tag{1-9}$$

A quad is the approximate energy used in 1 year by a metropolitan area such as San Francisco–Oakland.

Power is defined as the time rate at which work is done. Alternatively, power is the time rate of change of energy. Thus

$$p = \frac{dU}{dt} = \frac{dw}{dt} \tag{1-10}$$

where U represents work and w represents energy. The SI unit of power is the watt (W), which is work at a rate of 1 joule per second. Hence

$$1 \text{ W} = 1 \text{ J/s} \tag{1-11}$$

Multiples of the watt commonly used in power engineering are the kilowatt and the megawatt. Power ratings (or outputs) of electric motors are expressed in horsepower (hp), related to the watt as follows:

$$1 \text{ hp} = 745.7 \text{ W} \tag{1-12}$$

We illustrate these interrelationships between various units by the following examples.

*Some authors define 1 quad as 10^{18} Btu.

EXAMPLE 1-1

The net energy requirement for the United States for the year 1986 is estimated to be 2.82×10^6 GWh. What is the equivalent of this energy in Btu?

First we express GWh as

$$1 \text{ GWh} = 1 \text{ gigawatthour} = 10^9 \text{ watthour} = 10^6 \text{ kilowatthour} = 10^6 \text{ kWh}$$

$$2.82 \times 10^6 \text{ GWh} = 2.82 \times 10^{12} \text{ kWh}$$

Then from (1-4) we get

$$2.82 \times 10^{12} \text{ kWh} = 3.6 \times 2.82 \times 10^{12} \times 10^6 = 10.152 \times 10^{18} \text{ J}$$

From (1-9) we finally obtain

$$10.152 \times 10^{18} \text{ J} = \frac{10.152}{1.055} \times 10^{15} = 9.623 \times 10^{15} \text{ Btu}$$

$$= 9.623 \text{ quad}$$

■

EXAMPLE 1-2

Coal has an average energy content of 940 watt-years per ton and natural gas has an energy content of 0.036 watt-year per cubic foot. If 80 percent of the net energy requirement for the United States for 1986 is to be met by coal and 20 percent by gas, calculate the respective amounts of coal and gas required.

From Example 1-1,

$$2.82 \times 10^6 \text{ GWh} = \frac{2.82 \times 10^{15}}{365 \times 24} = 3.22 \times 10^{11} \text{ W-yr}$$

$$\text{energy to be supplied by coal} = 0.8 \times 3.22 \times 10^{11} = 2.576 \times 10^{11} \text{ W-yr}$$

$$\text{energy to be supplied by gas} = 0.2 \times 3.22 \times 10^{11} = 6.44 \times 10^{10} \text{ W-yr}$$

$$\text{amount of coal required} = \frac{2.576 \times 10^{11}}{940} = 2.74 \times 10^8 \text{ tons}$$

$$\text{amount of gas required} = \frac{6.44 \times 10^{11}}{0.036} = 1.79 \times 10^{13} \text{ ft}^3$$

■

1-3 GROWTH RATES

From Fig. 1-1 we observe that the rate of growth of energy demand from 1981 through 1990 is almost constant at 10^6 gigawatthours per year (GWh/yr). This has been essentially true for past periods also, although the rate was different [8]. In planning for the future it is important that we have a clear idea of the implication of constant growth rates. *Constant growth rate* of a quantity means that the increase is directly proportional to the quantity. Expressed mathematically, a constant growth rate of energy Q is given by

$$\frac{dQ}{dt} = aQ \tag{1-13}$$

where a is the constant of proportionality. The solution to (1-13) may be written as

$$Q = Q_0 e^{at} \tag{1-14}$$

where Q_0 is the value at $t = 0$. For two values of time, t_1 and t_2, the inverse ratio of the corresponding energies, Q_1 and Q_2, becomes

$$\frac{Q_2}{Q_1} = e^{a(t_2 - t_1)} \tag{1-15}$$

To find the doubling time, t_d, that is, for $Q_2 = 2Q_1$, we may write $t_2 - t_1 = t_d$ in (1-15) and obtain

$$t_d = \frac{0.693}{a} \tag{1-16}$$

A plot of doubling times for different yearly growth rates, a, is shown in Figure 1-3, from which it is clear that at a growth rate of 5 percent per year ($a = 0.05$), the energy consumption (Q) will be doubled in about 14 years.

A knowledge of growth rate is important in planning for the future. Equally important is an understanding of the significance of the peak power demand. Figure 1-4 shows the summer–winter reserve, peak electrical power demand, and net capability of power generation for the period 1977–1980 and projected through 1991 for the contiguous United States. It is clear that the peak demand in the summer is greater than that in the winter. As shown in the figure, the net capability of power generation is the sum of the peak demand and reserve power. In order to understand the significance of the area under the peak power demand curve in Fig. 1-4, we obtain the peak power demand for the 1977–1990 period, shown in Fig. 1-5. We approximate this curve by

$$P = P_0 e^{bt} \tag{1-17}$$

where P_0 is the power at $t = 0$ and b is the growth rate. The area under this curve over a given period is a measure of the energy consumed for that period.

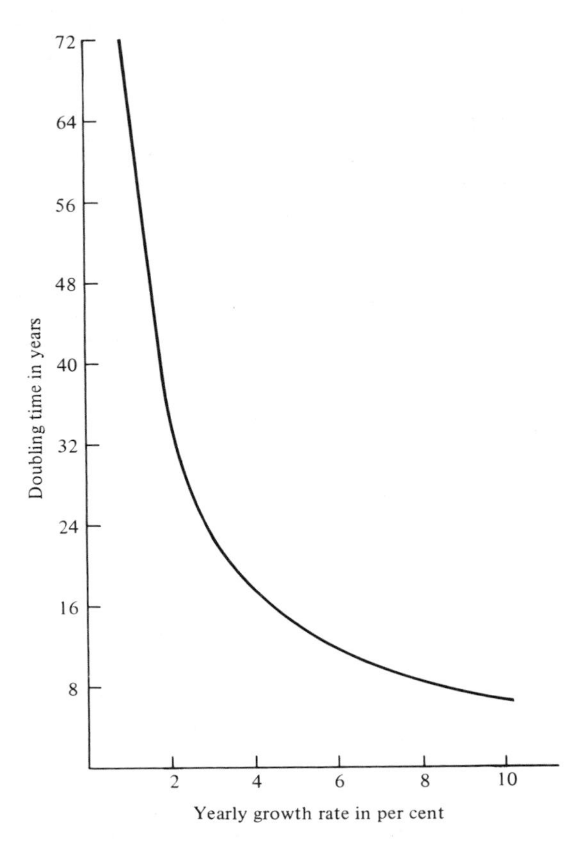

FIGURE 1-3. Doubling time versus growth rate.

Evaluating the energy consumed, Q_1, up to a certain time, t_1, and the energy consumed Q_2, for the doubling time, $t_d = (t_2 - t_1)$, we obtain

$$Q_1 = \int_{-\infty}^{t_1} P_0 e^{bt}\, dt = \frac{P_0}{b} e^{bt_1} \tag{1-18}$$

and

$$Q_2 = \int_{t_1}^{t_2} P_0 e^{bt}\, dt = \frac{P_0}{b} (e^{bt_d} - 1) e^{bt_1} \tag{1-19}$$

Since t_d is the doubling time and is given by (1-16) as $t_d = 0.693/b$, (1-19) becomes

$$Q_2 = \frac{P_0}{b} (2 - 1) e^{bt_1} = \frac{P_0}{b} e^{bt_1} \tag{1-20}$$

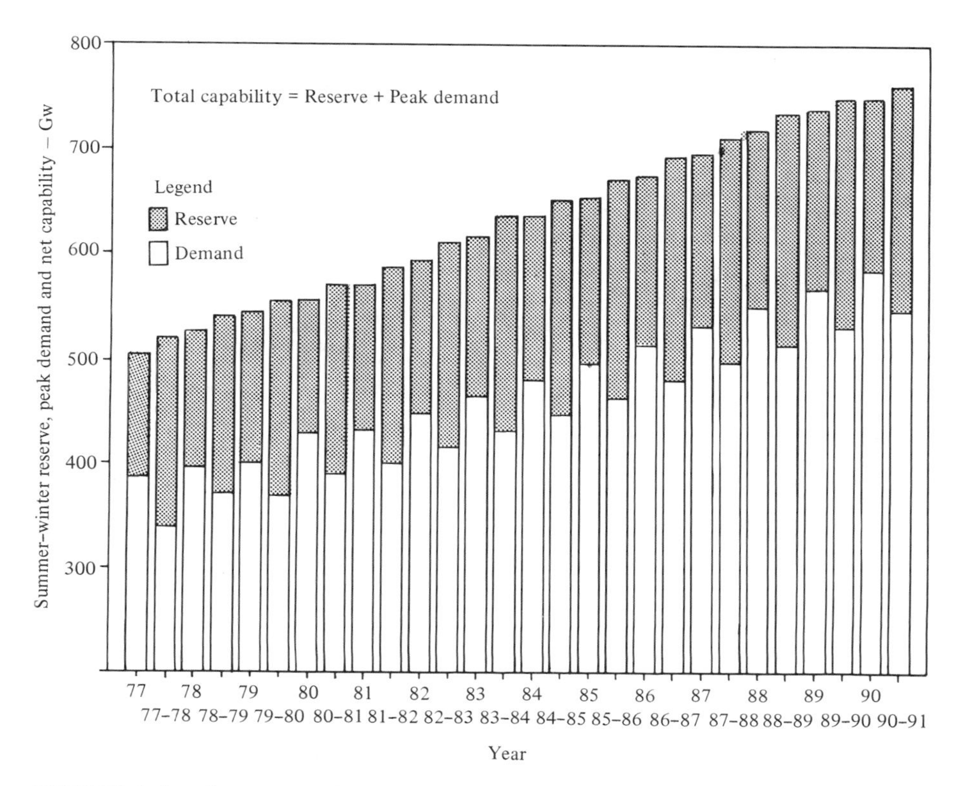

FIGURE 1-4. Summer–winter reserve, peak demand, and total capability for the contiguous United States. [From Ref. 2]

Comparing (1-18) and (1-20), we observe that if the energy consumption rate is constant, the energy consumed in one doubling period equals the energy consumed for the entire time prior to that doubling period. This conclusion is of great significance in estimating fuel consumption.

EXAMPLE 1-3

The present estimate of solid coal reserves in the United States is about 1.5×10^9 tons, having an energy content of 940 watt-years per ton. If the energy consumption rate increase is 3.38 percent, approximately how long will the coal reserves last? Assume that all the energy will be supplied by coal and that the present peak power demand is 425 GW.

If T is the time when the total consumption equals the reserve and Q_T is the total energy reserve, then, from (1-19), we obtain (with $t_1 = 0$)

$$Q_T = \frac{P_0}{b}(e^{bT} - 1)$$

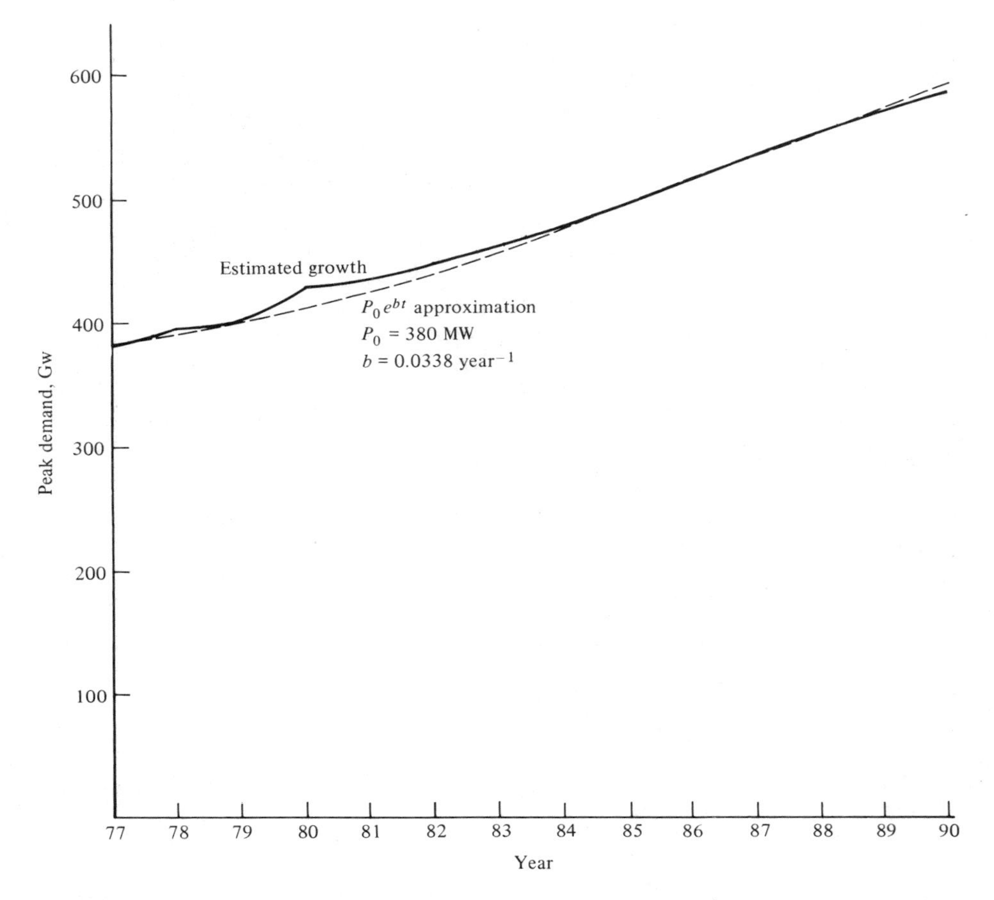

FIGURE 1-5. Peak demand and its approximation by P_0e^{bt}.

which may also be written as

$$e^{bT} = \frac{bQ_T}{P_0} + 1$$

or

$$T = \frac{1}{b} \ln \left(\frac{bQ_T}{P_0} + 1 \right)$$

From the numerical values given, we have

$$Q_T = 1.5 \times 10^9 \times 940 = 1.41 \times 10^{12} \text{ W-yr}$$

$$P_0 = 425 \times 10^9 \text{ W}$$

$$b = 0.0338$$

Hence

$$T = \frac{1}{0.0338} \ln \left(\frac{0.0338 \times 1.41 \times 10^{12}}{425 \times 10^9} + 1 \right)$$
$$= 3.14 \text{ years}$$

■

1-4 AVAILABLE MAJOR ENERGY RESOURCES

In Section 1-3 we have seen the effect of growth rate on doubling time. The consumption of energy for industrial purposes for the entire world is approximately doubling every 10 years. With such a growth rate, the world's fossil-fuel supply—coal, petroleum, and natural gas—will be depleted in a few hundred years unless we exploit other sources of energy. The other major source of energy on the earth is solar radiation. This energy may be used directly as intercepted solar radiation, or indirectly as wind and hydropower. Other significant forms of energy are tidal energy, geothermal energy, and nuclear energy.

Considering fossil fuels first, the estimate of reserves [9] and their approximate energy contents for the world are given in Table 1-1. It may be verified that if all the energy needs of the world were to be supplied by coal and the doubling time was about 14 years, the entire coal reserves will be depleted in about 79 years.* Even though the reserves of fossil fuels may be large, their ultimate production depends, in somewhat unpredictable ways on economic, political, and environmental considerations. Ultimately, we will need alternative energy sources in addition to using fossil fuels more efficiently than we do at present. A schematic representation of a coal-fired electric power plant providing electricity to a customer is given in Fig. 1-6. We will present an interesting summary of these efficiencies in Chapter 2, but first let us consider some of the alternative energy sources.

We mentioned earlier in this section that solar radiation is a major source of

TABLE 1-1 Estimate of Fossil-Fuel Reserves

Fuel	Estimated Reserves	Approximate Energy Content (watt-years)
Coal	7.6×10^{12} metric tons	937 per ton
Petroleum	2×10^{12} barrels	168 per barrel
Natural gas	10^{16} ft^3	0.036 per ft^3

*We have made the unrealistic assumption here that the growth rate of 5% (equivalent to a doubling time of 14 years) will remain unchanged for the entire 79 years.

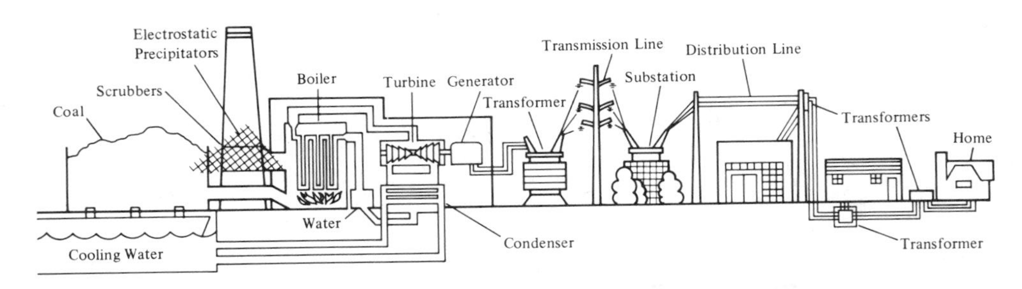

FIGURE 1-6. A coal-fired steam generating station supplying a home.

energy. It has been shown that the average worldwide incident power at the earth's surface is 182 W/m^2, which corresponds to a daily average energy of 4.4 kWh/m^2. The potential for saving fuel by use of solar energy in buildings in the United States has been estimated at 0.2 to 0.3 quadrillion Btu in 1985 [10]. The direct use of solar power is either the active type involving solar cells (or photovoltaic cells) or the passive type utilizing radiation to heat solar collectors. Obviously, the most favorable sites for the production of solar power are desert areas. At present, large-scale utilization of solar energy is limited by the cost of solar cells and solar collector–heat exchanger systems and by the requirement of an adequate energy storage system to smooth out the daily variation. Furthermore, the solar cells generate direct current (dc), and hence inverting and related equipment will be needed to obtain the desired alternating current (ac) for most large-scale operations. An active solar cell system supplying a building is shown in Fig. 1-7.

Indirect use of solar energy is manifested in wind energy. Turbine-type wind-energy generators transform the kinetic energy of the wind into rotary-shaft motion and, in turn, to electrical energy. The choice of wind as a source of energy depends on the following characteristics of wind [11]:

1. Wind is low-density fluid. Thus to convert its kinetic energy into a usable form, we must use blades of large dimensions. It has been found that the power that could be extracted from wind is given approximately by

$$P = 2.46 \times 10^{-3} D^2 u^3 \qquad \text{W} \tag{1-21}$$

where D is the blade diameter in feet and u is the wind velocity in miles per hour. When converted into SI units, (1-21) becomes

$$P = 2.0417 \times 10^{-5} D^2 u^3 \qquad \text{W} \tag{1-22}$$

where D is in meters and u is in meters per second.

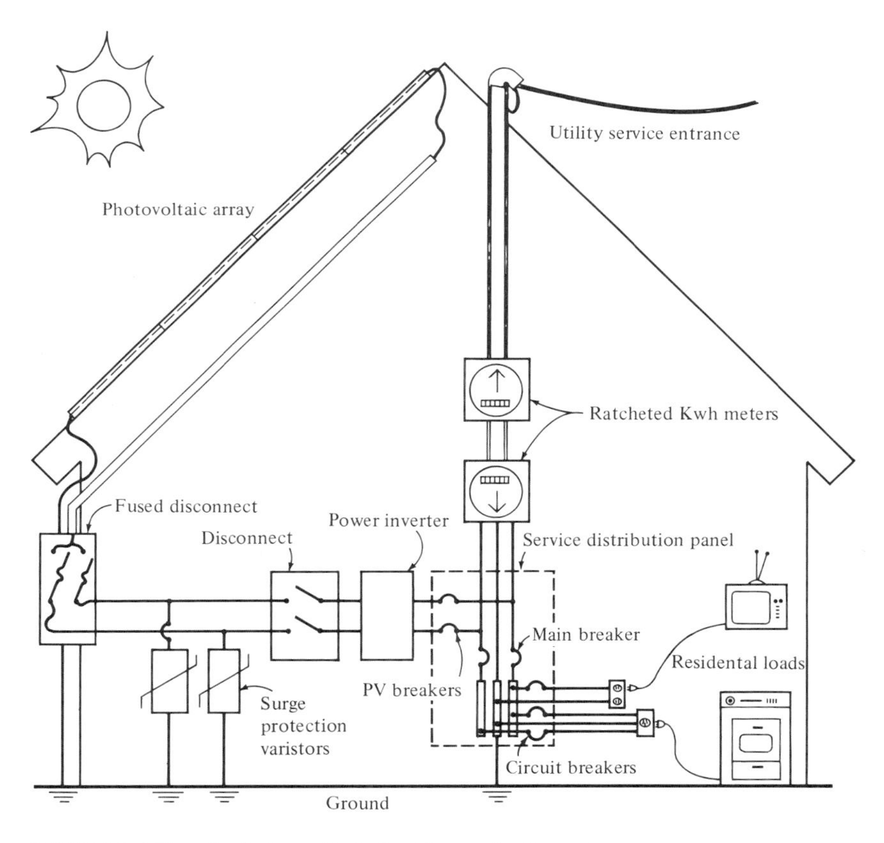

FIGURE 1-7. A photovoltaic (active solar) system.

TABLE 1-2 Comparative Costs[a] of Electric Power Generation by Different Methods

	Small Wind Generator[b]	Large Wind Generator[c]	Large Coal-Fired Plant	Light-Water Fission Reactor
Capital cost	$177/m^2$ [d]	$425/m^2$ [e]	$1107/kW	$1458/kW
Annual power production	613 kWh/m^2	1183 kWh/m^2	6132 kWh/kW	5256 kWh/kW
Heat rate	—	—	10.7 MJ/kWh	11.3 MJ/kWh
Fuel cost[f]	0	0	2.69¢/kWh	1.61¢/kWh
Cost of operation and maintenance[g]	0.72¢/kWh	0.90¢/kWh	0.57¢/kWh	0.21¢/kWh
Total generating cost[h]	2.8¢/kWh	3.5¢/kWh[e]	4.8¢/kWh	3.9¢/kWh

[a]All costs are estimated on the basis of mid-1981 prices.
[b]Power rating of 55 kW, rotor diameter 15 m.
[c]Power rating of 600 kW to 2.5 MW, rotor diameter 40 to 91 m.
[d]List price of manufacturer, including all service during first 2 years. Cost is given per square meter swept by the rotor.
[e]Estimated mass-production price.
[f]Averaged over the assumed 25-year depreciation period, using an annual cost escalation rate of 1.6% for coal and 3% for nuclear fuel.
[g]Operation and maintenance costs for wind generators, which are poorly known, have been put at 2.5% of the capital cost.
[h]All figures reflect 1981 prices. The assumptions are: depreciation time 25 years, annual market interest rate 18%, annual inflation 12%. The resulting capital charge rate in 1981 prices is 7.35% per year. It should be kept in mind that all comparisons of this kind are strongly dependent on the assumptions made.
Source: Ref. 12.

2. The directions and magnitudes of wind during a given day are random. Hence it is necessary to provide for the storage of energy, to smooth out these variations. Otherwise, the wind generator must operate in parallel with the utility lines. A typical small generator output may vary from 0.7 to 7.0 kW for wind speeds ranging from 20 to 50 km/h.

3. Wind velocity increases with height above the ground surface, and because the available power is proportional to the cube of the velocity, the choice of site of the plant is very critical.

Among the advantages of wind energy are its widespread availability, limited environmental impact, and the relatively small amount of land that is needed for the plant. The comparative costs of electric power generation by wind generators, coal-fired plants, and fission reactors are shown in Table 1-2.

A wind-generator system is shown in Fig. 1-8.

Another indirect means of using solar power is the stream-flow part of the hydrological cycle. In hydropower conversion, the potential energy of a mass of water at a hydraulic head is converted into the kinetic energy of the hydraulic turbine, which drives the electric generator. Such a system is represented sche-

FIGURE 1-8. A wind generator.

matically in Fig. 1-9. Since the potential energy is given by (1-6), the potential energy of 1000 kg of water for a head of 100 m is 9.8×10^5 J. Alternatively, a flow rate of 1 m^3/s, for a head of 100 m has a hydraulic power of $(9.8 \times 10^3 \times \text{head}) = 9.8 \times 10^3 \times 100 = 9.8 \times 10^5$ W of power. It has been estimated that the total water-power capacity of the world is about 3×10^{12} W, but only 8.5 percent of this capacity has been developed, because many of the regions having the greatest potential have economic problems.

The next form of energy we will discuss is *tidal energy,* which is obtained by filling a bay, closed by a dam, during periods of high tide, and emptying it during low-tide periods. Tidal power has been in use for centuries. A simple schematic of the use of tidal power to drive a turbine is shown in Fig. 1-10. The turbine is reversible, so that the flow of water during both the filling and

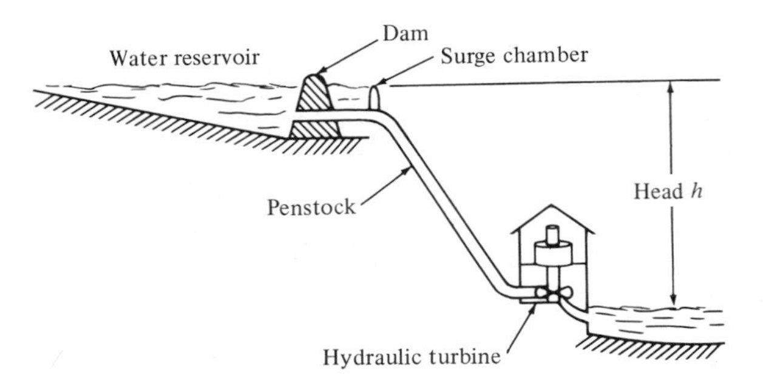

FIGURE 1-9. Schematic of a hydroelectric power station.

emptying of the bay may be utilized. Thus tidal power is available twice during each tidal period of 12 h and 25 min. The average tidal power per unit area of the tidal bay for a maximum tidal head, H, is given approximately by [8].

$$P_{av} = 0.219H^2 \qquad \text{MW/km}^2 \tag{1-23}$$

where the maximum tidal head is measured in meters and is shown in Fig. 1-10. This corresponds to energy of 2.45×10^{11} J/km^2 for a maximum tidal head of 5 m and for a 12 h 25 min tidal period. The total worldwide tidal-power potential is estimated at about 64 billion watts. However, large (of the order of gigawatts) tidal-power plants are not feasible, either economically or technically.

It has been found [8] that the average outward thermal power density at the earth's surface is 0.063 W/m^2 (or 1.5 μcal/cm^2-s). *Geothermal power* is obtained by extracting the heat that is stored temporarily in the earth. It is estimated [9] that the total stored thermal energy in major geothermal areas is

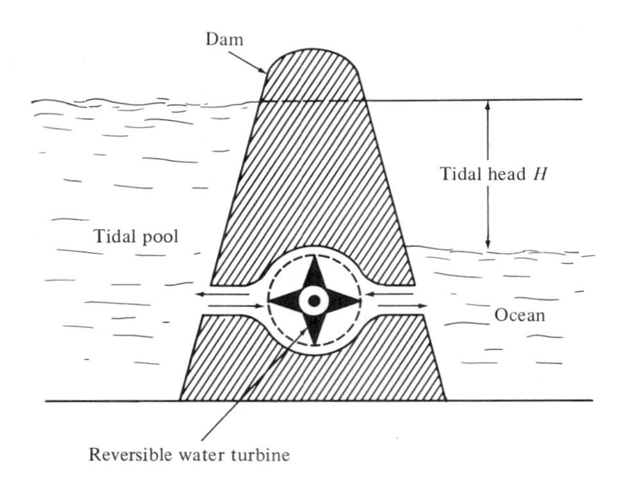

FIGURE 1-10. Tital power turbine, dam, and pool.

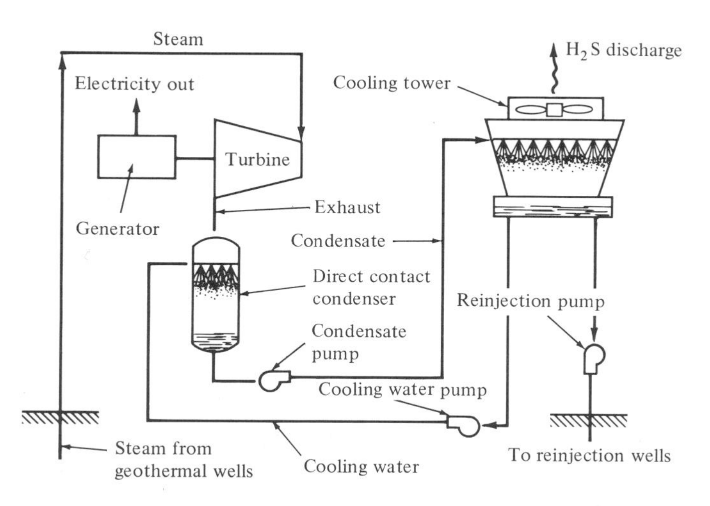

FIGURE 1-11. Flow diagram for a steam-dominated geothermal power plant using a direct-contact condenser. All plants installed at The Geysers site in California used some form of this arrangement until 1979.

4×10^{20} J. With a 25 percent conversion factor, this corresponds to electrical energy of 3×10^6 megawatt-years. A geothermal power plant is represented schematically in Fig. 1-11. The plant utilizes the steam produced in underground geothermal wells. The wells may be 1 to 2 km deep. The steam from the wells contains liquids and suspended solids, which are removed by a centrifugal separator. From the separator, the steam enters a turbine at a high temperature (e.g., 177°C for one well near San Francisco). The steam exits from the turbine at a reduced temperature (e.g., 50°C) by using a condenser having an internal pressure less than atmospheric pressure. Dry steam at a high temperature is ideal for turbine-driven electric generators. But hot water at about 125°C, available from some geothermal wells, can be used for operating a steam turbine by utilizing a secondary closed thermodynamic cycle with a suitable working fluid.

We consider finally the two forms of nuclear energy: fission and fusion. These two nuclear reactions are illustrated in Fig. 1-12. In fission, nuclei of a heavy element, such as uranium 235 (^{235}U) split, whereas fusion involves the combi-

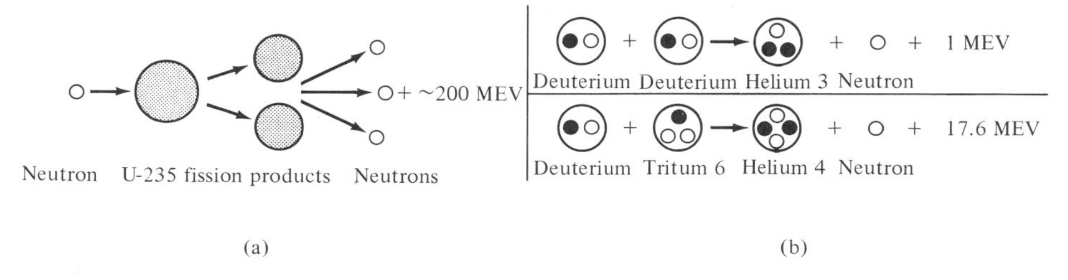

FIGURE 1-12. (a) A fission reaction; (b) a fusion reaction.

nation of light nuclei such as deuterium. Among fissionable materials, ^{235}U is the most suitable from an environmental standpoint. But ^{235}U is a rare isotope (each 100,000 atoms of uranium contain 711 atoms of ^{235}U). Thus naturally available nuclear fuel is in very limited supply. This difficulty may be overcome by the use of breeder reactors, in which uranium 238 is transformed into fissionable plutonium 239 by absorbing neutrons. Similarly, thorium 232 is transformed into uranium 233, which is fissionable. In terms of equivalent energy, with the breeder reactor 1 g of uranium 238 will produce 8.1×10^{10} J of heat, which is approximately equivalent to the heat produced by 2.7 metric tons of coal. The cost of production of electrical energy from nuclear fuels is approximately the same as that from fossil fuels (see Table 1-2).

Fusion power is scientifically feasible, but the engineering problems have not all been resolved. Thus no fusion power plants exist in the world. Engineering implementation of a practical fusion-power generating station does not seem feasible in the near future (say, by the year 2000). The hydrogen isotopes deuterium and tritium are suitable as fusion fuels. Deuterium is abundant (one atom to each 6,700 atoms of hydrogen), and the energy cost of separating it would be almost negligible compared with the amount of energy released by fusion. Tritium, however, exists only in tiny amounts in nature and must be bred in a reactor. For fusion to occur, the nuclei must form a plasma (atoms heated to such a high temperature that they are stripped of their electrons). The plasma must be contained or confined in a region of space such that the plasma density is high. Furthermore the plasma must be contained long enough for the fusion process to take place. Heating and plasma confinement are major engineering problems. [13]

1-5 ENVIRONMENTAL CONSIDERATIONS AND OTHER LIMITING FACTORS

From the discussion of Section 1-4 it is clear that some resources for energy are potentially unlimited. However, unlimited resources do not imply that the increase in energy conversion and consumption could be unlimited. The earth cannot sustain physical growth for more than a few tens of successive doublings. Besides constraints on the availability of fuels, and technical limitations (discussed for the various methods of energy conversion in Chapter 2), factors limiting energy conversion are biophysical, social, ethical, economic, and political in nature. [14]

Biophysical limitation is due to the degradation of all energy to low-temperature heat. All energy obtained from any resource, by an appropriate process of energy conversion, must end up as low-temperature heat added to the surface of the earth, causing the earth's surface temperature to increase. Also, the extreme heat affects climatic events which are rather complex and not fully understood. The burning of fossil fuels produces carbon dioxide and certain solid particles which tend to change the optical properties of air and have intricate effects on the climate. It is generally felt that "global climatic problems could be expected fairly early in the next century if energy use were to continue growing at historical rates" [14]. The adverse effects on public health of oxides of sulfur and nitrogen are among the biophysical limitations on energy conversion.

The impact of energy conversion on land, water, wilderness, and ecological systems constitutes a social and ethical limit. Ill effects of energy conversion and utilization include those related to strip mining, water pollution, oil spills, and acid rainfall. Since the Three Mile Island incident [15], safe use of nuclear energy poses an ever-increasing ethical problem.

High economic and technical risks also impose a limit on the sustained growth of energy conversion. It has been estimated that the cost of the 1976–1985 phase of Project Independence will equal about one year's GNP (at the 1976 level) compared with the energy sector's requirement for one-fourth of that investment. Complex technology is involved in the utilization of nuclear fuels for power plants. A good portion (about two-thirds [14]) of the nuclear power output may be needed to maintain its own growth, and the cumulative energy deficit may become insurmountable.

Public acceptance of complex and new energy technology—such as the use of nuclear fuels—presents another limit to the growth of energy systems. Because of "the susceptibility of most modern energy technologies to commercial monopoly or technical dependence, all producing inequity," [14] political forces are also considered a constraint to energy growth.

Emissions from thermal power plants include oxides of sulfur and nitrogen (SO_2 and NO_x), carbon monoxide, hydrocarbons, and particulates. Although means are taken to control air-polluting emissions, these emissions and waste-heat removal do present severe problems. In studying the growth of energy conversion processes we must consider the many limiting factors discussed above.

1-6 PRIME MOVERS FOR ELECTRIC GENERATORS

Most electric power is obtained by electromechanical energy conversion, whereby mechanical energy is converted into electrical energy by means of electric generators. The source of the mechanical energy is the prime mover, which is directly coupled to the generator, as illustrated in Fig. 1-13. Among the broad class of prime movers are thermal-, hydro-, and wind-driven prime movers. Certain interesting data pertinent to these prime movers, compared to the horse, ox, and human being, are presented in Fig. 1-14. Notice that the thermal class

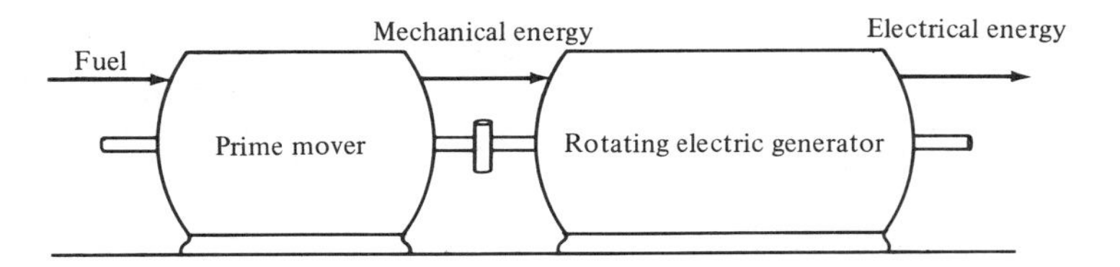

FIGURE 1-13. Prime mover driving an electric generator.

of prime movers is subdivided into the steam engine, the steam turbine, the gas turbine, and the internal-combustion engine. Steam turbines and internal-combustion engines (such as a diesel engine) are the most common prime movers for electric generators, the former being suitable for high-speed (3600-r/min) large-rating generators, whereas the latter is used for medium speed (1200-r/min) medium- and low-power generators. Water turbines are low- to medium-speed machines and are used to drive medium-size to large generators. A high-speed water turbine is the Pelton wheel, but this is not a very commonly used prime motor. Windmills and wind turbines are used to drive a large range of ratings of electric generators, depending on the potential application.

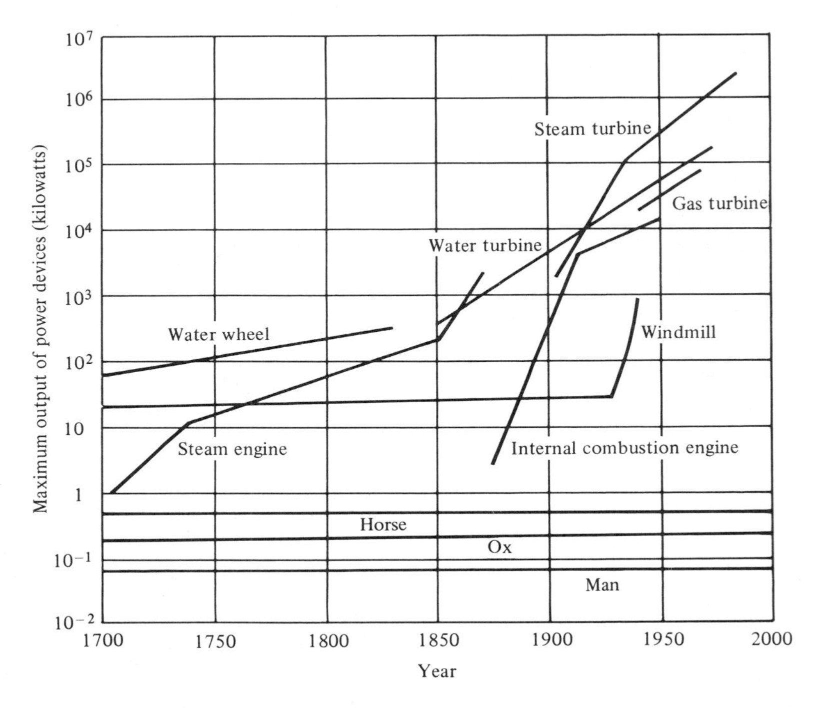

FIGURE 1-14. Prime movers for electric generators compared to horse, ox, and man. [From Ref. 1]

REFERENCES

1. **C. Starr,** "Energy and Power," *Scientific American,* September 1971, pp. 3–18.
2. "Electric Power Supply and Demand for the Contiguous United States 1981–1990," *U.S. Dept. of Energy Report DOE/EP-0022,* July 1981.
3. *The New York Times,* August 5, 1969, p. 1.
4. *The New York Times,* July 6, 1971, p. 1.
5. *The New York Times,* July 7, 1971, p. 1.
6. *The New York Times,* July 8, 1971, p. 1.
7. *The New York Times,* December 30, 1971, p. 8.
8. **J. H. Krenz,** *Energy Conversion and Utilization,* (second edition) Allyn and Bacon, Boston, 1984.
9. **M. K. Hubbert,** "The Energy Resources of the Earth," *Scientific American,* September 1971, pp. 31–40.
10. *Solar Age,* February 1981, p. 4.
11. **D. F. Warne** and **P. G. Calman,** "Generation of Electricity from the Wind," *IEE Reviews* (IEE, London), vol. 124, no. 11R, November 1977, pp. 963–985.
12. **B. Sorensen,** "Turning to the Wind," *American Scientist,* vol. 69, September–October 1981, pp. 500–506.
13. **G. H. Miley,** *Fusion Energy Conversion,* American Nuclear Society, La Grange Park, Ill., 1976.
14. **A. Lovins,** "Limits to Energy Conversion: The Case for Decentralized Technologies," in *Alternatives to Growth,* Vol. 1, D. L. Meadows, ed. Ballinger, Cambridge, Mass., 1977, pp. 59–76.
15. *Time,* "Nuclear Nightmare" (cover story), April 9, 1979, pp. 8–19.
16. **T. Collins,** "Solar Energy: Four Sites Demonstrate Potential," *IEEE Spectrum,* April 1979, pp. 60–65.
17. **M. N. John,** "Electricity Supply-Problems and Possibilities," *Electronics and Power,* October 1983, pp. 702–704.
18. **J. S. Forrest,** "Electricity Supply: Present and Future," *Electronics and Power,* November/December 1983, pp. 796–800.
19. **F. Kreith** and **R. T. Meyer,** "Large Scale Use of Solar Energy with Central Receivers," *American Scientist,* November/December, 1983, pp. 598–605.
20. **M. A. Fischetti,** "Power and Energy," *IEEE Spectrum,* January 1984, pp. 72–77.
21. *IEE News,* "California Steams Ahead," January 1984, p. 3.
22. **M. E. McCormick,** *Ocean Wave Energy Conversion,* Wiley-Interscience, New York, 1981.
23. **B. M. Count,** "Exploiting Wave Power," *IEEE Spectrum,* September, 1979, pp. 42–49.

PROBLEMS

1-1 One million cubic meters of water is stored in a reservoir feeding a water turbine. If the height between the turbine and the center of mass of the water is 50 m, how much total energy in megawatthours will be produced if all the water is utilized and the losses are negligible? The density of water is 993 kg/m^3.

1-2 In 1981, the U.S. consumption of fuels, in quads, was as follows: coal, 16.1; oil 32.1; gas, 20.2; hydro, 2.9; and nuclear, 2.9. Calculate the total energy produced during the year in gigawatthours, assuming an average overall power plant efficiency of 10 percent. Define efficiency as the ratio of output energy to input energy.

1-3 The average heat content of natural gas is 1.05 Btu/ft^3 and that of bituminous coal is 14,000 Btu/lb. From the data of Problem 1-2, determine the amounts of natural gas and coal consumed in the United States in 1981.

1-4 Suppose that the consumption of energy in a certain country has a growth rate of 4 percent per year. In how many years will the energy consumption be tripled?

1-5 At what growth rate will the energy consumption be doubled in 10 years?

1-6 The coal reserves in the eastern United States are estimated to contain 2250 quads of energy. If the energy content of this coal is 11,500 Btu/lb, determine the approximate weight of the coal reserve.

1-7 A power plant consumes 3600 tons of coal per day. If the coal has an average energy content of 10,000 Btu/lb, what is the plant's output rating? Assume an overall power efficiency of 15 percent.

1-8 The present recoverable natural gas reserve in the United States is estimated to contain 452 quads of energy. The present peak power demand is 450 GW. If the energy consumption rate increases at 6.5 percent per year, and 22 percent of the energy is to be supplied by natural gas, approximately how long will the natural gas reserve last?

1-9 Estimate the average power output from a wind turbine having a blade diameter of 35 ft. The wind velocity ranges from 10 to 30 mi/hr.

1-10 The maximum tidal head available for a proposed tidal-power station is 6 m. What must be the area of the tidal bay to generate 1000 MW of power?

CHAPTER 2

Electric Energy Conversion

2-1 BASIC ENERGY CONVERSION PROCESSES

In Chapter 1 we discussed various resources available for obtaining electrical energy. The three basic sources of energy are solar, chemical, and nuclear; and the two basic processes of conversion to electrical energy are *direct* and *electromechanical*. The interrelationships among these processes and various fuels are illustrated in Fig. 2-1. The utilization of these fuels involves a number of stages, except the fuel cell, which is a true direct energy conversion device in that it converts chemical energy directly into electrical energy. In electromechanical energy conversion, the fuel is transformed into heat, which may produce steam to run either a steam turbine or an internal-combustion engine. Thus the energy of the fuel is transformed into mechanical energy via heat energy. The mechanical energy of the steam turbine (or the internal-combustion energy) is converted into electrical energy by a rotary electric generator. On the other hand, solar energy, which produces rainfall and thus hydropower, is used indirectly to drive a hydraulic turbine, which in turn drives a rotary electric generator. Processes involving mechanical motion, such as in rotary electric generators, result in electromechanical energy conversion.

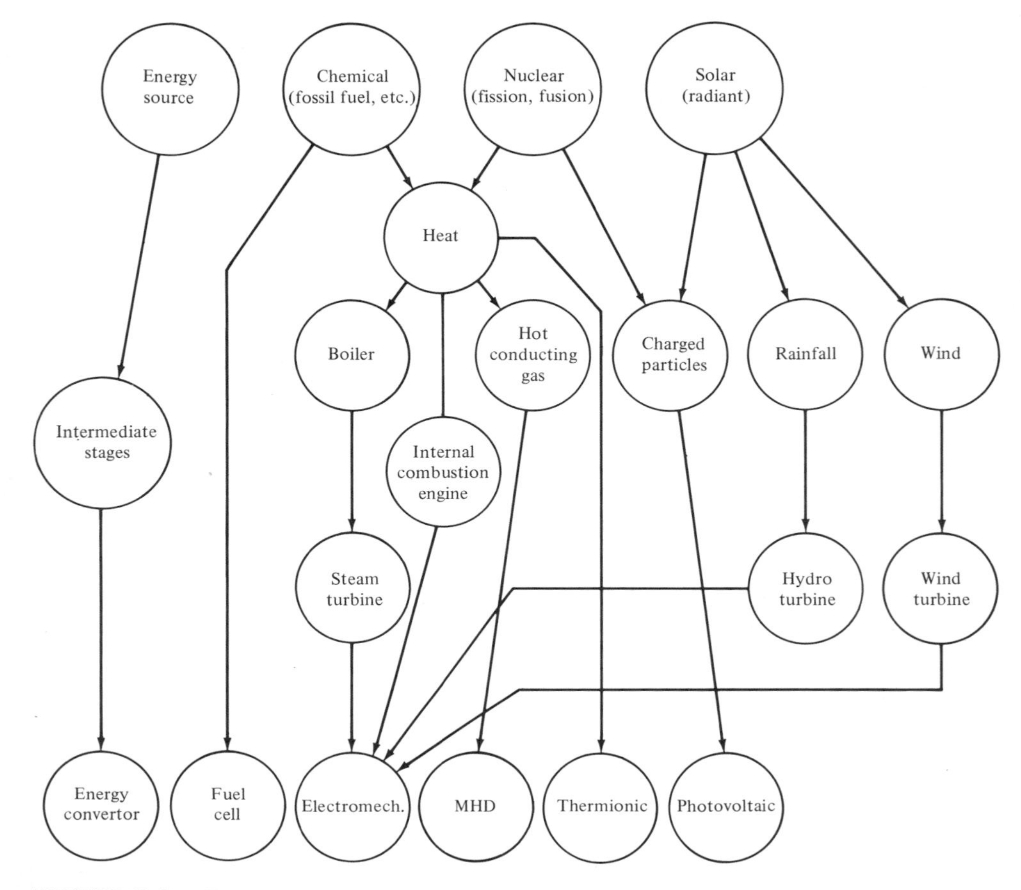

FIGURE 2-1. Energy sources, energy converters, and their interrelationships.

Although a fuel cell is the only device that accomplishes energy conversion directly, other devices not involving mechanical rotation are grouped in the class of direct energy conversion devices. Thus direct energy conversion devices are those which convert solar, thermal, chemical, and nuclear energies into electrical energy without involving a rotary or reciprocating mechanical prime mover. The various direct energy converters, illustrated in Fig. 2-1, are as follows:

1. Fuel cells.
2. Thermionic, thermoelectric, and ferroelectric generators.
3. Photovoltaic, photoelectric, and electrostatic generators.
4. Magnetohydrodynamic generators.
5. Piezoelectric generators.

In addition to those listed above, there are several other devices which are capable of direct energy conversion. But these, and some listed above, are insignificant for bulk power generation. In the following discussion we consider only those direct energy converters which are relevant to the present applications and have future prospects of bulk power generation. A general theory of direct energy converters has been developed on the basis of irreversible thermodynamics [1–3]. However, here we will take a simplified approach and consider the devices individually. Even for a simplified treatment, an understanding of the principles of thermodynamics is essential.

2-2 REVIEW OF LAWS OF THERMODYNAMICS

The *first law* of thermodynamics states that energy is conserved. Thus if heat ΔQ is added to a system to increase its energy by an amount ΔU resulting in mechanical output ΔW, then

$$\Delta Q = \Delta U + \Delta W \tag{2-1}$$

Now, if we consider a steady-flow system, such as the one shown in Fig. 2-2, where there is a change in the pressure, p, and volume, V, of the system, then (2-1) modifies

$$\Delta Q = \Delta U + \Delta W + \Delta(pV) \tag{2-2}$$

From Fig. 2-2, $\Delta Q = Q_1 - Q_2$, $\Delta U = U_1 - U_2$, and $\Delta(pV) = p_1V_1 - p_2V_2$. The quantity $\Delta(pV)$ is externally stored energy in that it reflects the difference in work required to move a unit mass into and out of the system. Both ΔU and $\Delta(pV)$ are unique forms of stored energy in that they depend on the state variables p, V, and T (T being the temperature) of the system. Usually, the sum $(U + pV)$ is considered as a single property H, called the *enthalpy* of the system; that is,

$$H = U + pV \tag{2-3}$$

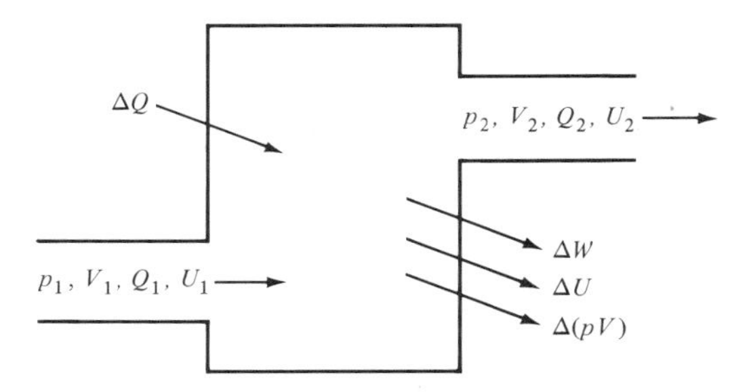

FIGURE 2-2. Illustration of a thermodynamic process.

According to the *second law* of thermodynamics, continuous conversion of heat into work by a device receiving heat from a source at a temperature T is possible only when part of the heat is rejected into a sink at a lower temperature, T_2 ($T_2 < T_1$). As a consequence, an energy conversion process involving heat must have an efficiency of less than 100 percent, even without loss. This ideal or maximum attainable efficiency is called the *Carnot efficiency,* η_C, and is given by

$$\eta_{\text{ideal}} = \eta_C = \frac{T_1 - T_2}{T_1} \tag{2-4}$$

where T_1 and T_2 are absolute temperatures for a reversible cycle. Accordingly, the Carnot efficiency of a modern large fuel-fired steam-turbine generating unit is about 61 percent. In an electric power station the mechanical energy of the turbine is converted to electrical energy by an electric generator, and the overall efficiency from fuel to electrical power can scarcely exceed 40 percent. Such an efficiency limitation exists on all closed-cycle heat engines. On the other hand, a fuel cell, which converts chemical energy into electrical energy, is not a heat engine and is not subject to the Carnot efficiency limitation. Rather, as we will see later, the efficiency of a fuel cell is limited by the thermodynamic quantities related to the fuel oxidation reaction. Fuel cells have efficiencies in the range 80 to 99 percent.

We realize that heat flow is a function of temperature difference. If a quantity of heat is divided by its absolute temperature, the quotient is called the *entropy* of the system. Entropy is a measure of degradation of energy through usage. This definition of entropy was proposed by Clausius, who visualized an ultimate "heat death" of the universe, implying a continual decrease in energy potential, eventually going down to zero. If heat ΔQ enters a system at an absolute temperature T, then the change in entropy of the system for a reversible process is given by

$$\Delta S = \frac{\Delta Q}{T} \tag{2-5}$$

In a reversible process, the state variables p, V, T, and U are well defined such that if the process is reversed, these state variables follow the same values in reverse order. Thus reversibility is measured by entropy S. The second law of thermodynamics can be expressed as

$$\Delta S \geq \frac{\Delta Q}{T} \tag{2-6}$$

where "$>$" is for an irreversible process. For a reversible cycle the net change in entropy is zero.

We now consider the thermodynamic limitation of energy conversion in an oxidation reaction, such as that in a fuel cell. We define *Gibbs free energy, G,* by

$$G = H - TS \tag{2-7}$$

where H is the enthalpy of the system at an absolute temperature T and entropy S. The maximum amount of electrical energy obtainable from an oxidation reaction is the change in the Gibbs free energy, ΔG, for the reaction. Thus from (2-7) for an isothermal reaction, we have

$$\Delta G = \Delta H - T\,\Delta S \tag{2-8}$$

where $T\,\Delta S$ is a measure of the heat absorbed by the system during a reversible change, and is the unavailable energy.

With this brief review of the principles of thermodynamics, we proceed to consider some direct energy conversion devices.

2-3 FUEL CELLS

A fuel cell is an electrochemical device which directly converts the chemical energy (of a fuel oxidation reaction) into electrical energy. In a fuel cell, energy conversion occurs by a process that is the reverse of electrolysis. Two distinct features of fuel cells are that the chemical energy is stored outside the cell, and the reaction products are commonly rejected from (although they may be stored in) the cells. A fuel cell is shown schematically in Fig. 2-3. Unlike the lead–acid storage battery, which is a secondary cell, the fuel cell is a primary cell. The hydrogen–oxygen fuel cell is one of the most highly developed fuel cells. In a hydrogen–oxygen fuel cell, hydrogen is the fuel and oxygen is the oxidant (Fig. 2-3). The electrodes are porous and are connected to the load. The hydrogen breaks into positive hydrogen ions and electrons upon passing through the anode. This reaction is expressed as

$$2H_2 \rightarrow 4H^+ + 4e^- \tag{2-9}$$

The flow of hydrogen ions and electrons is shown in Fig. 2-3. Upon flowing through the load circuit, the electrons and the hydrogen ions combine with oxygen at the cathode, where the reaction is given by

$$4H^+ + 4e^- + O_2 \rightarrow 2H_2O \tag{2-10}$$

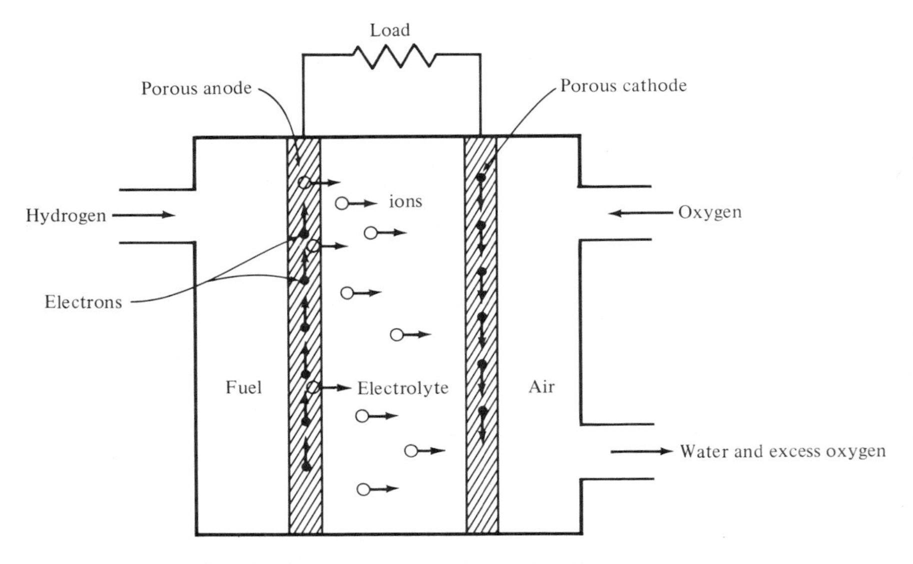

FIGURE 2-3. Schematic of a hydrogen–oxygen fuel cell.

The overall reaction is

$$2H_2 + O_2 \rightarrow 2H_2O \tag{2-11}$$

Obviously, the oxidation product is water, which is removed from the cell. Reactions (2-9) through (2-11) are controlled continuous chemical reactions in the hydrogen–oxygen fuel cell. Notice that reaction (2-11) is the reverse of the electrolysis of water. Recalling (2-8), ΔG, the Gibbs free-energy change, is the energy change theoretically available from the fuel cell.

The ideal electromotive force (emf) E of a fuel cell is given by

$$E = -\frac{\Delta G}{23.06n} \quad \text{V} \tag{2-12}$$

where the minus sign is an arbitrary sign convention, ΔG is the change in Gibbs free energy in kilocalories per gram mole, and n is the number of electrons transferred per molecule of fuel oxidized. In a fuel cell, the net amount of energy liberated (that is, the change in enthalpy ΔH) is the difference between the enthalpy of formation of the reactants and that of the products. Therefore,

$$\Delta H = \Sigma\, \Delta H_{\text{reactants}} - \Sigma\, \Delta H_{\text{products}} \tag{2-13}$$

Similarly, the change in Gibbs free energy can be expressed as

$$\Delta G = \Sigma\, \Delta G_{\text{reactants}} - \Sigma\, \Delta G_{\text{products}} \tag{2-14}$$

Values for the enthalpy of formation and Gibbs free energy of compounds and ions, at a given temperature and pressure, are available in the literature. From these values some characteristics of fuel cells may be determined.

EXAMPLE 2-1

For the reaction (2-11), $\Delta G = -56.7$ kcal at 25°C and $\Delta H = -68.3$ kcal. Determine the ideal emf and efficiency of the cell.

From (2-9) or (2-10), four electrons are transferred for two molecules of water. Thus $n = 2$, and (2-12) yields the ideal emf $E = -(-56.7)/(23.06 \times 2) = 1.23$ V.

The ideal efficiency is defined as the useful work per unit enthalpy change at a constant temperature and pressure. But useful work equals Gibbs free energy. Hence the ideal efficiency is given by

$$\eta = \frac{\Delta G}{\Delta H} = \frac{-56.7}{-68.3} = 83 \text{ percent}$$

The efficiency of the hydrogen–oxygen fuel cell decreases as the temperature increases. Hence this fuel cell is operated near ambient temperature. ■

Numerous other types of fuel cells exist and are under development. Although fuel cells have high efficiencies, their application for bulk power generation is unlikely in the near future because fuel cells are about twice as expensive as other fossil-fuel electrical plants on a per kilowatt basis. Furthermore, the life of fuel cells is rather low (about 16,000 h). However, some experimental large (4.5-MW) fuel cells are reported to have been built [4].

2-4 THERMOELECTRIC ENERGY CONVERSION

Thermoelectric energy converters are based on a combination of the Seebeck, Peltier, and Thomson effects. According to the *Seebeck effect,* an electromotive force is produced in a loop of two different materials if their junctions are at different temperatures. Consequently, a current flows through the loop. On the other hand, if a current is passed through a loop of two different materials, one junction becomes hot and the other cold. This phenomenon is known as the *Peltier effect*. The *Thomson effect* is due to Lord Kelvin. He discovered that when a current flows through a conductor in which a temperature gradient exists, heat is either liberated or absorbed, depending on the direction of flow of current. These three effects are reversible and are interrelated. The interrelationship can be analyzed by the principles of irreversible thermodynamics [1].

Schematically, the three effects are shown in Figs. 2-4, 2-5, and 2-6. Refer-

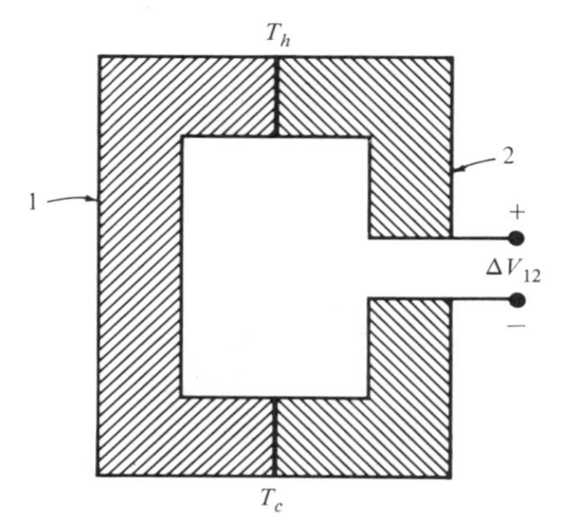

FIGURE 2-4. Junction of two dissimilar metals at temperatures T_h and T_c ($T_h > T_c$) resulting in a voltage ΔV_{12} as a consequence of the Seebeck effect.

ring first to Fig. 2-4, which shows the Seebeck effect, if T_h and T_c are the hot and cold junction temperatures, respectively, the open-circuit voltage ΔV_{12} is related to the temperature difference $(T_h - T_c)$ by the Seebeck coefficient α_S such that

$$\alpha_S = \lim_{(T_h - T_c) \to 0} \frac{\Delta V_{12}}{T_h - T_c} = \frac{dV_{12}}{dT} \tag{2-15}$$

Figure 2-5 shows the Peltier effect. The ratio of the heat change at the junction to the current flow is defined as the Peltier coefficient α_P; that is,

$$\alpha_P = \frac{Q_{12}}{T} \tag{2-16}$$

where I is the current and Q_{12} is the heat.

Finally, Fig. 2-6 shows the Thomson effect. In this case the ratio of heat change per unit of current flow to the local temperature is the Thomson coefficient α_T. From the figure, then,

$$\alpha_T = \frac{\Delta Q/\Delta x}{I(\Delta T/\Delta x)} \tag{2-17}$$

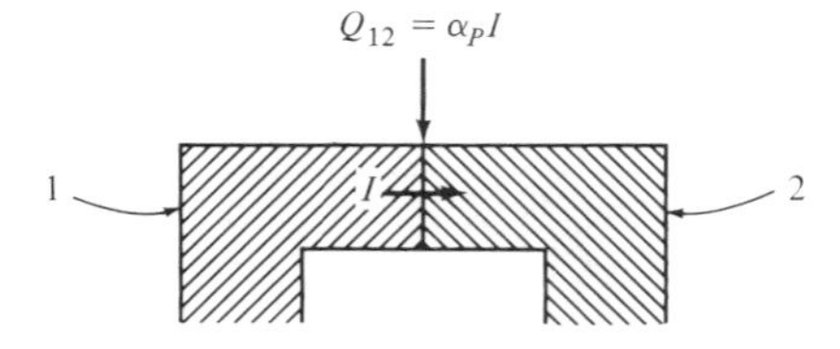

FIGURE 2-5. Schematic representation of the Peltier effect.

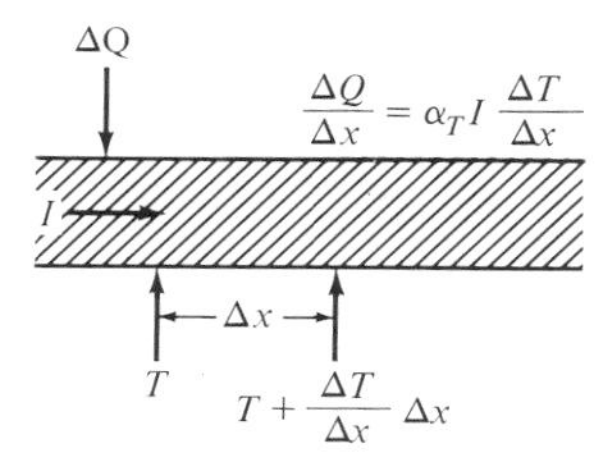

FIGURE 2-6. Schematic defining Thomson coefficient.

Thermoelectric generators can operate on solar heat, fossil-fuel heat, or heat from a nuclear reactor. However, like the fuel cell, the thermoelectric generator is not practical for bulk power generation.

2-5 THERMIONIC ENERGY CONVERSION

In a thermionic converter, heat energy is converted to electrical energy by thermionic emission, whereby electrons are emitted from the surface of certain metals, when sufficiently heated. Schematically, a thermionic energy converter is shown in Fig. 2-7. The three major components of a basic thermionic energy converter are the thermionic emitter, the collector, and the working fluid, which may be an electron gas or a partially ionized plasma. The emitter is heated by the input heat, Q_i, and the electrons are emitted. Some of these electrons are received by the cold collector at an output heat, Q_o. The difference $(Q_i - Q_o)$ is the energy that drives the electrons through the external circuit, and appears as electrical energy. The collector is cooled to remove the output heat.

Vacuum converters, which have electron gas as their operating fluid, operate at low temperatures, in the range 1200 to 1400 K. Their typical operating characteristics are: output power, 1 W/cm^2; efficiency 5 percent; life, tens of hours [3].

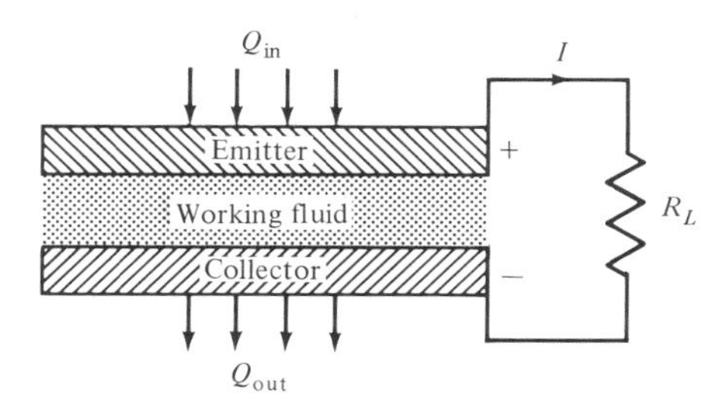

FIGURE 2-7. A thermionic energy converter.

Low-pressure converters have produced a power density of about 10 W/cm^2 at an efficiency of 10 percent and at emitter temperatures up to 2300 K. High-pressure converters have delivered about 40 W/cm^2 at efficiencies of 20 percent and at emitter temperatures of 2200 K.

The principal applications of thermionic energy converters are in regions not easily accessible, such as outer space, undersea, and polar regions. The two important heat sources are the sun and nuclear reactors.

2-6 PHOTOVOLTAIC ENERGY CONVERSION

A photovoltaic energy converter, commonly known as a *solar cell,* converts solar radiation directly into electrical energy. A high-efficiency silicon solar cell was developed by Bell Telephone Laboratories in 1954. At that time, solar cells were heavy and expensive, and since then there has been a concerted effort to develop solar cells suitable for practical applications. In 1959, the approximate cost of silicon solar cells was $2000 per watt, which was reduced to $21 per watt in 1976 and to $10 per watt in 1978. The goal for the mid-eighties has been set at 50 cents per watt. In terms of energy costs, 50 cents per watt would result in an energy cost of 6 to 8 cents per kilowatthour, which would be competitive with utility rates.

Solar cells are made from semiconductors, such as silicon or germanium, by doping them to obtain n-type or p-type layers. To obtain an n-type semiconductor, a small amount of phosphorus is added to silicon while the crystal is being formed. It is called n-type because the material has some free electrons which do not have definite places in the crystal structure. These electrons, which can move about freely, are called *n-type carriers*. On the other hand, if boron is added to silicon at the time of crystal formation, the result is a deficiency of electrons in the crystal structure. There are places in the crystal structure where there would be electrons. These voids in the crystal are known as holes, which can easily move within the crystal structure. The motion of holes occurs because electrons will fill one of the holes, but creating another hole, which is filled by another electron, and so on. Each hole may be considered to have a positive charge, and the holes are known as *p-type carriers*. A silicon crystal doped with boron is called p-type. Figures 2-8(a) and (b), respectively, show current flow in an n-type material by electron movement, and current flow in a p-type material where the majority current flow is hole movement.

The situation at a pn junction of two semiconductors is depicted in Fig. 2-8(a). From the n-type material free electrons will move into the region to the left, and the free holes will move to the right as positive charge carriers from the p-type material. Because of this diffusion, there is an excess of negative charges on the left and an excess of positive charges on the right, as shown in

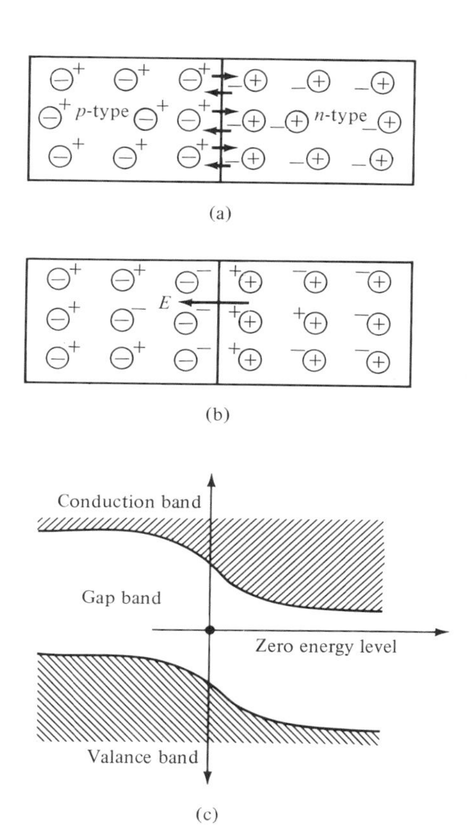

FIGURE 2-8. (a) A *pn*-junction; (b) diffusion of electrons to hole creating the electric field *E*; (c) potential energy of electrons and holes at a *pn*-junction.

Fig. 2-8(b). Thus an electric field is established in the interface region, when the diffusion process reaches an equilibrium. The electric potentials as seen by an electron and a hole are shown in Fig. 2-8(c). Electrons and holes will have a tendency to move in the directions of decreasing potential energies, implying that the electrons will tend to move to the right and holes to the left. Notice that the curves denoting the potentials are separated by a band gap. Electrons having energies greater than those corresponding to the electron potential energy curve are free to move and are in the conduction band. Similarly, holes with energies greater than those corresponding to the hole potential energy curve are free to move and are considered in the valence band. The minimum energy required to move an electron from the conduction band to the valence, or a hole from the valence band to the conduction band, is the gap energy as measured by the gap band.

When light, or a photon, hits a *pn* junction with an energy greater than the gap energy, electron–hole pairs are created. They tend to move in their respective directions. If this separation of electron–hole pairs is sustained, charge

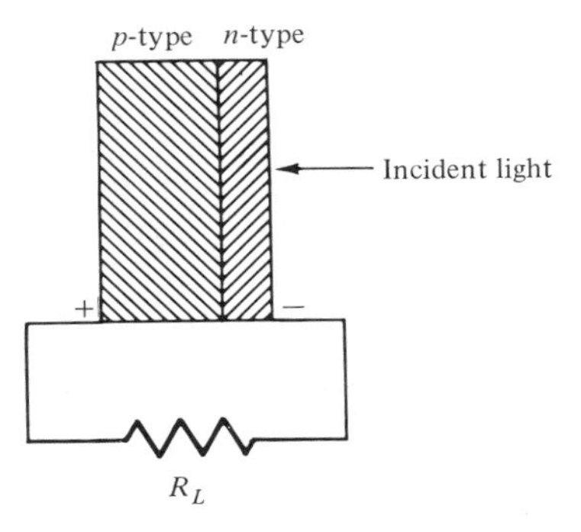

FIGURE 2-9. A solar cell supplying a load.

separation occurs and electric current flows through the load, as indicated in Fig. 2-9. Under open circuit, the voltage is determined by the production of electron–hole pairs and is limited by the charge separation and accumulation at the terminals. Theoretically, the greater the gap band, the greater is the open-circuit voltage.

The efficiency of solar cells is between 10 and 12 percent. For maximum utilization of the semiconductor material, it must have a low reflectivity and a high photon absorption coefficient. Whereas a typical commercial solar cell may be rated at about 0.45 V, 0.8 A, the current–density–voltage variation of a silicon cell, for an incident energy of 0.1 W/m^2, is shown in Fig. 2-10. From this curve, the output power per unit area may be obtained. If the target set for the future, especially the cost at 50 cents per watt and an efficiency of over 15 percent can be accomplished, solar cells are more likely to play an important role in electric power generation.

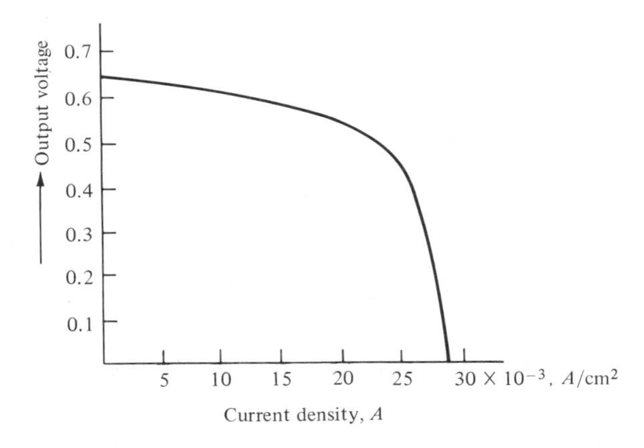

FIGURE 2-10. Output characteristic of a silicon solar cell.

MAGNETOHYDRODYNAMIC POWER CONVERSION

A magnetohydrodynamic (MHD) generator converts the internal energy of an ionized (electrically conducting) gas into electrical energy. [5] The principle of operation of a dc MHD generator is the same as that of a conventional rotary dc generator (see Chapter 5). The major difference between the two, however, arises from the fact that in an MHD generator an ionized gas, or plasma, serves the purpose of the conductor. Also, whereas for a dc generator we need a prime mover to drive, in an MHD generator the flow of ionized gas replaces the prime mover. The ionized gas is driven into the MHD generator by creating a pressure difference, as we will see later.

To understand the basic principle of operation of a dc MHD generator, we first state *Faraday's law* of electromagnetic induction for moving conductors: An emf is induced in a conductor "cutting" magnetic lines of flux. Specifically, if a conductor of length, l, moves with a velocity, u, in a uniform magnetic field, B, such that l, u, and B are mutually perpendicular, as illustrated in Fig. 2-11, then the emf, E induced in the conductor is given by

$$E = Blu \tag{2-18}$$

We now refer to Fig. 2-12, which shows the elements of a dc MHD generator. Notice that the effective "length" of the ionized gas, which is the conductor, is a. Hence the emf available at the electrodes is, according to (2-18),

$$E = Bau \tag{2-19}$$

where B is the uniform dc field and u is the constant velocity of the conductor. If J is the current density (assumed uniform) then, from Fig. 2-12, the load current, I, is

$$I = Jbc \tag{2-20}$$

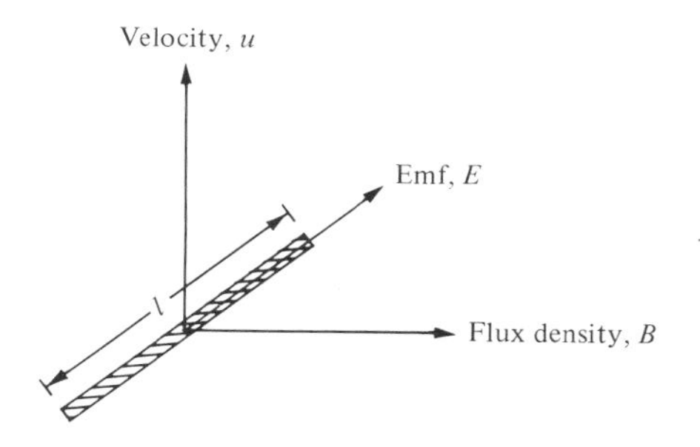

FIGURE 2-11. Relative directions of *E, u,* and *B*.

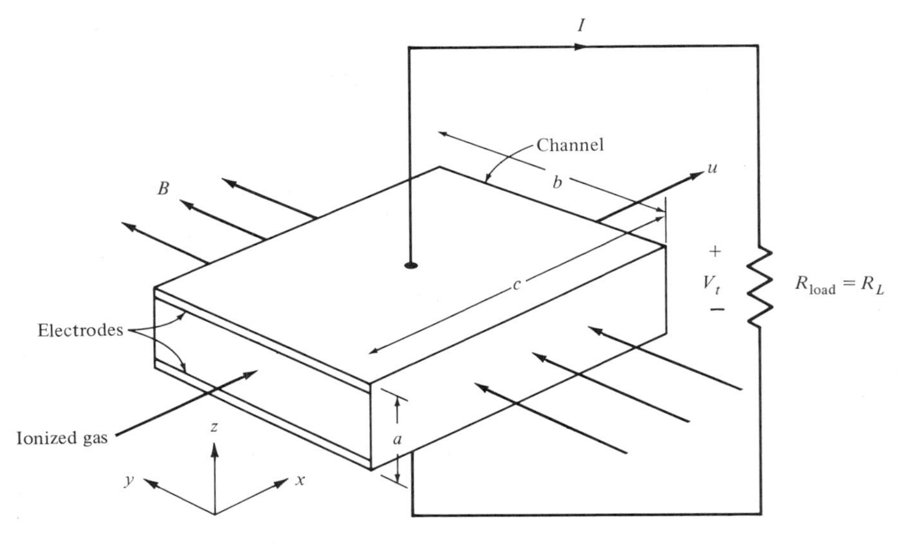

FIGURE 2-12. An elementary MHD generator.

If σ is the electrical conductivity of the ionized gas in the channel, then the resistance, R_G, between the electrodes is simply

$$R_G = \frac{a}{\sigma bc} \tag{2-21}$$

Consequently, the terminal voltage, V_t, across the load can be written as

$$V_t = E - IR_G \tag{2-22}$$

Substituting (2-19)–(2-21) in (2-22) yields

$$V_t = Bau - \frac{Ja}{\sigma} \tag{2-23}$$

Or the current density J may be written as

$$J = \sigma\left(Bu - \frac{V_t}{a}\right) = \sigma Bu\left(1 - \frac{V_t}{Bau}\right) \tag{2-24}$$

Defining the *loading factor*, K, as the ratio

$$K = \frac{V_t}{Bau} \tag{2-25}$$

we may rewrite (2-25) as

$$J = \sigma Bu(1 - K) = \sigma(E - uB) \tag{2-26}$$

The output power per unit volume (power density, P_0) of the gas can be written as

$$P_o = \frac{V_t I}{abc} \tag{2-27}$$

Substituting (2-20) and (2-25) in (2-27), we get

$$P_o = \frac{KBau}{abc} Jbc = KBuJ \tag{2-28}$$

Finally, (2-26) and (2-28) yield

$$P_o = \sigma u^2 B^2 K(1 - K) \tag{2-29}$$

For maximum output-power density, it can be readily verified from (2-29) that $K = \frac{1}{2}$. This corresponds to the matched-resistance condition for which the load resistance equals the internal resistance of the generator for maximum power transfer. The constant K can thus be interpreted as the ratio of the load resistance to the total resistance. Since the electrode area is bc, the load current I is, from (2-26),

$$I = \sigma(1 - K)uBbc \tag{2-30}$$

The load resistance R_L is

$$R_L = \frac{V_t}{I} = \frac{Ka}{\sigma(1 - K)bc} \tag{2-31}$$

But the internal resistance of the generator is $R_G = a/\sigma bc$. Thus (2-31) can also be expressed as

$$R_L = \frac{K}{1 - K} R_G \tag{2-32}$$

or

$$K = \frac{R_L}{R_T} \tag{2-33}$$

where $R_T = R_L + R_G$ = total resistance, which further validates the statement that K is the ratio of load resistance to total resistance.

Further analysis of this generator can be made by considering the fact that the output-power density P_o, as given by (2-28) and (2-29), is obtained by the work done by ionized gas as it flows against the body force per unit volume. Consequently, the force density JB and the work done JBu are, respectively, given by

$$JB = (1 - K)\sigma u B^2 \tag{2-34}$$

$$JBu = (1 - K)\sigma u^2 B^2 \tag{2-35}$$

The ohmic dissipation in gas is, from (2-29) and (2-35), the difference between the work done by the gas and the output power; or

$$P_{\text{ohmic}} = (1 - K)^2 \sigma u^2 B^2 \tag{2-36}$$

In the analysis above, fluid-flow equations have not been considered. Nevertheless, certain important features of the dc MHD generator emerge from the equations derived above. For example, from (2-29) we see that for maximum output density, $K = \frac{1}{2}$. But this is not the only consideration for the operation of an MHD generator. Equations (2-29) and (2-35) show that only a fraction K of the work done by the gas appears as electrical power, and the remaining $(1 - K)$ dissipates as ohmic heating in the gas. This ohmic heating does not represent a loss of energy from the system. Rather, it denotes an increase in entropy, a degradation of energy and a reduction in its availability for conversion. The factor K is analogous to the isentropic efficiency factor in a steam or gas turbine. Therefore, the balance between maximum output-power density and adequate isentropic efficiency leads to values of K between 0.7 and 0.8 for MHD generators.

Apart from the loading factor K there are a number of other thermodynamic, hydraulic, and electrical factors which affect the performance of MHD generators. The first two factors are beyond the scope of this book and only a few of the electrical parameters are considered here.

MHD generators have been proposed for topping in conventional steam-turbine power plants. An arrangement is illustrated in Fig. 2-13. Although MHD generators are exclusively suited, in principle, for very larger power generation (of the order of hundreds of megawatts), there are a number of practical limitations to their applications. Some of these constraints are: production of high magnetic fields (of the order of several tesla); rather low conductivity of the plasma (in the range 1 to 60 S/m at about 3000 K for a seeded carbon monoxide–oxygen flame at 1 atm pressure, compared to the conductivity of copper, which is 6×10^7 S/m); duct (or channel) material to withstand high combustion

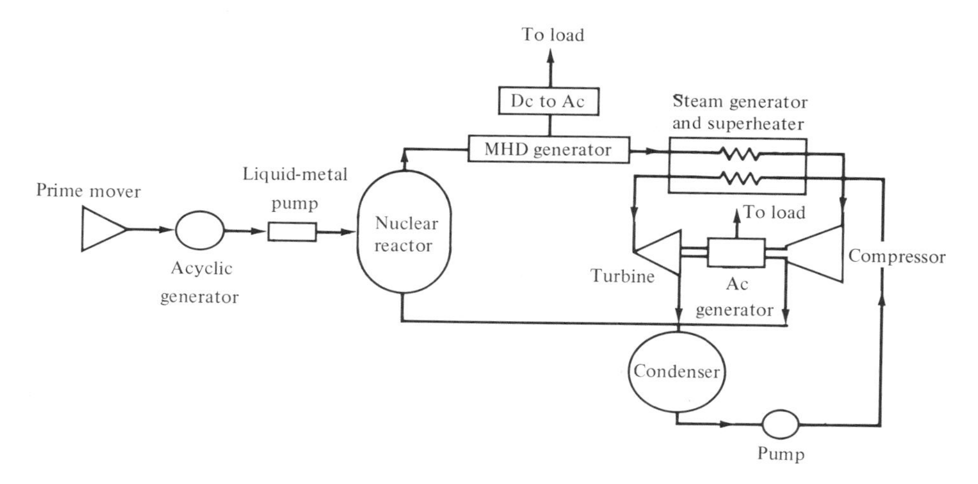

FIGURE 2-13. Schematic of a nuclear-steam-MHD power plant.

temperatures (of several thousand degrees kelvin); plasma wall losses; and economic viability. In short, MHD generators are feasible in principle, but technically realizable only in the distant future.

2-8

ELECTROMECHANICAL ENERGY CONVERSION

In the preceding we have considered various direct energy conversion devices. We pointed out that most of them at present are not developed to a stage where they could be used for bulk power generation. In fact, as most of us know, power is generated in electric power stations by rotary generators. These generators belong to a large class of devices known as *electromechanical energy converters*. From an application's viewpoint, electromechanical energy converters, commonly known as *electric machines*, are much more important than the direct energy converters presented earlier. We discuss the basic three types of electric machines in detail in later chapters. For the present, we include an overview of different types of electric machines.

As the name implies, an electromechanical energy converter converts mechanical energy into electrical energy, and vice versa. A device that converts energy from mechanical to electrical form, and modulates in response to an electrical signal, is a *generator*. When the conversion involved is from electrical to mechanical energy, and the modulating signal is electrical in nature, the device accomplishing such conversion is a *motor*. Incremental-motion electromechanical energy converters, whose main function is to process energy, are called *transducers*. For example, a microphone or an electric strain gauge are considered as transducers. Rotating electric machines operate on the principle of electromechanical power equivalence expressed as

$$T\omega_m = vi \tag{2-37}$$

where T is the mechanical torque, N·m; ω_m is the mechanical angular velocity, rad/s; v is the instantaneous electrical voltage, volts; and i is the instantaneous electrical current, amperes. We will reconsider (2-37) in Chapter 5, where we discuss its significance in detail. For the present it will suffice to say that electric generators are governed by Faraday's law of electromagnetic induction, stated earlier in connection with MHD generators. In practice, a mechanical prime mover is coupled to the generator (as mentioned in Chapter 1) and rotates the electrical conductors, constituting the generator windings, in a magnetic field, thereby inducing a voltage in the generator windings. These windings supply the electrical load on the generator. On the other hand, if a current-carrying conductor is placed in a magnetic field, the conductor experiences a mechanical force according to Ampère's law (or the Lorentz force equation). A large class of electric motors operate on this basis. From (2-37) and the preceding discussion we see that electric machines, in general are reversible in that they are capable of operating as generators as well as motors.

There are three major classes of rotating electric machines: dc commutator, induction, and synchronous machines. Several other types of machines exist, but the do not fit conveniently into any of these classifications. Some of the latter include stepper motors, which are synchronous machines operated in a digital manner; torque motors, which are either dc commutator or brushless synchronous machines operated in the torque (zero- or low-speed) mode; homopolar machines, which are a variation of the Faraday disk generator principle and which are used to supply low voltage and high current for plating loads; and electrostatic machines, which fall into a different category of theory and practice from the electromagnetic machines to be discussed in this book.

As mentioned earlier, we will discuss the major types of electric machines in detail in later chapters; in the following we present a brief qualitative description of these machines. This brief description of the various machines will aid in appreciating some of the qualitative features of these machines.

1. *Dc commutator machines.* These are commonly referred to just as "dc machines" and are distinguished by the mechanical switching device known as the commutator. They are widely used in traction and industrial applications and are discussed in Chapter 5.

2. *Induction machines.* The induction motor is the "workhorse" of industry, but it is also the principal appliance motor in homes and offices. It is simple, rugged, durable, and long-lived, which accounts for its widespread acceptability in almost all aspects of technology. It can be operated as a generator and is so

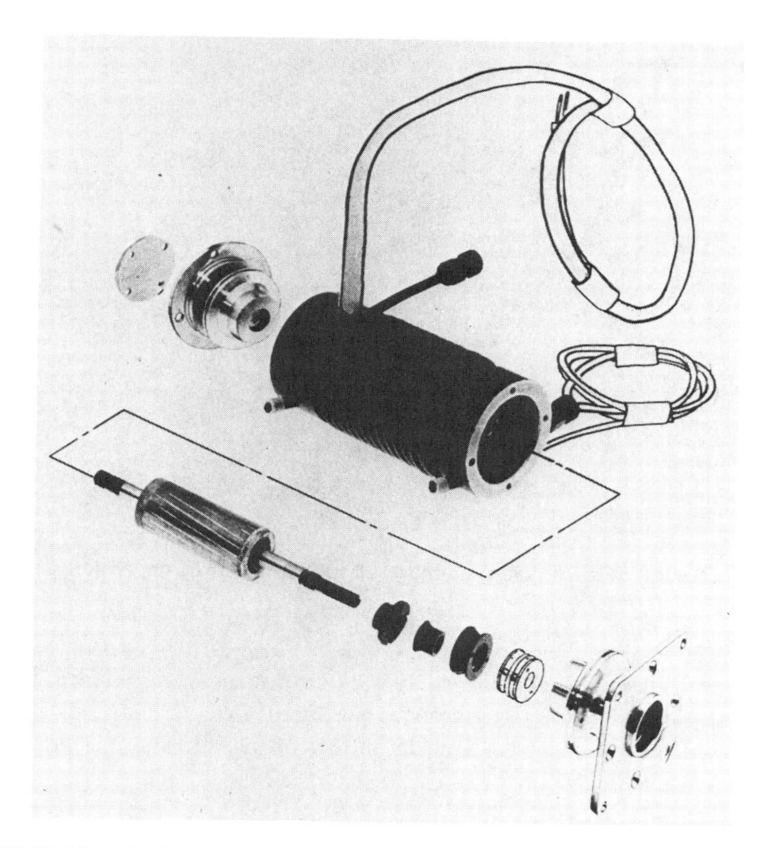

FIGURE 2-14. An induction motor for use in the aerospace industry, which operates at speeds of 64,000 r/min.

used in various aerospace and hydroelectric applications. Induction motors, because of their simple rotor structure, can operate at very high speeds. Figure 2-14 pictures an aerospace induction motor which operates at speeds near 64,000 r/min when driven from a source of 3200 Hz.

3. *Synchronous machines*. The synchronous machine is probably the most diversified machine configuration, and it is often difficult to recognize the many variations that this class of machines can have. The term *synchronous* refers to the relationship between the speed and frequency in this class of machine, which is given as

$$\text{r/min} = 120 \frac{f}{p} \tag{2-38}$$

where r/min = machine speed
f = frequency of applied source, Hz
p = number of poles on the machine

FIGURE 2-15. Rotor of a permanent-magnet motor used in an aerospace application.

A synchronous machine operates only at a synchronous speed, whereas induction machines, often termed asynchronous machines, operate at speeds somewhat below synchronous speeds. A wide variety of synchronous machines are in common use today:

(a) *Conventional*. This is the standard synchronous machine (discussed in Chapter 6). It is the machine used in most central-station electrical generating plants (as a generator), and in many motor applications for pumps, compressors, and so forth. A cutaway of a central-station generator is shown in Chapter 3.

(b) *Reluctance*. It is one of the simplest machine configurations and has recently been used in applications conventionally supplied by induction motors. In very small power ratings, it is used for electric clocks, timers, and recording applications.

(c) *Hysteresis*. This configuration, like the reluctance configuration, requires only one electrical input (singly excited). The rotor of a hysteresis motor is a solid cylinder constructed of permanent-magnet materials. Hysteresis motors are used in electric clocks, phonograph turntables, and other constant-speed applications. Recently, hysteresis motors have been used in applications requiring larger power output, such as centrifuge drives.

(d) *Permanent magnet*. This is a conventional synchronous machine in which the field excitation is supplied by a permanent magnet (PM) instead of by a source of electrical energy. It has the potential for very high energy efficiency since there are no field losses and can, in general, be constructed at a low cost.

An example of a rotor or a permanent-magnet machine used in an aerospace application is shown in Fig. 2-15. High efficiency is a characteristic of PM machines. However, to achieve relatively large power levels, PM machines require the use of permanent magnets of a type that are, at present, relatively costly—such as cobalt–platinum, and the cobalt–rare earth alloys. Also, the fixed field excitation of a PM machine eliminates one element of control that is a principal advantage of synchronous machines over induction machines—the field control.

2-9 EFFICIENCY, LOSSES, AND SOME PREDICTIONS

An important factor in the applications of energy conversion devices of all types is the efficiency of the device. Efficiency can have different meanings in different types of physical systems. In fact, it can have a fairly general meaning that is used in everyday conversation, which is "how well a specific job is done." In mechanical systems, use is made of thermal efficiency and mechanical efficiency, which describe the efficiency of two phases of a given process and also "ideal" efficiencies. In the electrical systems that will be discussed, efficiency is defined as

$$\eta = \frac{\text{output power or energy}}{\text{input power or energy}} \tag{2-39}$$

This can also be expressed in terms of mechanical and electrical losses in either energy or power terms as

$$\eta = \frac{\text{output}}{\text{output} + \text{losses}} = \frac{\text{input} - \text{losses}}{\text{input}} \tag{2-40}$$

As mentioned in Chapter 1, the SI units of power are watts, abbreviated W; SI units of energy are joules, J, or watt-seconds, W-s, or watthours, Wh.

The energy use or efficiency of an electric machine is becoming increasingly significant and is one of the more important design criteria today. Approximate maximum efficiencies of various types of energy converters, discussed previously, in relation to come common energy converter such as the automobile engine, are shown in Fig. 2-16.

It is projected that by the turn of the century, of all the direct energy converters, only solar power and fuel cells will contribute to the production of any

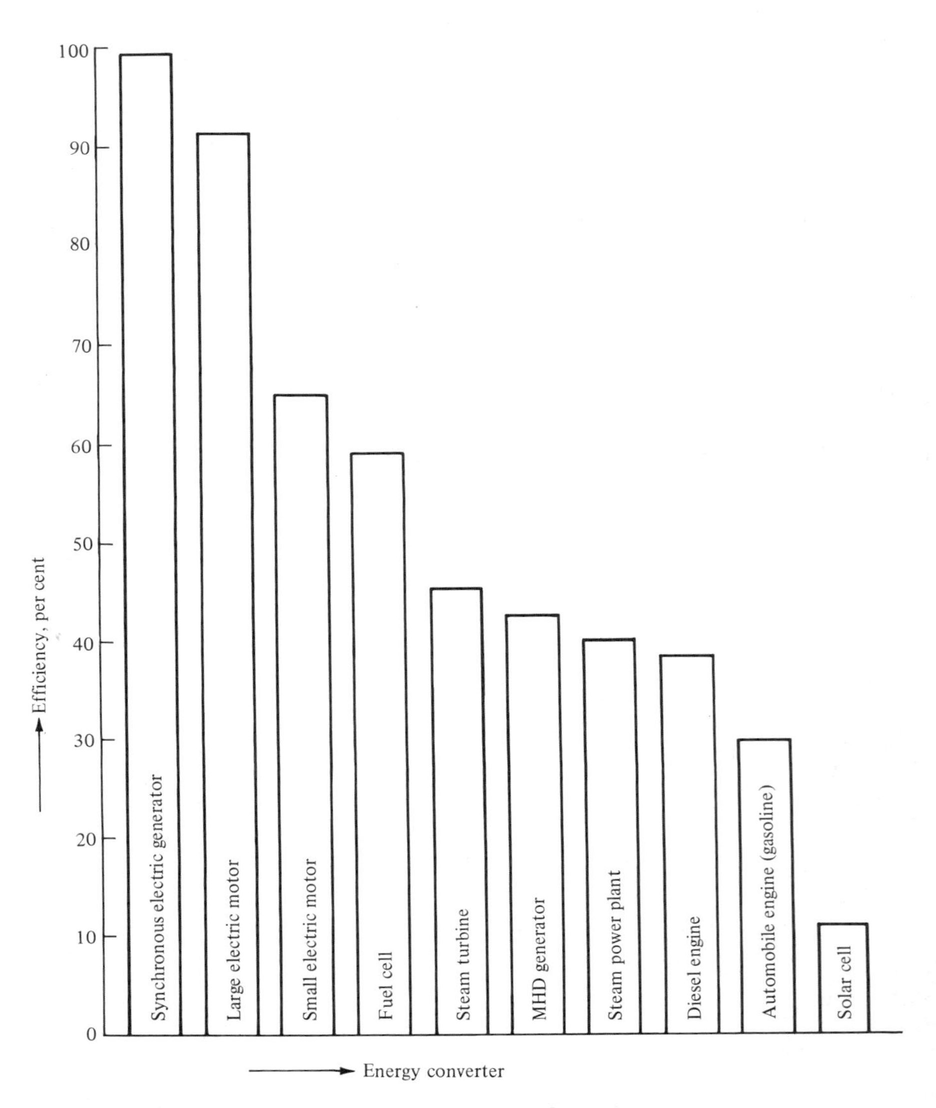

FIGURE 2-16. Approximate maximum efficiencies of energy converters.

significant amount of electrical power. For the decade 1981–1990, in the United States, considering wind, solar, geothermal, cogeneration*, solid waste, wood waste, and fuel cell as alternatives to conventional methods of generating electrical power, no more than 12 percent is expected from solar and fuel cells in terms of installed capacity.

*A cogeneration facility is a facility that produces (1) electric energy and (2) steam or other forms of useful energy (such as heat) which are used for industrial, commercial, heating, or cooling purposes.

PRIME MOVERS

From the discussion of the preceding sections it is clear that, for major applications, electromechanical energy converters will dominate for decades to come. Simply stated, most of the electrical power will be generated by synchronous electric generators. Mechanical power is provided to these generators, for converting to electrical power, by means of prime movers. Energy sources for the prime movers are thermal, hydro, and wind. A detailed discussion of the prime movers is beyond the scope of this book. We will, however, review very briefly some of the more common prime movers.

Thermal Prime Movers

Thermal prime movers include steam turbines, gas turbines, gasoline engines, and diesel engines. To produce the steam required to drive steam turbines, fuel such as coal, gas, or oil is burned in boilers. For a maximum efficiency of operation of the turbine, steam at maximum possible pressure and temperature is used. From cost considerations, the larger the size of the turbine, the less the capital cost. Consequently, units of rating of 500 MW and above are common. Steam at a pressure of over 30 MPa and temperature exceeding 650°C are used in such turbines. Respective efficiencies of steam turbines and coal-fired generating station are given in Fig. 2-16.

A gas turbine is driven by hot gases. These gases are produced in a combustion chamber, in which a continuous combustion of injected fuel oil occurs in the presence of compressed air simultaneously delivered to the chamber. In this regard, a gas turbine is an internal combustion engine. The main advantage of a gas turbine is its capability of starting quickly and taking up the load. Such a feature is required to meet sudden peak loads. Under continuous normal operating conditions, a gas turbine is less economical than a steam turbine (for the same rating).

Of the other internal-combustion engines, gasoline engines are seldom used to drive large (say 50 kW or more) electric generators, and diesel engines are used for isolated units rather than in central stations. The diesel engine differs from the gasoline engine mainly in the mechanism of combustion. In a diesel engine heat is generated in the cylinder to ignite the fuel. On the other hand, electric spark ignites the air–fuel mixture in the cylinder of a gasoline engine. Consequently, a diesel engine has a higher compression and is bulkier, heavier, and more expensive than a gasoline engine. But a diesel engine is relatively less expensive to operate.

Hydraulic Prime Movers

Hydraulic prime movers, commonly known as hydraulic turbines, are used in hydroelectric and tidal power stations (see Chapter 1). Whereas tidal energy conversion is technically and economically not feasible, numerous hydroelectric power stations have been in existence for a long time. Water is collected and stored at a high elevation and led through penstocks (or pipelines) to the power station, where the hydraulic turbine is installed, at a lower elevation. The power available is given by

$$P = 9.81HW \quad \text{kW} \tag{2-41}$$

where H is the water head in meters, and W is the flow rate, in cubic meters per second, of water through the turbine. Thus, for a given flow rate, the power developed is proportional to the working head. Roughly speaking, heads are classified into low, medium, and high heads, respectively, having values of 6 to 30 m, 30 to 200 m, and above 220 m. Low-head plants utilize dams, and high-head plants use penstocks for transmitting water downstream.

There are three types of hydraulic turbines suitable for three different heads: Kaplan turbines are used for heads up to 60 m; Francis turbines, for heads from 30 to 300 m; and Pelton wheels, for heads between 90 and 900 m. Kaplan and Francis turbines belong to the class of *reaction turbines,* in which the blades travel at about the same velocity as the water flow. Pelton wheels are *impulse turbines,* in which the turbine blades reverse the direction of the water flow.

Maximum efficiencies of hydraulic turbines are between 85 and 95 percent. Francis turbines of ratings exceeding 500 MW have been built and Pelton wheels of ratings of 40 MW are in use. Hydraulic turbines can be started almost instantaneously from rest, and have the obvious advantage that no losses are incurred when at a standstill. Thus, working in parallel with thermal power stations, hydroelectric plants can meet peak loads at minimum operating cost.

Wind Turbines

The two major class of wind turbines are the horizontal-axis and the vertical-axis turbine. In a horizontal-axis turbine, each blade section experiences a constant angle of attack during one revolution, under steady-state conditions. On the other hand, in a vertical-axis turbine, the angle of attack experienced by each blade section varies continuously through one revolution. Also, the vertical-axis turbine is a high-speed machine compared to a horizontal-axis turbine. Horizontal-axis rotor may be either *upwind* or *downwind* type, shown in Fig. 2-17(a) and (b), respectively. Notice the location of blades with respect to the supporting tower. Small generators use upwind horizontal-axis turbines because the tail vane keeps the blades pointed into the wind, and the protection mechanism in case of high winds can easily be designed. Downwind turbines are suitable for large generators. Small downwind turbines have a natural tendency to turn and align with the flow of wind. Large downwind machines are steered by pilot wind vanes. For protection against high winds, blades of downwind machines are designed to cone, as illustrated in Fig. 2-17(b). In a downwind

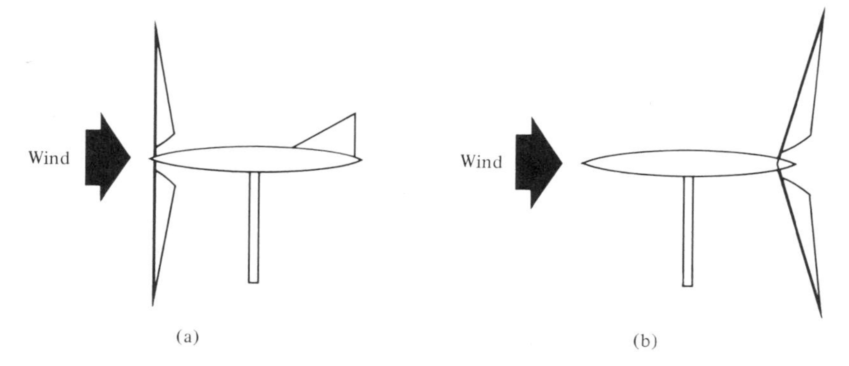

FIGURE 2-17. (a) Upwind and (b) downwind horizontal-axis turbines.

machine, the tower acts as a barrier to the windstream. Each time a blade passes the tower, the blade undergoes stress transients. This is a distinct disadvantage of the downwind turbine.

The main advantage of the vertical-axis machine is that it does not have to yaw (or turn) with changing wind directions. Two of the well-known vertical-axis turbines are the Savonius and the Darrieus turbines, shown in Fig. 2-18(a) and (b). The latter are not self-starting and must be started by an auxiliary means such as an electric motor.

The effectiveness of a wind turbine is measured by its power coefficient, C_p, defined as

$$C_p = \frac{\text{power delivered by the rotor}}{\text{power in the wind striking the area swept by the rotor}} \tag{2-42}$$

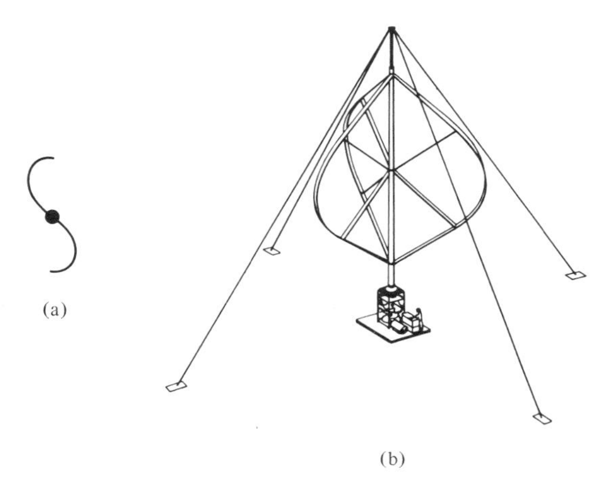

FIGURE 2-18. Vertical-axis turbines: (a) Savonius; (b) Darrieus.

The value of C_p depends on the ratio of the blade tip speed to the wind speed. The value of C_p for an ideal wind rotor is 0.59, and for actual rotors, C_p ranges from 0.15 to 0.45.

Power is transmitted from the wind turbines to the generators by gearboxes, belts and pulleys, roller chains, or by hydraulic transmissions. Direct drives are used on very small machines.

2-11 ENERGY STORAGE

In the preceding sections we have briefly reviewed various methods of electric energy conversion. In some cases, such as in solar and wind (and even hydro) energy conversion, the available power from the ''fuel'' will obviously fluctuate during a 24-h period. In such situations, the power generating system must have a provision for energy storage. Some of the options available for large-scale energy storage are storage batteries, inertial storage, pumped storage, hydrogen gas, compressed air, and superconducting coils.

The first and foremost of the characteristics of a *storage* (or *secondary*) *battery* with regard to energy storage is the energy density. Energy density refers to the volume or mass density of an energy source. The latter is the more common usage, and we will use the term *energy density* to refer to source energy divided by source weight (including ancillary subsystems weights), in kWh/kg, the energy or capacity of batteries may vary considerably with the power level at which the energy is used. Therefore, it is usually necessary to specify energy density at a specific power level. Table 2-1 summarizes the energy densities of a number of fuels and energy sources. These values are typical values; certain types of fuels or components in each category may vary appreciably.

TABLE 2-1 Nominal Energy Densities of Certain Energy Sources

Energy Source	Nominal Energy Density (Wh/kg)
Gasoline	12,300
Natural gas	9,350
Methanol	6,200
Hydrogen	28,000
Coal (bituminous)	8,200
Lead–acid battery	35
Sodium–sulfur battery	150–300
Flywheel (steel)	12–30

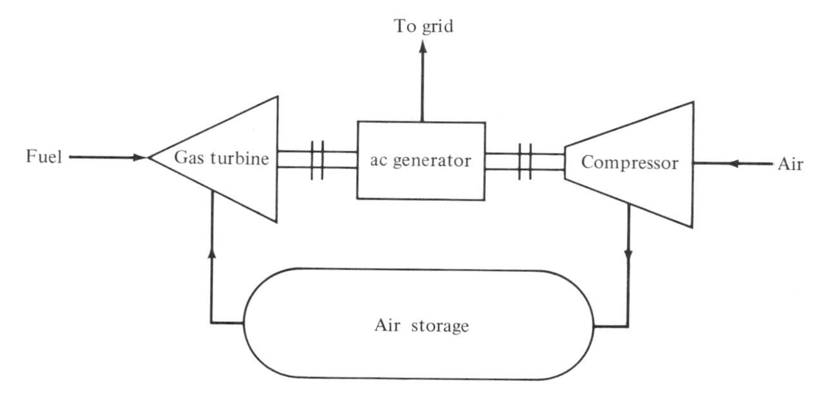

FIGURE 2-19. Compressed air storage system.

From Table 2-1 it is clear that secondary lead–acid batteries are not suitable for large-scale energy storage applications. Their main potential applications are in fluctuating energy sources such as solar and wind.

The pumped-storage method is applicable to hydroelectric plants. A pumped-storage system has an upper and a lower reservoir, a hydraulic turbine coupled to an electric generator. The upper reservoir has a capacity for about 8-h operation, and supplies the turbine during peak-load hours. During off-hours, the generator operates as a motor (from the interconnected power lines) and drives the turbine, which acts as a pump. Thus water from the lower reservoir is pumped back into the upper reservoir. Such a system has an approximate overall efficiency of 65 percent.

In the compressed-air storage system, air is pumped into a storage during the periods when the load on the system is light. During peak-load hours, the compressed air is used to drive a gas turbine coupled to an electric generator. The system is shown in its schematic form in Fig. 2-19. In a German system [6], the compressed-air storage system generates 580 MWh of electricity and takes 930 MWh of fuel and 480 MWh of off-peak electrical energy.

REFERENCES

1. **S. W. Angrist,** *Direct Energy Conversion,* Allyn and Bacon, Boston, 1965.
2. **K. H. Spring,** ed., *Direct Generation of Electricity,* Academic Press, New York, 1965.
3. **G. W. Sutton,** ed., *Direct Energy Conversion,* McGraw-Hill, New York, 1966.
4. **R. A. Bell** and **R. B. Hayman,** "The Electric Utility 4.5 MW Fuel Cell Power Plant—An Urban Demonstration," *IEEE Transactions on Power Apparatus and Systems,* vol. PAS-100, December 1981, pp. 4760–4764.
5. **R. J. Rosa,** *Magnetohydrodynamic Energy Conversion,* McGraw-Hill, New York, 1968.
6. **B. M. Weedy,** *Electric Power Systems,* 3rd ed., Wiley, New York, 1979.

PROBLEMS

2-1 In the thermodynamic process of an energy converter a portion of 350×10^6 cal of heat from a fuel is converted into electrical energy. The conversion process takes place at a constant volume of 2 m^3, but the pressure of the working fluid is reduced from 2 atm to 1 atm during the process. If the stored energy of the system is increased by 2×10^6 cal, calculate the energy converted to electrical form.

2-2 If initial stored energy of the system of Problem 2-1 is 0.5×10^6 cal, calculate the initial and final enthalpies of the system.

2-3 An energy conversion process takes place between 100 and 42°C. What is the maximum possible efficiency of the process?

2-4 In a fuel cell the chemical reaction takes place at 30°C. If the change in enthalpy is -70 kcal and the cell has an ideal efficiency of 84 percent, determine (a) the change in the Gibbs free energy and (b) the ideal emf of the cell.

2-5 In the fuel cell of Problem 2-4, calculate the change in enthalpy.

2-6 The following data pertain to a thermoelectric generator: hot junction temperature = 300°C; cold junction temperature = 25°C; open-circuit voltage = 0.13 V. What is the Seebeck coefficient?

2-7 The separation between the electrodes of a 50-cm-long rectangular-channel MHD generator is 20 cm. The channel is 20 cm wide. An ionized gas of conductivity 10 S/m traverses the channel at a velocity of 1000 m/s at 2500 K. The working flux density is 2.0 T. For a loading factor of 0.4, determine (a) the output voltage, (b) the output power, and (c) the ohmic loss of the generator.

2-8 For the generator of Problem 2-7, what is the loading factor for maximum power transfer to the load? For this condition calculate the output voltage, the current, and the ohmic loss of the generator.

2-9 For the generators of Problems 2-7 and 2-8, determine the force developed by the ionized gas.

2-10 Electromagnetic flowmeters are designed on the basis of MHD principles. Obtain an expression relating to the potential difference between two selected points in the fluid and the mean velocity of the fluid. Assume a configuration of a rectangular channel with a dc magnetic field at right angles to the direction of flow of the fluid.

2-11 An ideal rotary energy converter develops 65.5 kW of power at 1000 r/min. Calculate the electromagnetic torque developed.

2-12 An ideal energy converter develops 347.6 N·m of torque while running at 3600 r/min. If it takes 100 A of current while developing this torque, determine the input voltage.

2-13 A 60-Hz synchronous motor runs at 1200 r/min. How many poles does it have?

2-14 A hydroelectric power station is fed from a water reservoir of 60 million cubic meters, at a head of 160 m. Calculate the available electrical energy if the hydraulic turbine has an efficiency of 82 percent and the generator driven by the turbine has an efficiency of 92 percent.

2-15 Calculate the electrical energy spent in raising the temperature of 6 m^3 of water by 80°C if the efficiency of the water heater is 95 percent.

2-16 Determine the power available to a hydraulic turbine from a 100-m head of water flowing at the rate of 4 m^3/s.

CHAPTER 3

Components of Electric Energy Systems

In the preceding two chapters we have discussed energy resources, fundamentals of electric energy systems, and various methods of electric energy conversion with an emphasis on electromechanical energy conversion. We have considered the various energy converters as isolated entities. In practice, however, power generating stations all over the country are interconnected. In later chapters we shall study certain aspects of electric power systems and its major components in some detail. Before we consider the power system as a whole, it is worthwhile to familiarize ourselves with a brief historical development of electric central stations and review qualitatively some of the pertinent components that constitute an electric power system.

3-1 HISTORY OF CENTRAL-STATION ELECTRIC SERVICE

The concept of *central-station* electric power service was first applied at Thomas Edison's Pearl Street Station in New York City in 1882, when the first distribution line was strung along a few city blocks to provide lights in a few homes. The electric power industry in the United States has grown phenomenally from the novelty of one circuit with a few light bulbs to providing the main driving force of the greatest industrial nation on earth. All of this has taken place in the equivalent of one human life span. The high standard of living enjoyed in this country is closely tied to electric laborsaving appliances and tools, and to the high productivity made possible by the electric machinery of industry.

Central-station service, as opposed to individual generators in each home, possessed all the inherent advantages to make it a huge success. Economies of scale, convenience, and relative continuity combined to promote the growth of Edison's idea at a rapid rate. The first limiting factors encountered were voltage drop and resistance losses on the low-voltage *direct-current* (dc) distribution circuits. Distance of the customer from the generating station was severely limited as the problem of voltage regulation became more pronounced out toward the end of the line. The solution to these problems came with the introduction to this country of the *alternating-current* (ac) *transformer* by George Westinghouse in 1885. An ac distribution circuit was placed in service at Great Barrington, Massachusetts, by William Stanley in 1886, proving the feasibility of the technique. Power could be generated at low voltage levels by the simpler ac generator and stepped up to higher voltage for sending over long distances. Because the current required to deliver power at a given voltage is inversely proportional to the voltage, the current requirements and consequently the conductor size could be kept within practical limits and still deliver large amounts of power to distant areas. The first ac transmission line was put in service at Portland, Oregon, in 1890, carrying power 13 miles from a *hydro* generating station on the Willamette River.

During the next decade, *two-phase* and *three-phase* motors and generators were developed and were demonstrated to be superior to *single-phase* machines from the standpoint of size, weight, and efficiency. Three-phase transmission was shown to possess inherent advantages of conductor requirements and losses for a given power need. Consequently, by the turn of the century it was apparent that three-phase ac transmission systems would become standard. Three frequencies, 25, 50, and 60 Hz, battled for dominance, and it was several decades before the conversion to 60 Hz was complete in the United States.

3-2 ELECTRIC POWER GENERATORS

In order to visualize and discuss an entire functioning power system, it is necessary that we establish the necessary basic components of such a system. The first and most obvious is the three-phase ac generator or *alternator*. It must be driven mechanically by some sort of *prime mover* (outlined in Chapter 2). The early prime movers were primarily reciprocating engines and waterwheels. The simplest form of prime mover was the hydro station with a simple waterwheel.

(a)

(b)

FIGURE 3-1. (a) Stator of a steam-driven turbine generator; (b) turbine rotor with direct water cooling during the mounting of damper hollow bars. (Courtesy of Brown Boveri Company)

FIGURE 3-2. A hydroelectric generator. (Courtesy of Brown Boveri Company)

Once the original installation was made at a waterfall or dam, the fuel was free forever. For this reason, hydro stations are seldom retired from service. Thousands of tiny hydro stations are still in use today, often unattended and operated by remote control. As the more readily available sources of water power were developed, the emphasis shifted to the *steam turbine*. Fired by such fossil fuels such as coal, gas, and oil, steam turbines grew in numbers and unit size until they dominated the power generation industry. A stator of a typical steam-driven turbine-generator is shown in Fig. 3-1(a). The rotor of a turbo alternator is round or *cylindrical*, as shown in Fig. 3-1(b). The rotor of a hydroelectric generator is illustrated in Fig. 3-2. Such a rotor is known as a *salient-pole rotor*. Even when nuclear fuels began to command a sizable portion of the market, the nuclear reactor and its complex heat exchanger ultimately served only to make steam for driving a conventional steam turbine. The turbines driven by steam from nuclear reactors must be larger than fossil-fired units if they are to be economically feasible because of the huge fixed costs, which are independent of capability. Direct conversion of energy into its electrical form without the rotating prime mover and alternator offer great promise for the future, but is not yet economically competitive in commercial quantities, as discussed in Chapter 2. Conventional electric power generators are discussed in a subsequent chapter.

3-3 TRANSFORMERS

Practical design problems limit the voltage level at the terminals of the alternator to a relatively low value. The transformer is used to *step* the voltage up (as the current is proportionately reduced) to a much higher level so that power can be transmitted up to hundreds of miles while conductor size and losses are kept down within practical limits. Physically, a *power transformer* bank may consist of three single-phase transformers with appropriate electrical connections external to the cases, or a three-phase unit contained in a single tank (Fig. 3-3). The latter is predominant in the larger power ratings, for economic reasons. The windings usually are immersed in a special-purpose oil for insulation and cooling. Power transformers are discussed in some detail in Chapter 4.

FIGURE 3-3. A 3-phase 1300 MVA 345/24 kV power transformer. (Courtesy of Westinghouse Electric Corporation)

TRANSMISSION LINES

The *transmission line* usually consists of three *conductors* (either as three single wires or as bundles of wires) and one or more neutral conductors, although it is possible sometimes to omit the neutral conductor since it carries only the unbalanced return portion of the line current. A three-phase circuit with perfectly balanced phase currents has no neutral return current. In most instances the current is accurately balanced among the phases at transmission voltages, so that the neutral conductor may be much smaller. In some locations the soil conditions permit an effective neutral return current through the earth. The neutral conductors have another equally important function; they are installed above the phase conductors and provide an effective electrostatic shield against lightning (Fig. 3-4). Manufacturers of high-voltage equipment tend to standardize as much as possible on a few *nominal voltage classes*. The most common transmission line-to-line voltage classes in use within the United States are 115, 138, 230, 345, 500, and 765 kV. Developmental work is being done for utilizing voltages up to 2000 kV. Costs of line construction, switchgear, and transformers

FIGURE 3-4. A 345-kV transmission line, insulators, and supporting structure.

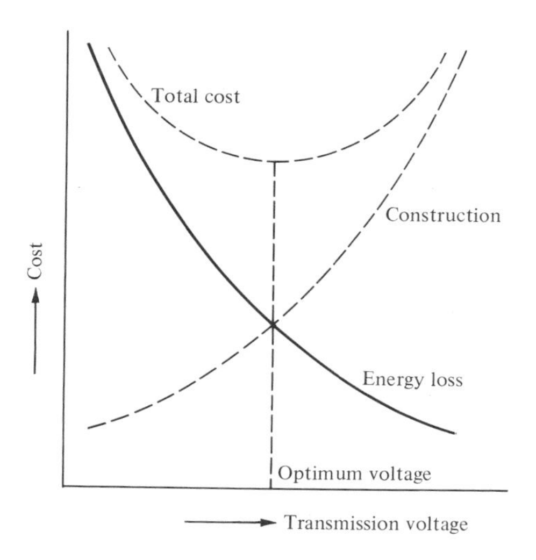

FIGURE 3-5. Determination of approximate optimum transmission voltage.

rise exponentially with voltage, leading to the use of the lowest voltage class capable of carrying the anticipated load over the required interval of time. However, the cost of energy loss is inversely proportional to the square of transmission voltage. The construction cost and the cost of energy loss are depicted in Fig. 3-5. The point of intersection of these two curves gives the theoretical optimum transmission voltage for a total minimum cost.

3-5 CIRCUIT BREAKERS AND DISCONNECT SWITCHES

Circuit breakers are large three-pole switches located at each end of every transmission-line section, and on either side of large transformers. The primary function of a circuit breaker is to open under the control of automatic *protective relays* in the event of a fault or short circuit in the protected equipment. The relays indicate the severity and probable location of the fault, and may contain sufficient electromechanical or solid-state logic circuitry to decide whether the line or transformer could be reenergized safely, and initiate reclosure of its circuit breakers. In very critical, extra-high-voltage (EHV)* switchyards or substations, a small digital computer may be used to analyze fault conditions and perform logical control functions. If the fault is a transient one from which the system may recover, such as a lightning stroke, the integrity of the network is

*Commonly accepted high voltage levels are:
high voltage (HV), 115 to 230 kV;
extra high voltage (EHV), 345 to 765 kV (up to 1000 kV);
ultra high voltage (UHV), 1000 kV and up.

FIGURE 3-6. Power circuit breakers.

FIGURE 3-7. A disconnect switch.

best served by restoring equipment to service automatically, preferably within a few cycles. If the fault is persistent, such as a conductor on the ground, the relays and circuit breakers will isolate the faulted section and allow the remainder of the system to continue in normal operation. The secondary function of a circuit breaker is that of a switch to be operated manually by a local or remote operator to deenergize an element of the network for maintenance. When a circuit breaker's contacts open under load, there is a strong tendency to arc across the contact gap as it separates. Various methods are used to suppress the arc, including submersion of the contact mechanism in oil or gas such as sulfur hexafluoride (SF_6). The highest voltage classes use a powerful air blast to quench the arc, using several interrupters or contact sets in series for each phase (Fig. 3-6).

One side of an open-circuit breaker generally remains energized. To completely isolate (or deenergize) a circuit breaker, a disconnect switch is placed in a series with the current breaker, as shown in Fig. 3-7.

3-6 VOLTAGE REGULATORS

When electric power has been transmitted into the area where it is to be used, it is necessary to transform it back down to a distribution level voltage which can be utilized locally. The step-down transformer bank may be very similar to the step-up bank at the generating station, but of a size to fill the needs only of the immediate area. To provide a constant voltage to the customer, a *voltage regulator* is usually connected to the output side of the step-down transformer. It is a special type of 1:1 transformer with several discrete taps of a fractional percent each over a voltage range of $\pm 10\%$. A voltage-sensing device and automatic control circuit will position the tap contacts automatically to compensate the low-side voltage for variations in transmission voltage. In many cases the same effect is accomplished by incorporating the regulator and its control circuitry into the step-down transformer, resulting in a combination device called a *load tap changer* (LTC), and the process is known as tap changing under load (TCUL).

3-7 SUBTRANSMISSION

Some systems have certain intermediate voltage classes which they consider *subtransmission*. Probably at the time it was installed it was considered transmission, but with rapid system growth and a subsequent overlay of higher-voltage transmission circuits, the earlier lines were tapped at intervals to serve more load centers and become local feeders. In most systems 23, 34.5, and 69 kV are considered of subtransmission type, and on some larger systems 138 kV may also be included in that category, depending on the application. As the frontiers of higher voltages are pushed back inexorably, succeeding higher voltage classes may be relegated to subtransmission service.

3-8 DISTRIBUTION SYSTEMS

A low-voltage *distribution system* is necessary for the practical distribution of power to numerous customers in a local area. A distribution system resembles a transmission system in miniature, having lines, circuit breakers, and transformers, but at lower voltage and power levels. Electrical theory and analytical methods are identical for both, since the distinction is purely arbitrary. Distribution voltages range from 2.3 to 35 kV, with 12.5 and 14.14 kV predominant. Such voltage levels are sometimes referred to as *primary* voltages of 240/120 V at which most customers are served. Single-phase distribution circuits are supplied from three-phase transformer banks, balancing the total load on each phase as nearly as possible. Three-phase distribution circuits are erected only to serve large industrial or motor loads. The ultimate transformer which steps voltage down from distribution to customer service level may be mounted on a pole for overhead distribution systems or on a pad or in a vault for underground distribution. Such transformers usually are protected by *fuses* or *fused cutouts*.

3-9 LOADS

Countless volumes have been written about the systems and techniques necessary for the production and delivery of electric power, but very little has been recorded about *loads,* for which all the other components exist. Perhaps the main reason is that loads are so varied in nature as to defy comprehensive classification. In the simplest concept, any device that utilizes electric power can be said to impose a load on the system. Viewed from the source, all loads can be classed as resistive, inductive, capacitive, or some combination of them. Loads may also be time variant, from a slow random swing to rapid cyclic pulses which cause distracting flicker in the lights of customers nearby. The composite load on a system has a predominant resistive component and a small net inductive component. Inductive loads such as induction motors are far more prevalent than capacitive loads. Consequently, to keep the resultant current as small as possible, *capacitors* are usually installed in quantities adequate to balance most of the

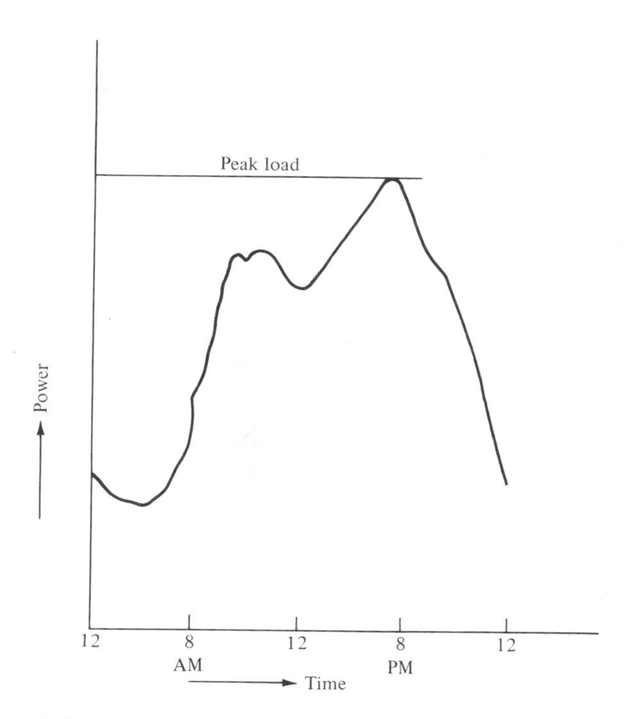

FIGURE 3-8. A typical load curve on a power system.

inductive current. It has been shown that the power consumed by the composite load on a power system varies with system frequency. This effect is imperceptible to the customer in the range of normal operating frequencies (± 0.02 Hz), but can make an important contribution to the control of systems operating in synchronism. System load also varies through daily and annual cycles, creating difficult operating problems. A typical load curve is shown in Fig. 3-8.

3-10 CAPACITORS

When applied on a power system for the reduction if inductive current (*power factor correction*), capacitors can be grouped into either transmission or distribution classes. In either case, they should be installed electrically as near to the load as possible for maximum effectiveness. When applied properly, capacitors balance out most of the inductive component of current to the load, leaving essentially a unity power factor load. The result is a reduction in size of the conductor required to serve a given load and a reduction in I^2R losses. *Static capacitors* may be used at any voltage, but practical considerations impose an upper limit of a few kilovolts per unit, therefore, high-voltage banks must be composed of many units connected in series and parallel. High-capacity transmission capacitor banks should be protected by a high-side circuit breaker and its associated protective relays. Small distribution capacitors may be vault- or pole-top-mounted and protected by fuses.

Industrial loads occasionally require very large amounts of power factor correction, varying with time and the industrial process cycle. The *synchronous condenser* is ideally suited to such an application. Its contribution of either capacitive or inductive current can be controlled very rapidly over a wide range, using automatic controls to vary the excitation current. Physically, it is very similar to a synchronous generator operating at a leading power factor, except that it has no prime mover. The synchronous condenser is started as a motor and has its losses supplied by the system to which it supplies capacitive current. We will discuss the operation or synchronous condensers in a subsequent chapter.

3-11 REPRESENTATION OF AN ELECTRIC POWER SYSTEM

In the preceding ssections, we have mentioned the various basic components of an electric power system. These components include generators, tranformers, transmission lines, and loads. In this section we consider the representation of these components interconnected to constitute a power system. First, we review the graphical representation and one-line diagram of a power system. This is followed by the impedance diagrams obtained from the most commonly used equivalent circuits of the components. Finally, because the components have different voltage and kilovoltampere (kVA) ratings, we introduce the per unit quantities as a common basis for analyzing the interconnected components and systems.

Graphical Representation of Components. Figure 3-9 shows the symbols used to represent the typical components of a power system.

One-Line Diagrams. Using the symbols of Fig. 3-9, a system consisting of two generating stations interconnected by a transmission line is shown in Fig. 3-10, a one-line diagram. From even such an elementary network as this, it is easy to imagine the confusion that would result in making diagrams showing all three phases. The ratings of all the generators, transformers, and loads are specified and the voltage levels at the buses are assumed to be known. The advantage of such a one-line representation is rather obvious in that a complicated system can be represented simply. A concerted effort is made to keep the currents equal in each phase. Consequently, on a balanced system, one phase can represent all three by proper mathematical treatment. From the one-line

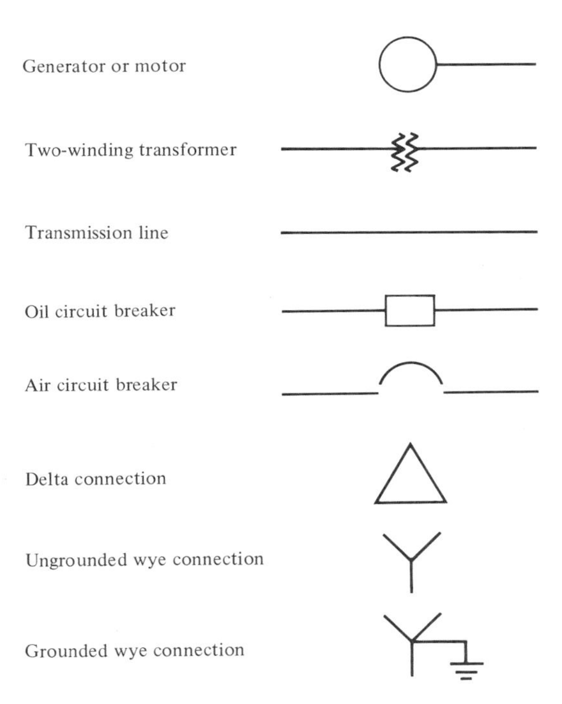

FIGURE 3-9. Symbolic representation of elements of a power system. Other legends are also used—see for instance Fig. 3-11.

diagram the impedance, or reactance, diagrams can be conveniently developed, as shown in the following section. A further advantage of the one-line diagram is in the power flow studies. The one-line diagram rather becomes second nature to power system engineers as they attempt to visualize a widespread complex network. A section of a one-line diagram of a typical power system is shown in Fig. 3-11.

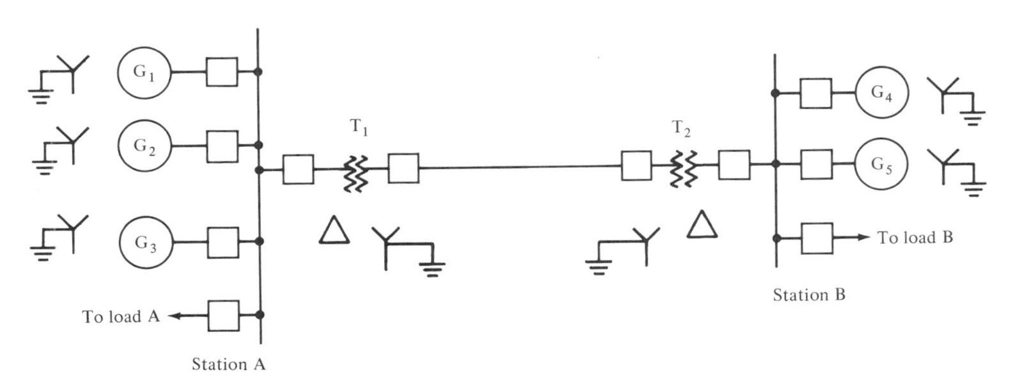

FIGURE 3-10. One-line diagram of a portion of a power system formed by interconnecting two stations.

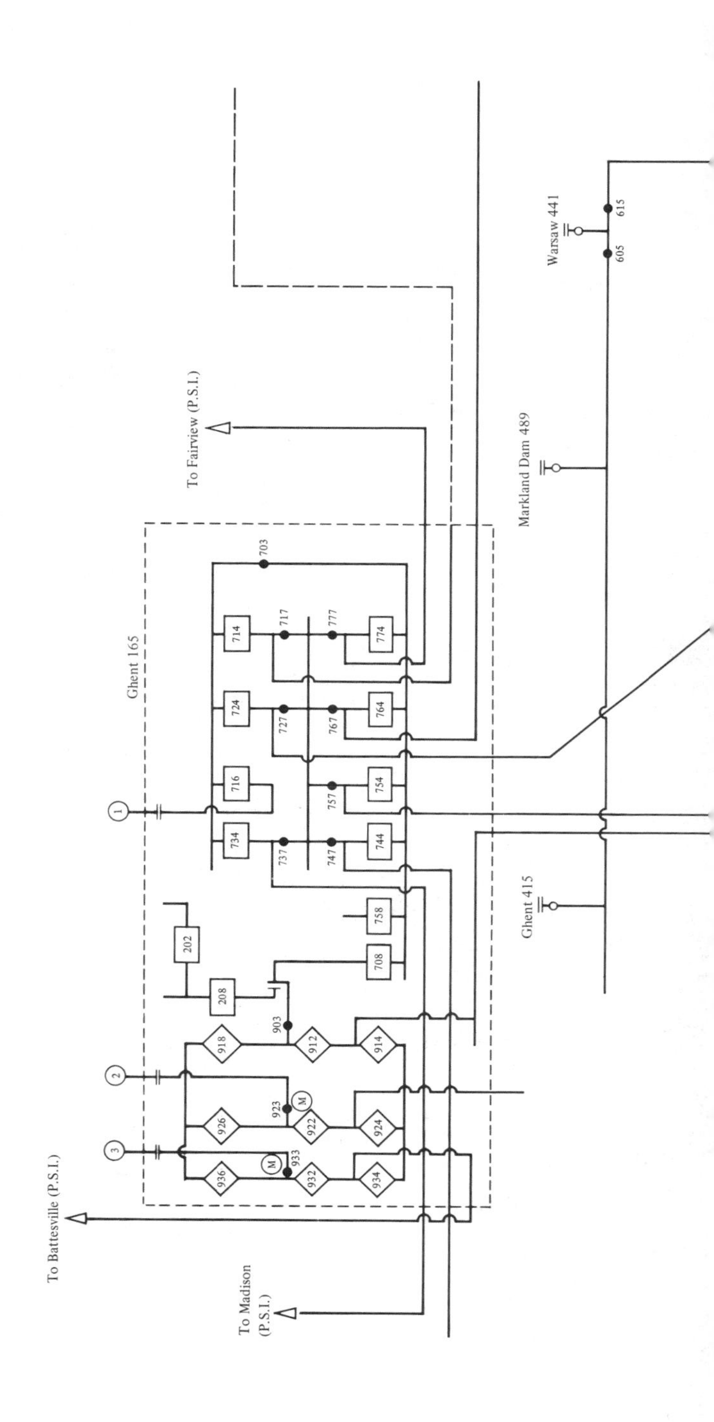

To Fairview (P.S.I.)
Warsaw 441
605
615
Markland Dam 489
Ghent 165
703
714
717
777
774
724
727
767
764
716
757
754
734
737
747
744
758
202
208
708
903
918
912
914
923
926
922
924
933
936
932
934
Ghent 415
To Battesville (P.S.I.)
To Madison (P.S.I.)
1
2
3
M

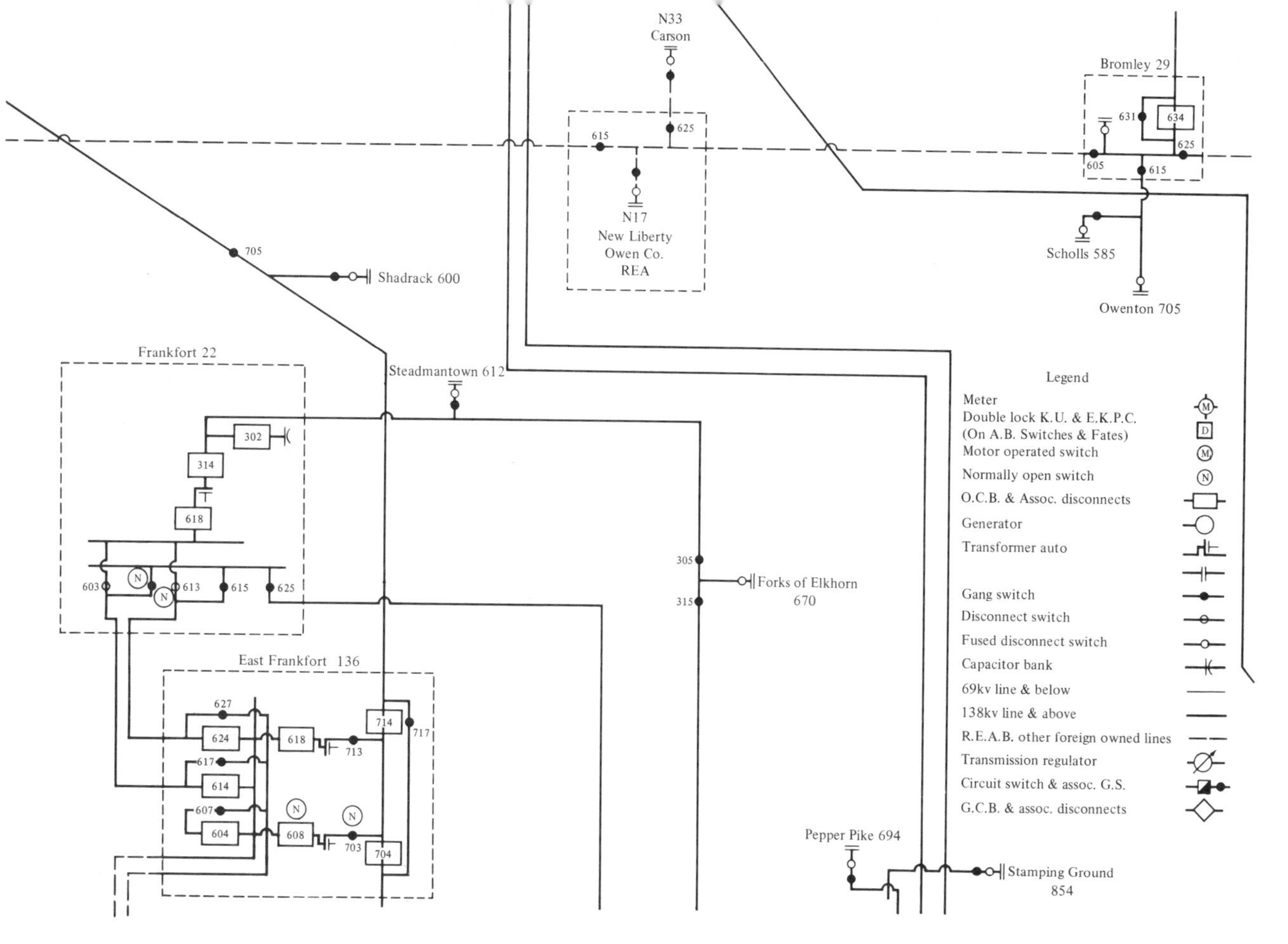

FIGURE 3-11. A one-line diagram of a power system.

3-12

EQUIVALENT CIRCUITS AND REACTANCE DIAGRAMS

We note from Fig. 3-11 that the power system consists of the components: generators, transformers, transmission lines, and loads. Equivalent circuits of these components may then be interconnected to obtain a circuit representation for the entire system. In other words, the one-line diagram may be replaced by an *impedance diagram* or a *reactance diagram* (if resistances are neglected). Thus, corresponding to Fig. 3-10, the impedance and reactance diagrams are shown in Fig. 3-12(a) and (b), respectively, on a per phase basis. In the equivalent circuits of the components in Fig. 3-12(a) we have made the following assumptions:

1. A generator can be represented by a voltage source in series with an inductive reactance. The internal resistance of the generator is negligible compared to the reactance.
2. The loads are inductive.
3. The transformer core is assumed ideal, and can be represented by a reactance.
4. The transmission line is of medium length and can be denoted by a T circuit. An alternate representation, such as by a Π circuit is equally applicable.
5. The delta/wye-connected transformer, T_1, is replaced by an equivalent wye/wye-connected transformer (by a delta-to-wye transformation) so that the impedance diagram may be drawn on a per phase basis. The exact nature and values of the impedances (or reactances) are determined in later chapters.

The reactance diagram, Fig. 3-12(b), is drawn by neglecting all the resistances, loads, and capacitances of the transmission line. For short-circuit calculations reactance diagrams are generally used, whereas the impedance diagram is used for power-flow studies.

3-13

PER UNIT REPRESENTATION

When making computations on a power system network having two or more voltage classes, it is very cumbersome to convert currents to a different voltage level at each point where they flow through a transformer, the change in current being inversely proportional to the transformer turns ratio. A much simplified system has been devised whereby a set of *base quantities* are assumed for each voltage class, and each parameter is expressed as a decimal fraction of its respective base. For convenience, base quantities are chosen such that they correspond rather closely to the range of values normally expected in the pa-

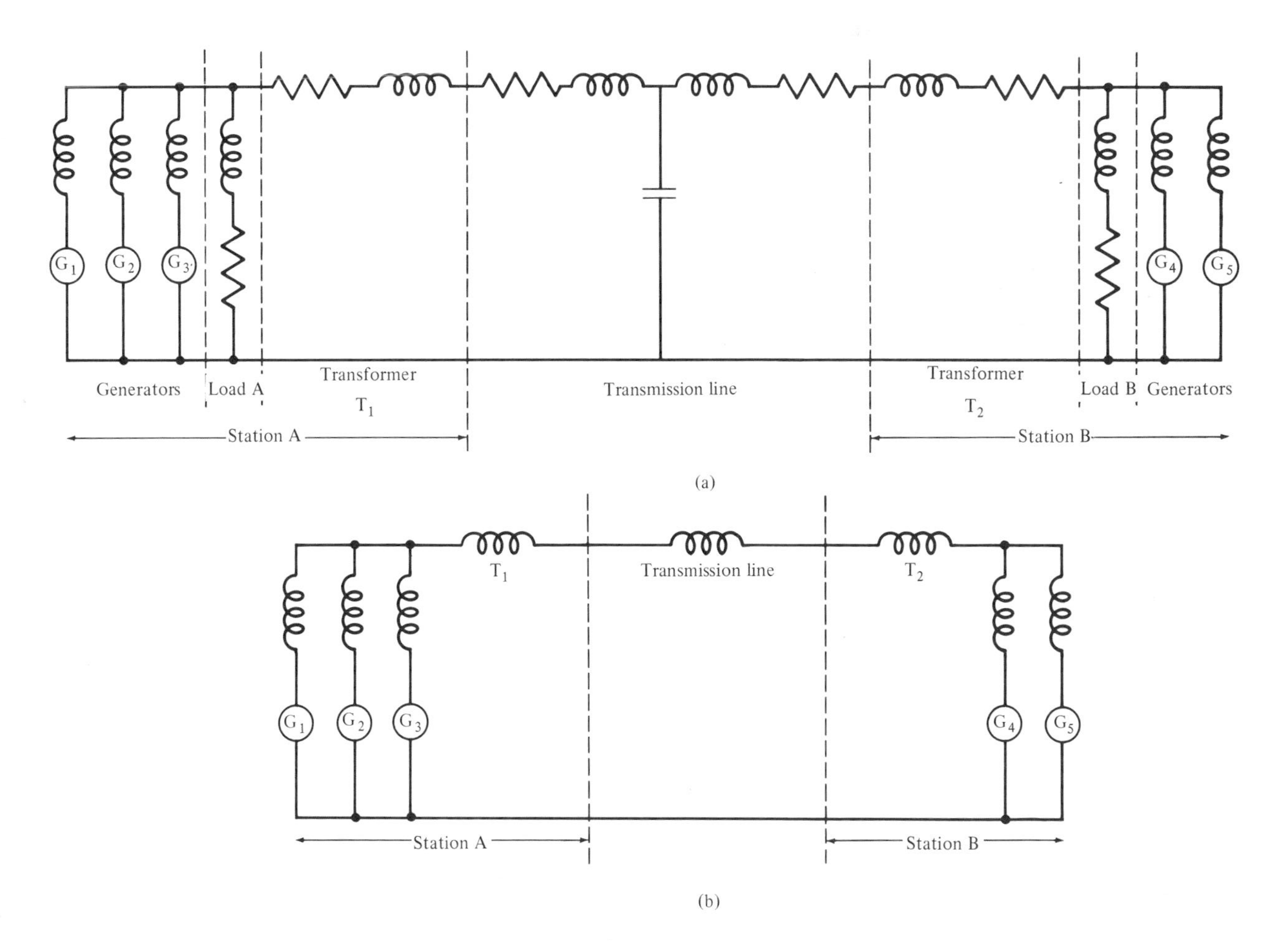

FIGURE 3-12. (a) Impedance diagram for Fig. 3-8; (b) corresponding reactance diagram, neglecting loads and resistances.

rameter. For instance, if a voltage base of 345 kV is chosen and under certain operating conditions the actual system voltage is 334 kV, the ratio of actual to base voltage is 0.97. This is expressed as 0.97 *per unit* volts. An equally common practice is to multiply each result by 100, under which the preceding would be expressed as 97 *percent* volts. With experience, this technique can give an excellent grasp of or feeling for the system conditions, whether normal or abnormal. It works equally well for single-phase or polyphase circuits. Per unit or percent quantities and their bases must obey the same relationships (such as Ohm's law and Kirchhoff's laws, etc.) to each other as with other systems of units. A minimum of four base quantities are required to define completely a per unit system: voltage, current, power, and impedance (or admittance). If two of these are chosen arbitrarily, for example, voltage and power, the others are fixed. Therefore, on a per phase basis the following relationships hold:

$$\text{base current} = \frac{\text{base voltamperes}}{\text{base voltage}} \quad \text{amperes} \tag{3-1}$$

$$\text{base impedance} = \frac{\text{base voltage}}{\text{base current}} \quad \text{ohms} \tag{3-2}$$

$$\text{per unit voltage} = \frac{\text{actual voltage}}{\text{base voltage}} \quad \text{pu} \tag{3-3}$$

$$\text{per unit amperage} = \frac{\text{actual current}}{\text{base current}} \quad \text{pu} \tag{3-4}$$

$$\text{per unit impedance} = \frac{\text{actual impedance}}{\text{base impedance}} \quad \text{pu} \tag{3-5}$$

The preceding equations are now applied to the following simple example.

EXAMPLE 3-1

A 10-kVA 1800-V 0.8 lagging power factor load is connected to a generator by a transmission line having an impedance of (12 + *j*20) ohms. Assume 10 kVA and 2000 V as base quantities, and calculate the per unit voltage at the load, per unit current, and per unit impedance of the line.

The following table identifies the per unit (pu) values and the base quantities:

Quantity	Per Unit Value	Unit Value or Base Quantity
Voltamperes	1.0	10 kVA (assumed)
Voltage	1.0	2000 V (assumed)
Current	1.0	10,000/2000 = 5 A
Impedance	1.0	2000/5 = 400 Ω
Admittance	1.0	5/2000 = 0.0025 Ω

■

In a three-phase system, the base kVA may either be chosen as the three-phase kVA and the base voltage as line-to-line voltage, or the base values may be taken as the phase quantities. In either case, the per unit three-phase kVA and voltage on the three-phase kVA base and the per unit per phase kVA and voltage on the kVA per phase base remain the same. This point is illustrated by the next example.

EXAMPLE 3-2

Consider a three-phase wye-connected 50,000-kVA 120-kV system. Express 40,000-kVA three-phase apparent power in per unit values on (a) the three-phase kVA base and (b) the per phase kVA base.

(a) Three-phase basis:

$$\text{base kVA} = 50{,}000 \text{ kVA} = 1 \text{ pu}$$

$$\text{base kV} = 120 \text{ kV (line to line)} = 1 \text{ pu}$$

$$\text{pu kVA} = \frac{40{,}000}{50{,}000} = 0.8 \text{ pu}$$

(b) Per phase basis:

$$\text{base kVA} = \frac{1}{3} \times 50{,}000 = 16{,}667 = 1 \text{ pu}$$

$$\text{base kV} = \frac{120}{\sqrt{3}} = 69.28 \text{ kV} = 1 \text{ pu}$$

$$\text{pu kVA} = \frac{1}{3} \times \frac{40{,}000}{16{,}667} = 0.8 \text{ pu}$$

■

We stated earlier that a power system consists of generators, transformers, transmission lines, and loads. The per unit impedances of generators and transformers, as supplied from tests by manufacturers, are generally based on their own ratings. However, these per unit values could be referred to a new voltampere base according to the following equation:

$$(\text{pu impedance})_{\text{new base}} = \frac{[(\text{VA})_{\text{new base}}][(\text{kV})^2_{\text{old base}}]}{[(\text{VA})_{\text{old base}}][(\text{kV})^2_{\text{new base}}]} (\text{pu impedance})_{\text{old base}} \quad (3\text{-}6)$$

If the old base and the new base voltages are the same, (3-6) simplifies to

$$(\text{pu impedance})_{\text{new base}} = \frac{\text{new base voltamperes}}{\text{old base voltamperes}} (\text{pu impedance})_{\text{old base}} \quad (3\text{-}7)$$

The impedances of transmission lines are expressed in ohms, which can be easily converted to the pu value on a given voltampere base.

EXAMPLE 3-3

A three-phase wye-connected 6.25-kVA 220-V synchronous generator has a reactance of 8.4 Ω per phase. Choose the rated kVA and voltage as base values, and determine the per unit reactance. Convert this per unit value to a 230-V 7.5-kVA base.

$$\text{base VA} = 6250 = 1 \text{ pu}$$

$$\text{base V} = \frac{220}{\sqrt{3}} = 127 \text{ pu}$$

$$\text{base A} = \frac{6250}{\sqrt{3} \times 220} = 16.4 = 1 \text{ pu}$$

$$\text{base } X = \frac{220\sqrt{3}}{16.4} = 7.75 = 1 \text{ pu}$$

$$\text{pu } X = \frac{8.4}{7.75} = 1.08 \text{ pu}$$

On a 230-V 7.5-kVA base, from (3-6) we obtain

$$\text{pu } X = 1.08 \left(\frac{220}{230}\right)^2 \left(\frac{7500}{6250}\right) = 1.18 \text{ pu}$$

■

EXAMPLE 3-4

A three-phase 13.2-kV transmission line delivers 8 MVA of load. The per phase impedance of the line is $(0.01 + j0.05)$ pu on a 13-kV 8-MVA base. How much voltage will be dropped across the line?

$$\text{base kVA} = 8000 = 1 \text{ pu}$$

$$\text{base kV} = 13 = 1 \text{ pu}$$

$$\text{base A} = \frac{8000}{13\sqrt{3}} = 355.3 = 1 \text{ pu}$$

$$\text{base } Z = \frac{13{,}000}{355.3} = 36.6 = 1 \text{ pu}$$

$$\text{impedance} = 36.69(0.01 + j0.05) = 0.366 + j1.83\ \Omega$$

$$\text{voltage drop} = 355.3(0.366 + j1.83) = 130 + j650 = 663 \text{ V}$$

■

The per unit system gives us a better feeling of relative magnitudes of various quantities, such as voltage, current, power, and impedance. The remarks in this section are meant to be introductory. We will have a better appreciation of the usefulness of the per unit concept in later chapters, particularly in fault calculations. Having briefly reviewed the major components of an electric power system, we now proceed to study some of these components in detail.

PROBLEMS

3-1 A system is rated at 300 kVA and 11 kV. Using these as base values, find the base current and base impedance.

3-2 Using 10 MVA and 345 kV as base values, express 138 kV, 6 MVA, 250 A, and 50 Ω in (a) per unit and (b) percent values.

3-3 For a system, use 50 Ω and 250 A as base impedance and base current, respectively. What is the base kVA and base voltage?

3-4 The per unit values of impedance, current, voltage, and VA of a system are 0.9, 0.3, 0.8, and 12, respectively. The base impedance is 35 Ω and the base current is 80 A. Determine the actual values of impedance, current, voltage, and voltamperes.

3-5 The per unit impedance of a system is 0.7. The base kVA is 300 kVA and the base voltage is 11 kV. What is the ohmic value of the impedance? Will this ohmic value change if 400 kVA and 38 kV are chosen as base values? What is the per unit impedance for the 400 kVA and 38 kV base values?

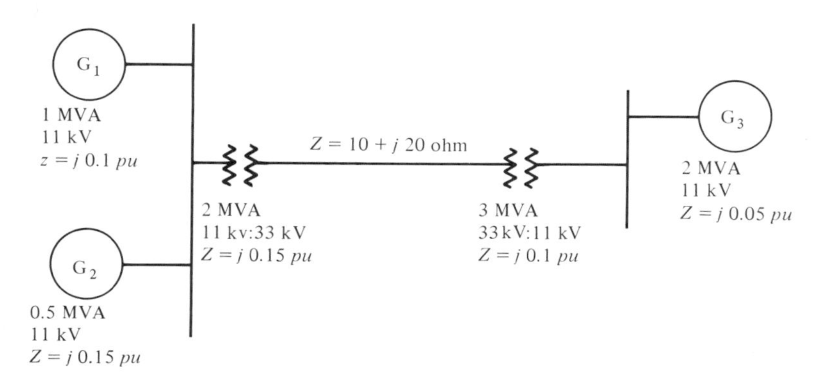

FIGURE 3P-7. PROBLEM 3-7.

3-6 A single-phase transmission line supplies a reactive load at a lagging power factor. The load draws 1.2 pu current at 0.6 pu voltage while taking 0.5 pu (true) power. If the base voltage is 20 kV and the base current is 160 A, calculate the power factor and the ohmic values of the resistance of the load.

3-7 A one-line diagram of a two-generator system is shown in Fig. 3P-7. Redraw the diagram to show all values in per unit on a 7000-kVA base.

3-8 Redraw the diagram of Fig. 3P-7 showing all values in ohms.

3-9 A 100-kVA 20/5-kV transformer has an equivalent impedance of 10 percent. Calculate the impedance of the transformer referred to (a) the 20-kV side and (b) the 5-kV side.

CHAPTER 4

Power Transformers

Ferromagnetic materials are used in the construction of transformers and electric machines so that the magnetic flux may be appropriately guided. Thus it is important that we acquire the basic knowledge pertinent to magnetic circuits to appreciate fully the working of transformers and electric machines.

4-1 MAGNETIC CIRCUITS

By a *magnetic circuit* we mean a path for magnetic flux, just as an electric circuit provides a path for the flow of electric current. Magnetic circuits are an integral part of transformers and electric machines. Here we present certain basic concepts relating to the analysis of magnetic circuits.

We define the magnetic flux density B by the force equation

$$F = BlI \tag{4-1}$$

where F is the force (newtons) experienced by an l (meters) long straight conductor carrying a current I (amperes) and oriented at right angles to a magnetic field of flux density B (tesla). In other words, if a conductor is 1 m long, carries 1 A of current and experiences 1 N of force when located at right angles to certain magnetic flux lines, the flux density is 1 T.

The magnetic flux, φ, through a given surface is defined by

$$\varphi = \int_s \mathbf{B} \cdot d\mathbf{s} \tag{4-2}$$

If B is uniform over an area A and is perpendicular to A, then

$$\varphi = BA \tag{4-3}$$

from which

$$B = \frac{\varphi}{A} \tag{4-4}$$

The unit of magnetic flux is the weber (Wb) and B may be expressed in $\mathrm{Wb/m^2} = 1\ \mathrm{T}$.

The source of magnetic flux is either a permanent magnet or an electric current. To measure the effectiveness of electric current in producing a magnetic field (or flux), we introduce the concept of magnetomotive force (or mmf), $\mathcal{F}$, defined as

$$\mathcal{F} \equiv NI \tag{4-5}$$

where I is the current (A) flowing in an N-turn coil. The unit of $\mathcal{F}$ is the ampere-turn (At). Schematically, a magnetic circuit with an mmf and magnetic flux are shown in Fig. 4-1.

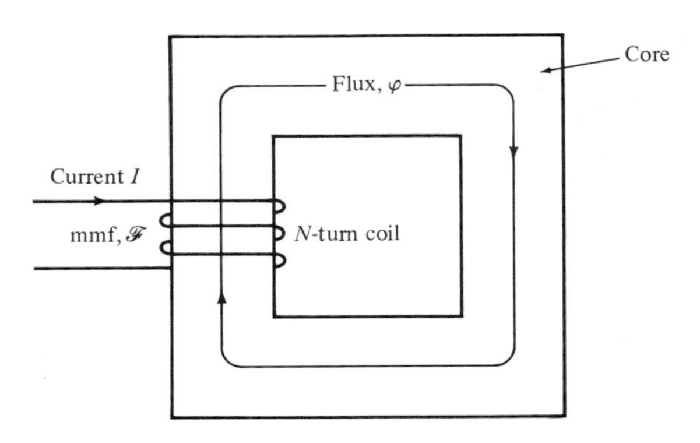

FIGURE 4-1. A magnetic circuit, showing mmf and flux.

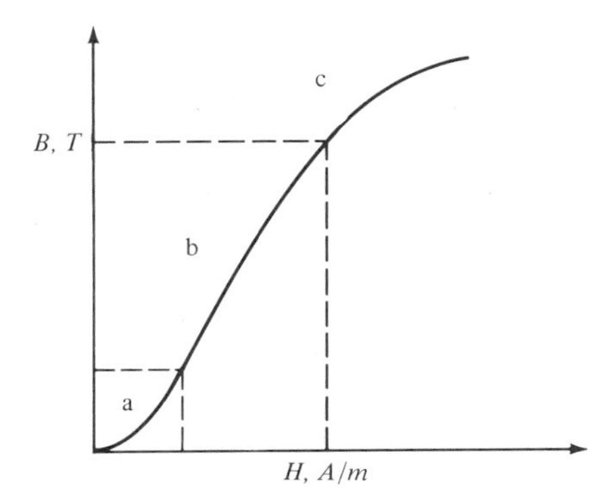

FIGURE 4-2. A *BH* curve.

Although we have defined mmf by (4-5), the mutual relationship between an electric current I and the corresponding *magnetic field intensity, H,* is given by *Ampère's circuital law,* expressed as

$$\oint \mathbf{H} \cdot d\mathbf{l} = I \tag{4-6}$$

When the closed path is threaded by the current N times, as in Fig. 4-1, (4-6) becomes

$$\oint \mathbf{H} \cdot d\mathbf{l} = NI \equiv \mathcal{F} \tag{4-7}$$

The core material of a magnetic circuit (constituting a transformer or an electric machine) is generally ferromagnetic, and the variation of B with H is depicted by the saturation curve of Fig. 4-2. The slope of the curve depends on the operating flux density, as classified in regions a, b, and c. For region b, which is of a constant slope, we may write

$$B = \mu H \tag{4-8}$$

where μ is defined as the *permeability* of the material and is measured in henries per meter (H/m). For free space (or air) we have $\mu = \mu_0 = 4\pi \times 10^{-7}$ H/m. In terms of μ_0, (4-8) is sometimes written as

$$B = \mu_r \mu_0 H \tag{4-9}$$

where $\mu_r = \mu/\mu_0$ and is called *relative permeability*.

Based on an analogy between a magnetic circuit and a dc resistive circuit, Table 4-1 summarizes the corresponding quantities. In the table, l is the length

TABLE 4-1 Analogy Between Magnetic and DC Resistive Circuits

DC Resistive Circuit	Magnetic Circuit
Current, I	Flux, φ
Voltage, V	mmf, $\mathcal{F}$
Conductivity, σ	Permeability, μ
Ohm's law, $I = V/R$	$\varphi = \mathcal{F}/\mathcal{R}$
Resistance, $R = l/\sigma A$	Reluctance, $\mathcal{R} = l/\mu A$
Conductance, $G = 1/R$	Permeance, $\mathcal{P} = 1/\mathcal{R}$

and A is the cross-sectional area of the path for the current in the electric circuit, or for the flux in the magnetic circuit. Based on the analogy above, the laws of resistances in series or parallel also hold for reluctances.

The differences between a dc resistive circuit and a magnetic circuit are:

1. We have an I^2R loss in a resistance but do not have a $\varphi^2\mathcal{R}$ loss in a reluctance.
2. Magnetic fluxes take *leakage* paths, as φ_l in Fig. 4-3, whereas electric currents (flowing through resistances) do not.
3. In magnetic circuits with air gaps we encounter *fringing* (Fig. 4-3) of flux lines, but we do not have fringing of currents in electric circuits. Fringing increases with the length of the air gap and increases the effective area of the air gap.

If the mmf acting in a magnetic circuit is ac, then the B–H curve takes the form shown in Fig. 4-4. The loop shown is known as hysteresis loop, and the area within the loop is proportional to the energy loss (as heat) per cycle. This energy loss is known as *hysteresis loss*. In addition, *eddy-current loss*, due to the eddy currents induced in the core material of a magnetic circuit excited by an ac mmf, is another feature of an ac-operated magnetic circuit. The losses

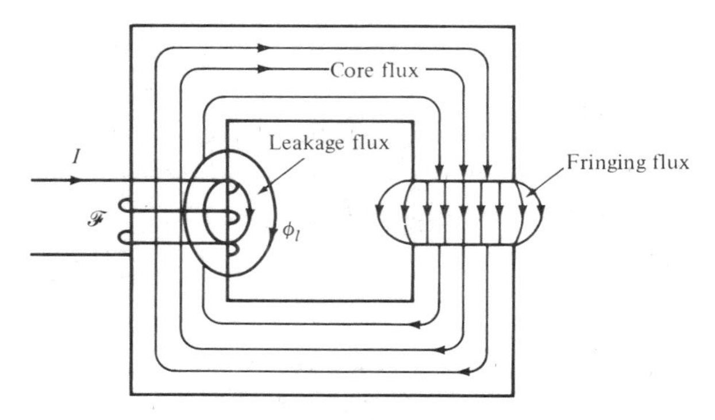

FIGURE 4-3. Leakage flux and fringing flux in a magnetic circuit.

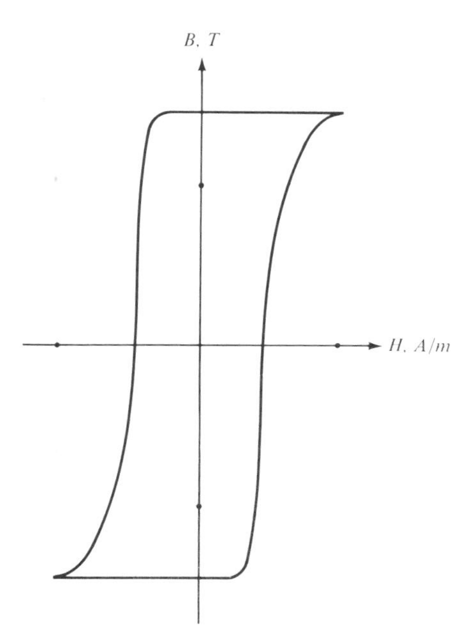

FIGURE 4-4. A hysteresis loop of a core material.

due to hysteresis and eddy currents, collectively known as *core losses* or *iron losses,* are approximately given by

$$\text{eddy-current loss:} \quad P_e = k_e f^2 B_m^2 \qquad \text{W/kg} \tag{4-10}$$

$$\text{hysteresis loss:} \quad P_h = k_h f B_m^{1.5 \text{to} 2.5} \qquad \text{W/kg} \tag{4-11}$$

where k_e is a constant depending on material conductivity and thickness; k_h is another constant, depending on the hysteresis loop of the material; B_m is the maximum core flux density; and f is the frequency of excitation.

The hysteresis-loss component of the core loss in a magnetic circuit is reduced by using "good"-quality electrical steel, having a narrow hysteresis loop, for the core material.

To reduce eddy-current loss, the core is made of laminations, or thin sheets, with very thin layers of insulation alternating with laminations. The laminations are oriented parallel to the direction of flux (Fig. 4-5). Laminating a core increases its cross-sectional area and hence the volume. The ratio of the volume actually occupied by the magnetic material to the total volume of the core is called the *stacking factor*. Table 4-2 gives some values for stacking factors. Typical magnetic characteristics and core losses of certain core materials are given in Fig. 4-6.

The magnetic circuit concepts developed so far are now illustrated by the following examples.

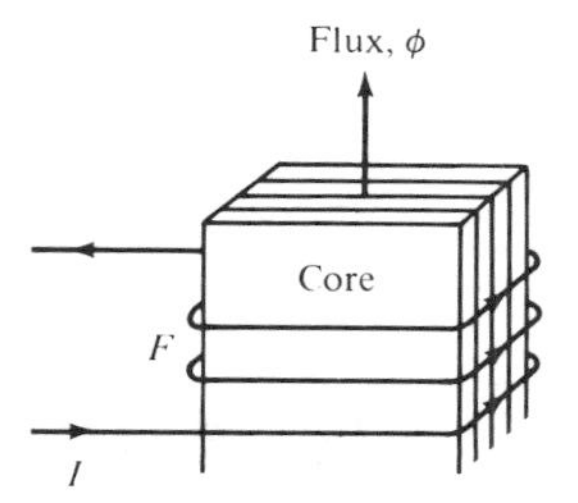

FIGURE 4-5. A laminated core.

EXAMPLE 4-1

The magnetic circuit shown in Fig. 4-6 carries a 50-turn coil on its core. The core is made of 0.15-mm-thick laminations. Total core length is 10 cm and the air-gap length is 0.1 mm. (a) Calculate the coil current to establish an air-gap flux density of 1 T. (b) What is the core flux if the core cross section is 2.5 cm by 2.5 cm? Neglect fringing and leakage. For the core, H = 130 A/m at B = 1.11 T.

(a) For the air gap:

$$H_g = \frac{B_g}{\mu_0} = \frac{1.0}{4\pi \times 10^{-7}} = 7.95 \times 10^5 \text{ A/m}$$

$$\mathcal{F}_g = H_g l_g = (7.95 \times 10^5)(10^{-4}) = 79.5 \text{ At}$$

For the core, from Table 4-2, for 0.15-mm-thick lamination, the stacking factor = 0.9 and

$$B_c = \frac{B_g}{0.9} = \frac{1}{0.9} = 1.11 \text{ T}$$

For the core at 1.11 T we have

$$H_c = 130 \text{ A/m}$$

$$\mathcal{F}_c = H_c l_c = 130 \times 0.1 = 13 \text{ At}$$

TABLE 4-2 Stacking Factor Values

Lamination Thickness (mm)	Stacking Factor
0.0127	0.50
0.0254	0.75
0.0508	0.85
0.10–0.25	0.90
0.27–0.36	0.95

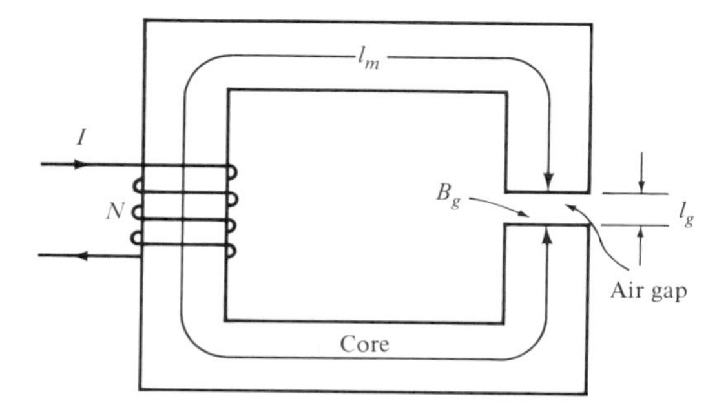

FIGURE 4.6.

For the entire magnetic circuit,

$$NI = \mathscr{F}_{\text{total}} = \mathscr{F}_g + \mathscr{F}_c = 79.5 + 13 = 92.5 \text{ At}$$

from which

$$I = \frac{92.5}{50} = 1.85 \text{ A}$$

(b) The core flux $= B_c A_c = 1.11 \times (2.5)^2 \times 10^{-4} = 0.694$ mWb. ■

EXAMPLE 4-2

Define flux linkage, λ, by $\lambda = N\varphi$, where φ is the magnetic flux threading an N-turn coil. Also, define inductance, L by $L = \lambda/i$ (or flux linkage per ampere), i being the current in the coil to produce the flux φ. Hence, (a) determine the inductance of the coil of Example 4-1. Also, (b) evaluate the energy stored in the coil. By computing the energy stored in the air gap, calculate the magnetic energy stored in the core of the magnetic circuit. Hence show that the energy density in the air gap is much greater than the energy density in the core:

(a) From Example 4.1, $\varphi = 6.94 \times 10^{-4}$ and $N = 50$. Thus

$$\lambda = N\varphi = 50 \times 6.94 \times 10^{-4} = 3.47 \times 10^{-2}$$

$$L = \frac{\lambda}{i} = \frac{3.47}{1.85} \times 10^{-2} = 18.76 \text{ mH}$$

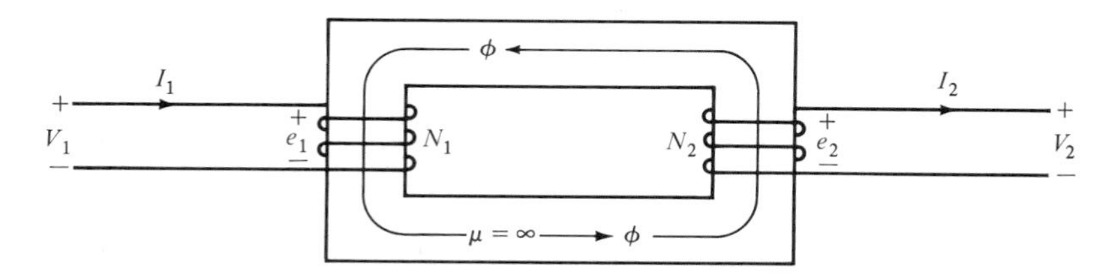

FIGURE 4-7. Schematic of a two-winding transformer.

(b) energy stored in L, $W_L = \frac{1}{2} Li^2 = \frac{1}{2} \times 18.76 \times 10^{-3} \times 1.85^2$

$$= 0.0321 \text{ J}$$

$$\text{energy stored in air gap, } W_g = \frac{B_g^2}{2\mu_0} \times \text{volume}_g$$

$$= \frac{1}{2 \times 4\pi \times 10^{-7}} \times 0.1 \times 10^{-3} \times 2.5^2 \times 10^{-4}$$

$$= 0.0249 \text{ J}$$

$$\text{energy stored in the core, } W_c = W_L - W_g = 0.0321 - 0.0249$$

$$= 0.0072 \text{ J}$$

$$\frac{\text{energy density in air}}{\text{energy density in core}} = \frac{W_g/\text{volume}_g}{W_c/\text{volume}_c}$$

$$= \frac{0.0249/(0.1 \times 10^{-3} \times 2.5^2 \times 10^{-4})}{0.0072/(10 \times 10^{-3} \times 2.5^2 \times 10^{-4})}$$

$$= 345.8$$

■

EXAMPLE 4-3

Using the definition of inductance given in Example 4-2, obtain an expression for inductance in terms of reluctance, $\mathcal{R}$, and number of turns, N. Hence determine the reluctance of the entire magnetic circuit of Example 4-1.

$$L = \frac{\lambda}{i} = \frac{N\varphi}{i} = \frac{N}{i}\frac{\mathcal{F}}{\mathcal{R}} = \frac{N}{i}\frac{Ni}{\mathcal{R}} = \frac{N^2}{\mathcal{R}}$$

Since $N^2/\mathcal{R} = 18.76 \times 10^{-3}$ (from Example 4-2), we obtain

$$\mathcal{R} = \frac{50 \times 50}{18.76} = 1.33 \times 10^5 \text{ H}^{-1}$$

■

4-2 PRINCIPLE OF OPERATION OF A TRANSFORMER

A *transformer* is an electromagnetic device having two or more mutually coupled windings. Figure 4-8 shows a two-winding ideal transformer. The transformer

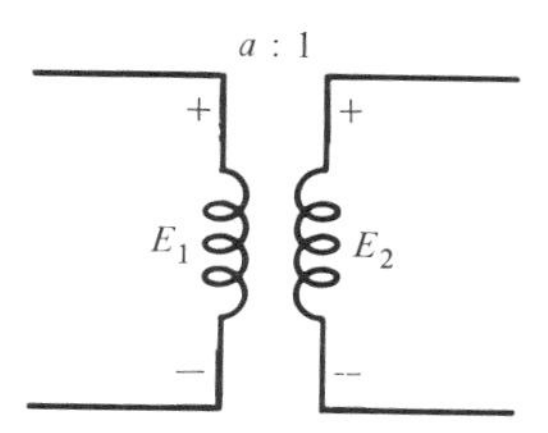

FIGURE 4-8. An ideal transformer.

is *ideal* in the sense that its core is lossless and is infinitely permeable, has no leakage fluxes, and the windings have no losses.

In Fig. 4-8, the basic components are the *core,* the *primary winding* N_1, and the *secondary winding* N_2. If φ is the mutual (or core) flux linking N_1 and N_2, then according to *Faraday's law* of electromagnetic induction, emf's e_1 and e_2 are induced in N_1 and N_2 due to a time rate of change of φ such that

$$e_1 = N_1 \frac{d\varphi}{dt} \tag{4-12}$$

and

$$e_2 = N_2 \frac{d\varphi}{dt} \tag{4-13}$$

The direction of e_1 is such as to produce a current that opposes the flux change, according to Lenz's law. The transformer being ideal, $e_1 = v_1$ (Fig. 4-8). From (4-12) and (4-13),

$$\frac{e_1}{e_2} = \frac{N_1}{N_2}$$

which may also be written in terms of root-mean-square (rms) values as

$$\frac{E_1}{E_2} = \frac{N_1}{N_2} = a \tag{4-14}$$

where a is known as the *turns ratio*.

Since $e_1 = v_1$ (and $e_2 = v_2$), the flux and voltage are related by

$$\varphi = \frac{1}{N_1}\int v_1\,dt = \frac{1}{N_2}\int v_2\,dt \tag{4-15}$$

If the flux varies sinusoidally such that

$$\varphi = \varphi_m \sin \omega t$$

then the corresponding induced voltage, e, linking an N-turn winding is given by

$$e = \omega N \varphi_m \cos \omega t \tag{4-16}$$

From (4-16), the rms value of the induced voltage is

$$E = \frac{\omega N \varphi_m}{\sqrt{2}} = 4.44 f N \varphi_m \tag{4-17}$$

which is known as the *emf equation*. In (4-17), $f = \omega/2\pi$ is the frequency in hertz.

4-3 VOLTAGE, CURRENT, AND IMPEDANCE TRANSFORMATIONS

Major applications of transformers are in voltage, current, and impedance transformations, and for providing isolation (that is, eliminating direct connections between electrical circuits). The voltage transformation property (mentioned in Section 4-2) of an ideal transformer is expressed as

$$\frac{V_1}{V_2} = \frac{E_1}{E_2} = a \tag{4-18}$$

where the subscripts 1 and 2 correspond to the primary and the secondary sides, respectively. This property of a transformer enables us to interconnect transmission and distribution systems of different voltage levels in an electric power system.

For an ideal transformer, the net mmf around its magnetic circuit must be zero, implying that

$$N_1 I_1 - N_2 I_2 = 0 \tag{4-19}$$

where I_1 and I_2 are the primary and the secondary currents, respectively. From (4-14) and (4-19) we get

$$\frac{I_2}{I_1} = \frac{N_1}{N_2} = a \tag{4-20}$$

From (4-18) and (4-20) it can be shown that if an impedance Z_2 is connected to the secondary, the impedance Z_1 seen at the primary satisfies

$$\frac{Z_1}{Z_2} = \left(\frac{N_1}{N_2}\right)^2 \equiv a^2 \tag{4-21}$$

EXAMPLE 4-4

How many turns must the primary and the secondary windings of a 220/110-V 60-Hz ideal transformer have if the core flux is not allowed to exceed 5 mWb?

From the emf equation (4-17), we have

$$N = \frac{E}{4.44 f \varphi_m}$$

Consequently,

$$N_1 = \frac{220}{4.44 \times 60 \times 5 \times 10^{-3}} \simeq 166 \text{ turns}$$

$$N_2 = \frac{1}{2} N_1 = 83 \text{ turns}$$

■

EXAMPLE 4-5

A 220/110-V 10-kVA transformer has a primary winding resistance of 0.25 Ω and a secondary winding resistance of 0.06 Ω. Determine (a) the primary and secondary currents on rated-load referred to the primary, and (b) the total winding resistance referred to the primary and referred to the secondary.

(a) transformation ratio, $a = \frac{220}{110} = 2$

primary current, $I_1 = \frac{10 \times 10^3}{220} = 45.45$ A

secondary current, $I_2 = aI_1 = 2 \times 45.45 = 90.9$ A

(b) secondary winding resistance referred to the primary

$$= a^2 R_2 = 2^2 \times 0.06 = 0.24\ \Omega$$

total resistance referred to the primary, R_e'

$$= R_1 + a^2 R_2 = 0.25 + 0.24 = 0.49\ \Omega$$

primary winding resistance referred to the secondary

$$= \frac{R_1}{a^2} = \frac{0.25}{4} = 0.0625\ \Omega$$

total resistance referred to the secondary, $R''_e = \frac{R_1}{a^2} + R_2$

$$= 0.0625 + 0.06 = 0.1225\ \Omega$$

■

EXAMPLE 4-6

Determine the I^2R loss in each winding of the transformer of Example 4-5, and thus find the total I^2R loss in the two windings. Verify that the same result can be obtained by using the equivalent resistance referred to the primary winding.

$$I_1^2R_1 \text{ loss} = (45.45)^2 \times 0.25 = 516.425 \text{ W}$$

$$I_2^2R_2 \text{ loss} = (90.9)^2 \times 0.06 = 495.768 \text{ W}$$

$$\text{total } I^2R \text{ loss} = 1012.19 \text{ W}$$

For the equivalent resistance,

$$I_1^2R'_e = (45.45)^2 \times 0.49 = 1012.19 \text{ W}$$

which is consistent with the preceding result. ■

4-4 NONIDEAL TRANSFORMER AND ITS EQUIVALENT CIRCUITS

A nonideal (or an actual) transformer differs from an ideal transformer in that the former has hysteresis and eddy-current (or core) losses, and has resistive (I^2R) losses in its primary and secondary windings. Furthermore, the core of a nonideal transformer is not perfectly permeable, and the transformer core requires a finite mmf for its magnetization. Also, not all fluxes link with the

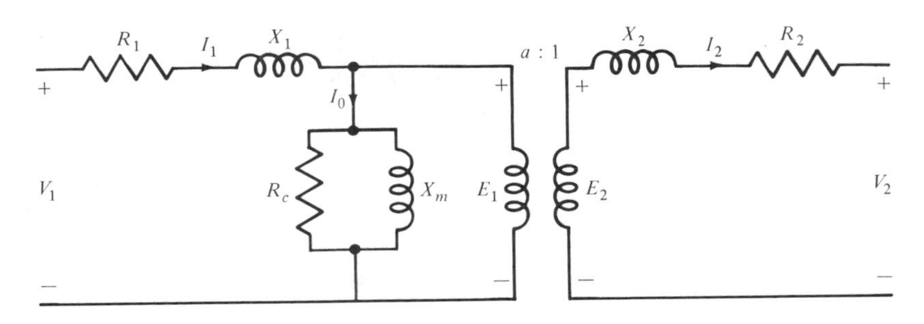

FIGURE 4-9. Nonideal transformer.

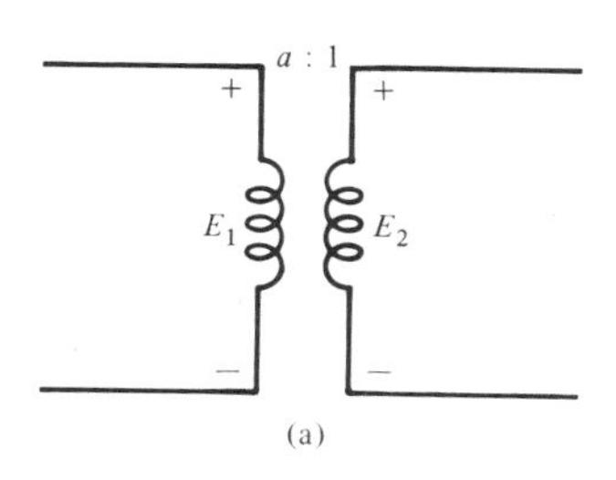

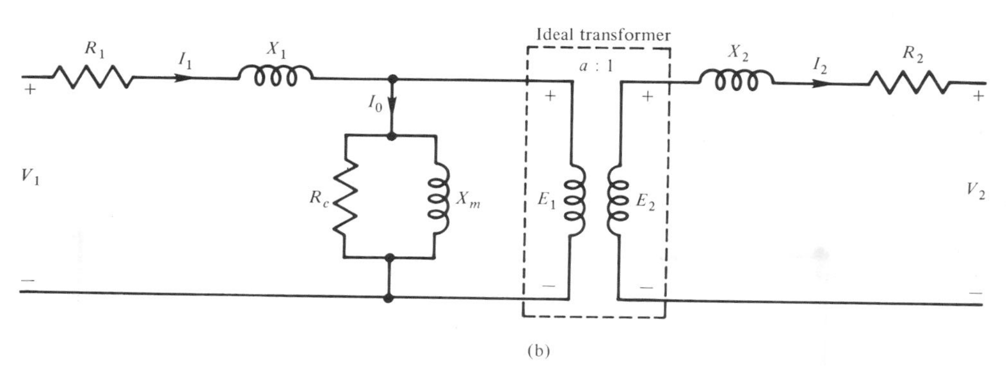

FIGURE 4-10. Equivalent circuits of (a) ideal and (b) non-ideal transformers.

primary and the secondary windings simultaneously in a nonideal transformer because of leakage.

An equivalent circuit of an ideal transformer is shown in Fig. 4-8. When the nonideal effects of winding resistances, leakage reactances, magnetizing reactance, and core losses are included, the circuit of Fig. 4-8 is modified to that of Fig. 4-10(b), where the primary and the secondary are coupled by an ideal transformer. By using (4-18), (4-20), and (4-21), the ideal transformer may be removed from Fig. 4-10(b) and the entire equivalent circuit may be referred either to the primary, as shown in Fig. 4-11, or to the secondary, as shown in Fig. 4-12.

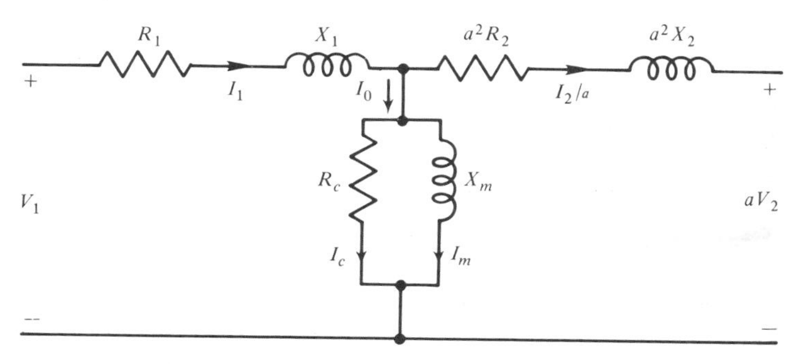

FIGURE 4-11. Equivalent circuit referred to primary.

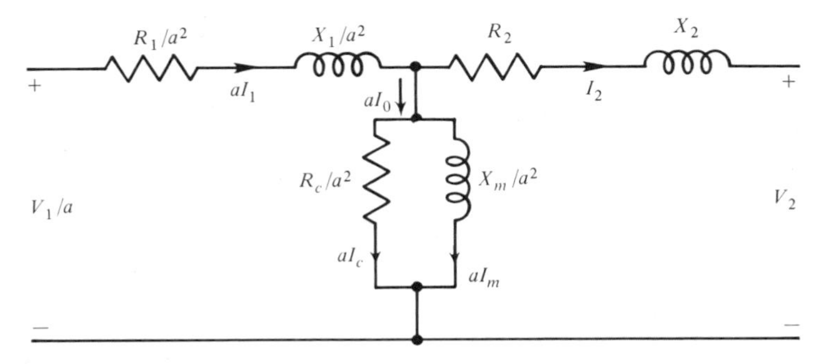

FIGURE 4-12. Equivalent circuit referred to secondary.

A phasor diagram for the circuit of Fig. 4-11, for lagging power factor, is shown in Fig. 4-13. In Figs. 4-9 to 4-13, the various symbols are

$a \equiv$ turns ratio

$\mathbf{E}_1 \equiv$ primary induced voltage

$\mathbf{E}_2 \equiv$ secondary induced voltage

$\mathbf{V}_1 \equiv$ primary terminal voltage

$\mathbf{V}_2 \equiv$ secondary terminal voltage

$\mathbf{I}_1 \equiv$ primary current

$\mathbf{I}_2 \equiv$ secondary current

$\mathbf{I}_0 \equiv$ no-load (primary) current

$R_1 \equiv$ resistance of the primary winding

$R_2 \equiv$ resistance of the secondary winding

$X_1 \equiv$ primary leakage reactance

$X_2 \equiv$ secondary leakage reactance

$\mathbf{I}_m, X_m \equiv$ magnetizing current and reactance

$\mathbf{I}_c, R_c \equiv$ current and resistance accounting for the core losses

The major use of the equivalent circuit of a transformer is in determining its characteristics. The characteristics of most interest to power engineers are *voltage regulation* and *efficiency*. Voltage regulation is a measure of the change in the terminal voltage of the transformer with load. From Fig 4-13 it is clear that

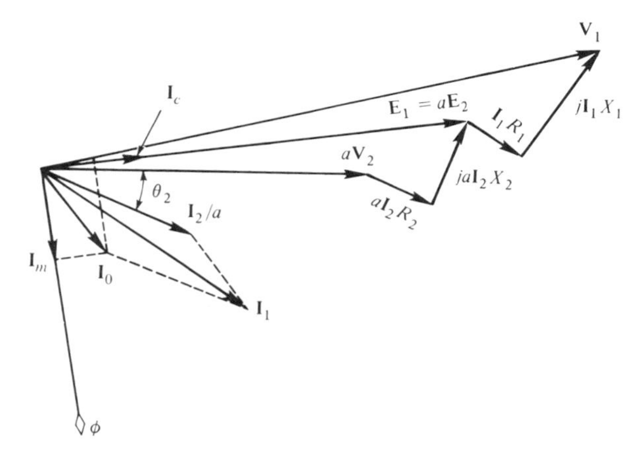

FIGURE 4-13. Phasor diagram corresponding to Fig. 4-10.

the terminal voltage V_1 is load dependent. Specifically, we define voltage regulation as

$$\text{percent regulation} = \frac{V_{\text{noload}} - V_{\text{load}}}{V_{\text{load}}} \times 100 \tag{4-22}$$

With reference to Fig. 4-13, we may rewrite (4-22) as

$$\text{percent regulation} = \frac{V_1 - aV_2}{aV_2} \times 100$$

for a given load. Notice that aV_2 is the terminal voltage referred to the primary. See Example 4-7 for a further explanation.

There are two kinds of efficiencies of transformers of interest to us, known as *power efficiency* and *energy efficiency*. These are defined as follows:

$$\text{power efficiency} = \frac{\text{output power}}{\text{input power}} \tag{4-23}$$

$$\text{energy efficiency} = \frac{\text{output energy for a given period}}{\text{input energy for the same period}} \tag{4-24}$$

Generally, energy efficiency is taken over a 24-h period and is called *all-day efficiency*. In such a case (4-24) becomes

$$\text{all-day efficiency} = \frac{\text{output for 24 h}}{\text{input for 24 h}} \tag{4-25}$$

For our purposes, we will use the term *efficiency* to mean power efficiency from now on. It is clear from our discussion of nonideal transformers that the output power is less than the input power because of losses, the losses being I^2R losses in the windings, and hysteresis and eddy-current losses in the core. Thus, in terms of these losses, (4-23) may be more meaningfully expressed as

$$\begin{aligned} \text{efficiency} &= \frac{\text{input power} - \text{losses}}{\text{input power}} \\ &= \frac{\text{output power}}{\text{output power} + \text{losses}} \\ &= \frac{\text{output power}}{\text{output power} + I^2R \text{ loss} + \text{core loss}} \end{aligned} \tag{4-26}$$

Obviously, I^2R loss is load dependent, whereas the core loss is constant and almost independent of the load on the transformer, as may be inferred from Fig. 4-9 or Fig. 4-11. The next examples show voltage regulation and efficiency calculations of a transformer.

EXAMPLE 4-7

A 150 kVA 2400/240-V transformer has the following parameters: $R_1 = 0.2\ \Omega$, $R_2 = 0.002\ \Omega$, $X_1 = 0.45\ \Omega$, $X_2 = 0.0045\ \Omega$, $R_c = 10{,}000\ \Omega$, and $X_m = 1550\ \Omega$, where the symbols are shown in Fig. 4-10. Refer the circuit to the primary. From this circuit, calculate the voltage regulation of the transformer at rated load with 0.8 lagging power factor.

The circuit referred to the primary is shown in Fig. 4-11. From the data given, we have $V_2 = 240$ V, $a = 10$, and $\theta_2 = -\cos^{-1} 0.8 = -36.8°$.

$$I_2 = \frac{150 \times 10^3}{240} = 625 \text{ A}$$

$$\frac{\mathbf{I}_2}{a} = \frac{I_2}{a}\underline{/\theta_2} = 62.5\,\underline{/-36.8} = 50 - j37.5 \text{ A}$$

$$a\mathbf{V}_2 = 2400\,\underline{/0°} = 2400 + j0 \text{ V}$$

$$a^2R_2 = 0.2\ \Omega \qquad \text{and} \qquad a^2X_2 = 0.45\ \Omega$$

Hence

$$\begin{aligned} \mathbf{E}_1 &= (2400 + j0) + (50 - j37.5)(0.2 + j0.45) \\ &= 2427 + j15 = 2427\,\underline{/0.35°} \text{ V} \end{aligned}$$

$$\mathbf{I}_m = \frac{2427 \angle 0.35^\circ}{1550 \angle 90^\circ} = 1.56 \angle -89.65 = 0.0095 - j1.56 \text{ A}$$

$$\mathbf{I}_c = \frac{2427 + j15}{10{,}000} \simeq 0.2427 + j0 \text{ A}$$

$$\mathbf{I}_0 = \mathbf{I}_c + \mathbf{I}_m = 0.25 - j1.56 \text{ A}$$

$$\mathbf{I}_1 = \mathbf{I}_0 + \frac{\mathbf{I}_2}{a} = 50.25 - j39.06 = 63.65 \angle -37.85$$

$$\mathbf{V}_1 = (2427 + j15) + (50.25 - j39.06)(0.2 + j0.45)$$

$$= 2455 + j30 = 2455 \angle 0.7^\circ \text{ V}$$

$$\text{percent regulation} = \frac{V_1 - aV_2}{aV_2} \times 100$$

$$= \frac{2455 - 2400}{2400} \times 100 = 2.3\%$$

■

EXAMPLE 4-8

Determine the efficiency of the transformer of Example 4-7, operating on rated load and 0.8 lagging power factor.

$$\text{output} = 150 \times 0.8 = 120 \text{ kW}$$

$$\text{losses} = I_1^2 R_1 + I_c^2 R_c + I_2^2 R_2$$

$$= (63.65)^2 \times 0.2 + (0.2427)^2 \times 10{,}000 + (625)^2 \times 0.002$$

$$= 2.18 \text{ kW}$$

$$\text{input} = 120 + 2.18 = 122.18 \text{ kW}$$

$$\text{efficiency} = \frac{120}{122.18} = 98.2\%$$

■

EXAMPLE 4-9

The transformer of Example 4-7 operates on full-load 0.8 lagging power factor for 12 h, on no-load for 4 h, and on half-full load unity power factor for 8 h. Calculate the all-day efficiency.

$$\text{output for 24 h} = (150 \times 0.8 \times 12) + (0 \times 4) + (150 \times \tfrac{1}{2} \times 8)$$

$$= 2040 \text{ kWh}$$

Losses for 24 h are:

$$\text{core loss} = (0.2427)^2 \times 10{,}000 \times 24 = 14.14 \text{ kWh}$$

I^2R loss on full load for 12 h

$$= 12[(63.65)^2 \times 0.2 + (625)^2 \times 0.002] = 19.1 \text{ kWh}$$

I^2R loss on half-full load for 8 h

$$= 8\left[\left(\frac{63.65}{2}\right)^2 \times 0.2 + \left(\frac{625}{2}\right)^2 \times 0.002\right] = 3.18 \text{ kWh}$$

$$\text{total losses for 24 h} = 14.14 + 19.1 + 3.18 = 36.42$$

$$\text{input for 24 h} = 2040 + 36.42 = 2076.42 \text{ kWh}$$

$$\text{all-day efficiency} = \frac{2040}{2076.42} = 98.2\%$$

■

4-5 TESTS ON TRANSFORMERS

Transformer performance characteristics can be obtained from the equivalent circuits of Section 4-4. The circuit parameters are determined either from design data or from test data. The two common tests are as follows.

Open-Circuit (or No-Load) Test

Here one winding is open-circuited and voltage—usually, rated voltage at rated frequency—is applied to the other winding. Voltage, current, and power at the terminals of this winding are measured. The open-circuit voltage of the second winding is also measured, and from this measurement a check on the turns ratio can be obtained. It is usually convenient to apply the test voltage to the winding that has a voltage rating equal to that of the available power source. In step-up voltage transformers, this means that the open-circuit voltage of the second winding will be higher than the applied voltage, sometimes much higher. Care must be exercised in guarding the terminals of this winding to ensure safety for test personnel and to prevent these terminals from getting close to other electrical circuits, instrumentation, grounds, and so forth.

In presenting the no-load parameters obtainable from test data, it is assumed that voltage is applied to the primary and the secondary is open-circuited. The no-load power loss is equal to the wattmeter reading in this test; core loss is found by subtracting the ohmic loss in the primary, which is usually small and may be neglected in some cases. Thus, if P_0, I_0, and V_0 are the input power, current, and voltage, the core loss is given by

$$P_c = P_0 - I_0^2 R_1 \tag{4-27}$$

The primary induced voltage is given in phasor form by

$$\mathbf{E}_1 = V_0 \angle 0^\circ - (I_0 \angle \theta_0)(R_1 + jX_1) \tag{4-28}$$

where $\theta_0 \equiv$ no-load power factor angle $= \cos^{-1}(P_0/V_0 I_0) < 0$. (*Note:* The determination of R and X is discussed under "Short-Circuit Test" below; see also Examples 4-10 and 4-11.)

$$R_c = \frac{E_1^2}{P_c}$$

$$I_c = \frac{P_c}{E_1}$$

$$I_m = \sqrt{I_0^2 - I_c^2}$$

$$X_m = \frac{E_1}{I_m}$$

$$a \approx \frac{V_0}{E_2}$$

Short-Circuit Test

In this test, one winding is short-circuited across its terminals, and a reduced voltage is applied to the other winding. This reduced voltage is of such a magnitude as to cause a specific value of current—usually, rated current—to flow in the short-circuited winding. Again, the choice of the winding to be short-circuited is usually determined by the measuring equipment available for us in the test. However, care must be taken to note which winding is short-circuited, for this determines the reference winding for expressing the impedance components, obtained by this test. Let the secondary be short-circuited and the reduced voltage be applied to the primary.

With a very low voltage applied to the primary winding, the core-loss current and magnetizing current become very small, and the equivalent circuit reduces to that of Fig. 4-14. Thus if P_s, I_s and V_s are the input power, current, and voltage under short circuit, then, referred to the primary,

$$Z_s = \frac{V_s}{I_s} \tag{4-29}$$

$$R_1 + a^2 R_2 \equiv R_s = \frac{P_s}{I_s^2} \tag{4-30}$$

$$X_1 + a^2 X_2 \equiv X_s = \sqrt{Z_s^2 - R_s^2} \tag{4-31}$$

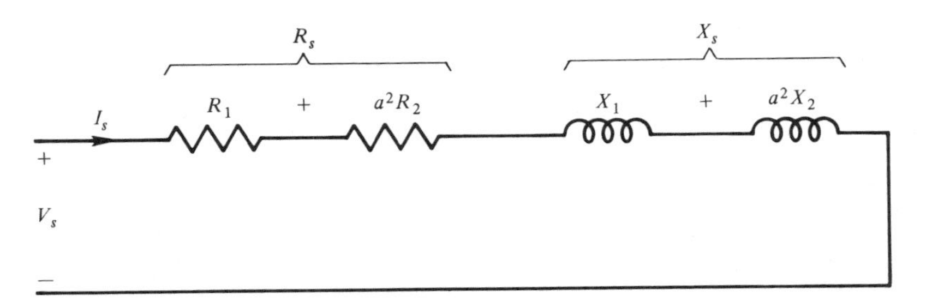

FIGURE 4-14. Equivalent circuit test for short circuit.

The primary resistance R_1 can be measured directly, and knowing a, R_2 can be found from (4-30). In (4-31) it is usually assumed that the leakage reactance is divided equally between the primary and the secondary; that is,

$$X_1 = a^2X_2 = \tfrac{1}{2}X_s$$

EXAMPLE 4-10

A certain transformer, with its secondary open, takes 80 W of power at 120 V and 1.4 A. The primary winding resistance is 0.25 Ω and the leakage reactance is 1.2 Ω. Evaluate the magnetizing reactance, X_m, and the core-loss equivalent resistance, R_c.

$$\text{no-load power factor angle, } \theta_0 = \cos^{-1}\frac{80}{1.4 \times 120} = -61.6°$$

$$\text{primary induced voltage } \mathbf{E}_1 = 120\underline{/0} - 1.4\underline{/-61.6}\,(0.25 + j1.2)$$

$$\simeq 118.29 \text{ V}$$

$$R_c = \frac{(118.29)^2}{79.5} = 176\ \Omega$$

$$I_c = \frac{118.29}{176} = 0.672 \text{ A}$$

$$I_m = \sqrt{(1.4)^2 - (0.672)^2} = 1.228 \text{ A}$$

$$X_m = \frac{118.29}{1.228} = 96.3\ \Omega$$

■

EXAMPLE 4-11

The results of open-circuit and short-circuit tests on a 25-kVA 440/220-V 60-Hz transformer are as follows:

Open-circuit test: primary open-circuited, with instrumentation on the low-voltage side. Input voltage, 220 V; input current, 9.6 A; input power, 710 W.

Short-circuit test: secondary short-circuited, with instrumentation on the high-voltage side. Input voltage, 42 V; input current, 57 A. Input pwoer = 1030 W. Obtain the parameters of the exact equivalent circuit (Fig. 4-11), referred to the high-voltage side. Assume that $R_1 = a^2R_1$ and $X_1 = a^2X_2$

From the short-circuit test:

$$Z_{s1} = \frac{42}{57} = 0.737\ \Omega$$

$$R_{s1} = \frac{1030}{(57)^2} = 0.317\ \Omega$$

$$X_{s1} = \sqrt{(0.737)^2 - (0.317)^2} = 0.665\ \Omega$$

where the subscript s corresponds to the short-circuit condition and the subscript 1 implies that the quantities are referred to the primary. Consequently,

$$R_1 = a^2R_2 = 0.158\ \Omega \qquad R_2 = 0.0395\ \Omega$$

$$X_1 = a^2X_2 = 0.333\ \Omega \qquad X_2 = 0.0832\ \Omega$$

From the open-circuit test:

$$\theta_0 = \cos^{-1}\frac{710}{(9.6)(220)} = \cos^{-1} 0.336 = -70°$$

$$\mathbf{E}_2 = 220\,\underline{/0°} - (9.6\,\underline{/-70°})(0.0395 + j0.0832) \approx 219\,\underline{/0°}\ \text{V}$$

$$P_{c2} = 710 - (9.6)^2(0.0395) \approx 710\ \text{W}$$

$$R_{c2} = \frac{(219)^2}{710} = 67.5\ \Omega$$

$$I_{c2} = \frac{219}{67.5} = 3.24\ \text{A}$$

$$I_{m2} = \sqrt{(9.6)^2 - (3.24)^2} = 9.03\ \text{A}$$

$$X_{m2} = \frac{219}{9.03} = 24.24\ \Omega$$

where the subscript 2 implies that the quantities are referred to the secondary.

$$X_{m1} = a^2X_{m2} = 97\ \Omega$$

$$R_{c1} = a^2R_{c2} = 270\ \Omega$$

■

TRANSFORMER POLARITY

Polarities of a transformer identify the relative directions of induced voltages in the two windings. The polarities result from the relative directions in which the two windings are wound on the core. For operating transformers in parallel it is necessary that we know the relative polarities. Polarities can be checked by a simple test, requiring only voltage measurements with the transformer on no-load. In this test, rated voltage is applied to one winding, and an electrical connection is made between one terminal from one winding and one from the other, as shown in Fig. 4-15. The voltage across the two remaining terminals (one from each winding) is then measured. If this measured voltage is *larger* than the input test voltage, the polarity is *additive;* if smaller, the polarity is *subtractive*.

A standard method of marking transformer terminals is as follows: The high-voltage terminals are marked H1, H2, H3, . . . , with H1 being on the right-hand side when facing the high-voltage side. The low-voltage terminals are designated X1, X2, X3, . . . , and X1 may be on either side, adjacent to H1 or diagonally opposite. The two possible locations of X1 with respect to H1 for additive and subtractive polarities are shown in Fig. 4-16. The numbers must be so arranged that the voltage difference between any two leads of the same set, taken in order from smaller to larger numbers, must be of the same sign as that between any other pair of the set taken in the same order. Furthermore, when the voltage is directed from H1 to H2, it must simultaneously be directed to X1 and X2. Additive polarities are required by the American National Standards Institute (ANSI) in large (>200 kVA) high-voltage (>8660 V) power transformers. Small transformers have subtractive polarities (which reduce the voltage stress between adjacent leads).

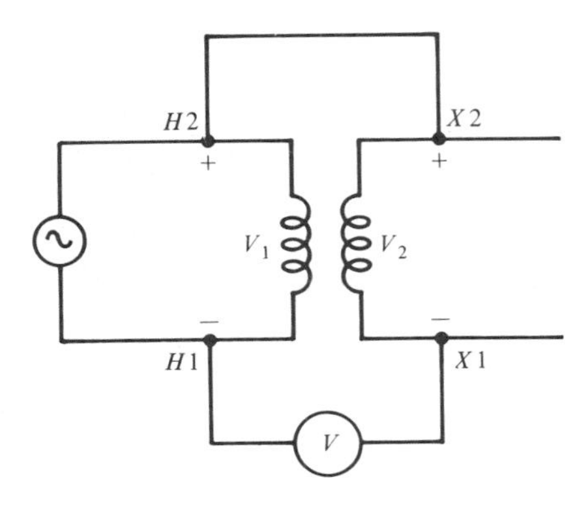

FIGURE 4-15. Polarity test on a transformer.

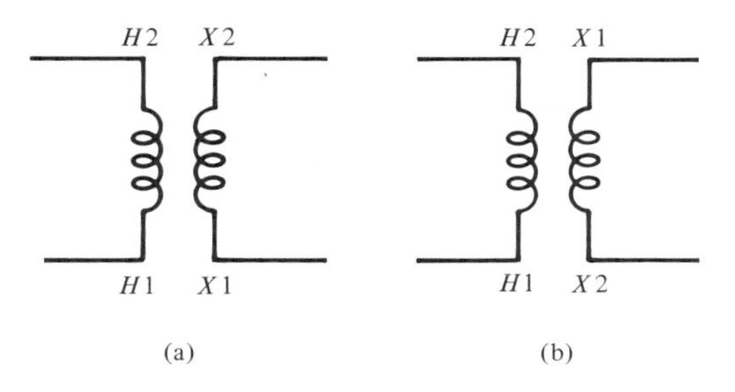

FIGURE 4-16. (a) Subtractive and (b) additive polarities of a transformer.

Knowing the polarities of a transformer aids in obtaining a combination of several voltages from the transformer. Certain single-phase connections of transformers are illustrated by the next example.

EXAMPLE 4-12

Two 1150/115-V transformers are given. Using appropriate polarity markings, show the interconnections of these transformers for (a) 2300/230-V operation and (b) 1150/230-V operation.

We mark the high-voltage terminals by H1 and H2 and the low-voltage terminals by X1 and X2 (as in Figs. 4-15 and 4-16). According to the ANSI convention, H1 is marked on the terminal on the right-hand side of the transformer case (or housing) when facing the high-voltage side. With this nomenclature, the connections desired are as shown in Fig. 4-17. ■

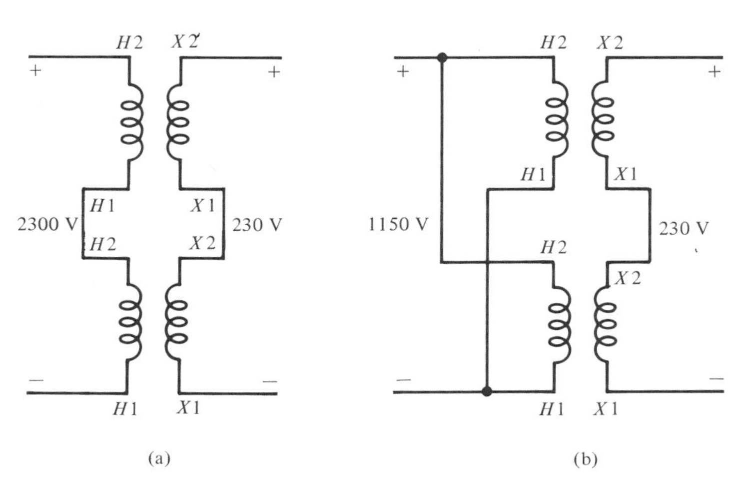

FIGURE 4-17. Example 4-12.

TRANSFORMERS IN PARALLEL

Transformers are connected in parallel to share loads exceeding the capacity of an individual transformer. Let us consider two transformers operating in parallel (with proper polarities) and supplying a common load, as given in Fig. 4-18. Neglecting the core losses and magnetizing currents, the equivalent circuit becomes as shown in Fig. 4-19. If the transformers have a' and a'' as their respective turns ratio, then from Fig. 4-19 we have

$$\mathbf{V}_1 = \mathbf{V}_1' = a'\mathbf{V}_t + \mathbf{I}_1'\mathbf{Z}_e' \tag{4-32}$$

$$\mathbf{V}_1 = \mathbf{V}_1'' = a''\mathbf{V}_t + \mathbf{I}_1''\mathbf{Z}_e'' \tag{4-33}$$

$$\mathbf{I}_1 = \mathbf{I}_1' + \mathbf{I}_1'' \tag{4-34}$$

Subtracting (4-33) from (4-32) and solving with (4-34) yields

$$\mathbf{I}_1' = \frac{-\mathbf{V}_t(a' - a'') + \mathbf{I}_1\mathbf{Z}_e''}{\mathbf{Z}_e' + \mathbf{Z}_e''} \tag{4-35}$$

$$\mathbf{I}_2'' = \frac{\mathbf{V}_t(a' - a'') + \mathbf{I}_1\mathbf{Z}_e'}{\mathbf{Z}_e' + \mathbf{Z}_e''} \tag{4-36}$$

Notice from (4-35) and (4-36) that the current through any one of the transformers consists of two components. One component is proportional to $\mathbf{I}_1$ or the load, whereas the other component varies with $\mathbf{V}_t(a' - a'')$. The latter component is constant, and gives rise to the local circulating current between

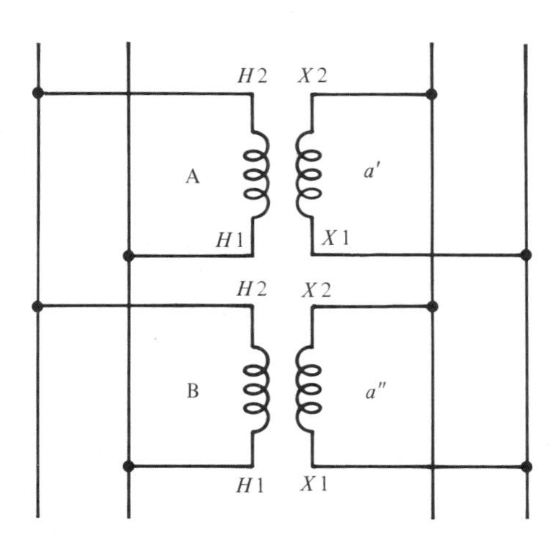

FIGURE 4-18. Two transformers in parallel.

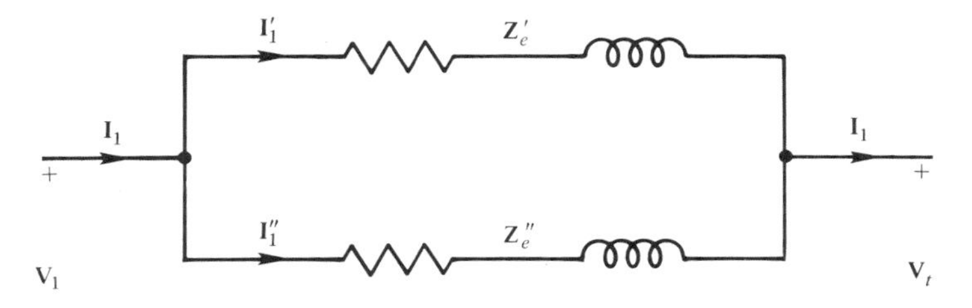

FIGURE 4-19. Equivalent circuit of transformers in parallel.

the two transformers. Clearly, with transformers having an equal turns ratio, $a' = a''$, there will be no circulating current.

EXAMPLE 4-13

A 2300/230-V transformer has an equivalent series impedance of $1.84 \angle 84.2^\circ$ Ω referred to the primary. Another transformer rated at 2300/230 V has an equivalent impedance of $0.77 \angle 82.5^\circ$ Ω. The two transformers are connected in parallel and supply a 230-V 400-kVA 0.8 lagging power factor load. Determine the power supplied by each transformer.

This problem involves an application of (4-35) and (4-36). However, since the load is specified on the secondary side, we must solve the problem in terms of secondary quantities. It may be shown that the secondary currents of the transformer, corresponding to (4-35) and (4-36) (which are the primary currents), are

$$\mathbf{I}_2' = \frac{-\mathbf{V}_t(a' - a'') + a''\mathbf{I}_2\mathbf{Z}_{e2}''}{a'\mathbf{Z}_{e2}' + a''\mathbf{Z}_{e2}''} \tag{4-37}$$

$$\mathbf{I}_2'' = \frac{\mathbf{V}_t(a' - a'') + a'\mathbf{I}_2\mathbf{Z}_{e2}'}{a'\mathbf{Z}_{e2}' + a''\mathbf{Z}_{e2}''} \tag{4-38}$$

where $\mathbf{Z}_{e2}'$ and $\mathbf{Z}_{e2}''$ are the equivalent impedances of the first and second transformers respectively referred to their secondaries.

We now substitute the following numerical values in (4-37) and (4-38):

$$a' = a'' = 10$$

$$\mathbf{Z}_{e2}' = \frac{1.84 \angle 84.2^\circ}{10^2} = 0.0184 \angle 84.2^\circ = (1.86 + j18.3) \times 10^{-3}$$

$$\mathbf{Z}_{e2}'' = \frac{0.77 \angle 82.5^\circ}{10^2} = 0.0077 \angle 82.5^\circ = (1.07 + j7.63) \times 10^{-3}$$

$$\mathbf{I}_2 = \frac{400{,}000}{230} = 1739 \angle -36.8^\circ \text{ A}$$

$$\mathbf{V}_t = 230 \angle 0^\circ \text{ V}$$

Thus we obtain

$$\mathbf{I}_2' = \frac{10 \times 1739 \angle -36.8^\circ \times 0.0077 \angle 82.5^\circ}{10(2.86 + j25.93) \times 10^{-3}} = 513.0 \angle -38.1^\circ \text{ A}$$

$$\mathbf{I}_2'' = \frac{10 \times 1739 \angle -36.8^\circ \times 0.0184 \angle 84.2^\circ}{10(2.86 + j25.93) \times 10^{-3}} = 1226.0 \angle -36.3^\circ \text{ A}$$

The power supplied by the first transformer is calculated as

$$230 \times 513 \times \cos 38.1 = 92.7 \text{ kW}$$

and the power supplied by the second transformer, as

$$230 \times 1226 \times \cos 36.3 = 227.3 \text{ kW}$$

Check: The sum of the power supplied by the two transformers = 92.7 + 227.3 = 320 kW = total load = 400 × 0.8 = 320 kW.

From (4-37) and (4-38) it follows that load sharing by tranformers in proportion to their ratings can be obtained by a proper choice of the respective $\mathbf{Z}_e$. ■

4-8 THREE-PHASE TRANSFORMER CONNECTIONS

Electric power systems are three-phase systems in that power is generated by three-phase generators and transmitted by three-phase transmission lines. Obviously, these generators and transmission lines must be linked by three-phase transformers. The primary and secondary windings of single-phase transformers may be interconnected to obtain three-phase transformer banks.

Some of the factors governing the choice of connections are as follows:

1. Availability of a neutral connection for grounding, protection, or load connections.
2. Insulation to ground and voltage stresses.
3. Availability of a path for the flow of third-harmonic (exciting) currents and zero-sequence (fault) currents.
4. Need for partial capacity with one unit out of service.
5. Parallel operation with other transformers.
6. Operation under fault conditions.
7. Economic considerations.

Keeping these factors in mind, we now consider some of the three-phase transformer connections.

Wye/Wye Connection

The wye/wye connection is shown in Fig. 4-20, where the terminal markings show subtractive polarities. Primary terminals are designated by *ABC*, whereas *abc* is used to indicate the secondary terminals. The phase relationships between the various voltages are shown in Fig. 4-20(b). The equilateral triangles super-

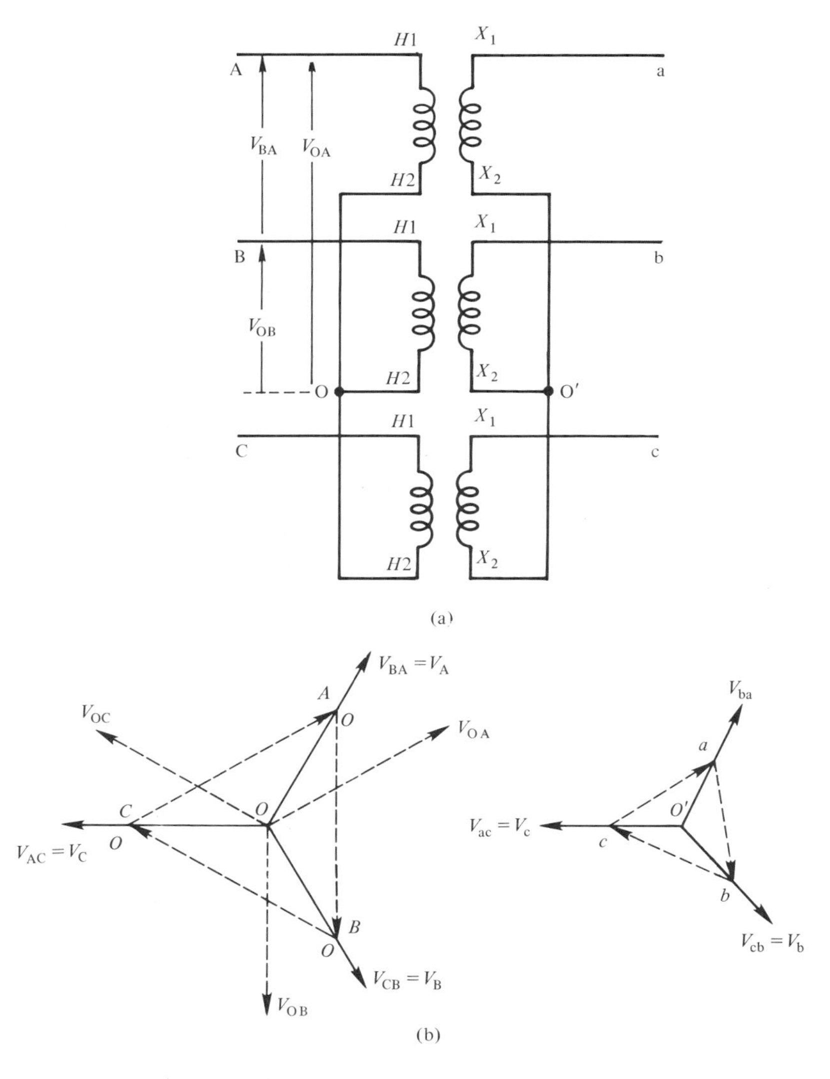

FIGURE 4-20. (a) Three-phase wye–wye connection; (b) phasor diagrams.

imposed on the phasor diagrams aid in the construction of the phasor diagrams. For instance, OA, OB, and OC denote the phase voltages of the primary and $O'a$, $O'b$, and $O'c$ correspond to the phase voltages of the secondary. To determine the time–phase relationships of line voltages, we observe that the primary phase sequence is ABC. The phase of the voltage between A and B is found by tracing the circuit AOB, where AO is traversed in the positive direction and OB in the negative direction. The phasor relationship between the voltages around the circuit AOB is

$$\mathbf{V}_{BA} + \mathbf{V}_{OB} - \mathbf{V}_{OA} = 0$$

or

$$\mathbf{V}_{BA} = \mathbf{V}_{OA} - \mathbf{V}_{OB} = \mathbf{V}_{A}$$

Thus we reverse OB and geometrically add it to OA to obtain BA, and hence $\mathbf{V}_{BA}$. Other voltages are determined in a similar fashion.

From Fig. 4-20(b) it is clear that there is no phase shift between the primary and secondary voltages. Furthermore, if V is the voltage between lines, then $V/\sqrt{3}$ is the voltage across the phase, on the terminals of a wye-connected transformer. Thus, compared to a delta connection (discussed later), the wye-connected transformer will have fewer turns, will require windings of larger cross section, and will have less dielectric stress on the insulation. On the other hand, the main disadvantage of wye/wye-connected transformers is that such transformers have *roving neutrals* when supplying unbalanced loads. By ''roving neutral'' we mean that the potential of point O in Fig. 4-20(b) is not fixed with respect to the lines, and may take any position within the triangle if the transformer supplies an unbalanced load. Figure 4-21 shows two positions of O under

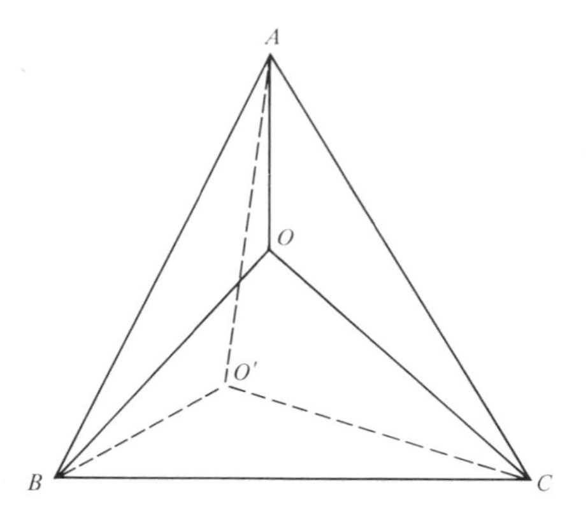

FIGURE 4-21. Roving neutral under unbalanced loads.

two unbalanced loading conditions. One way to prevent the shifting of the neutral is to connect the primary neutral of the transformer to the neutral of the generator. In such a case, however, if there is a third harmonic component in the generator voltage waveform, there will be a third harmonic in the secondary voltage, and there will be corresponding triple frequency currents in the secondary circuits.

Delta/Delta Connection

Figure 4-22 shows three tranformers connected to make a three-phase delta/delta system. The phasor diagram showing the phase relationships of various voltages is also included in the figure. The transformers have subtractive polarities. Clearly, in such a connection, the individual windings of the transformers must be designed for full line voltages. This arrangement, however, has the advantage that

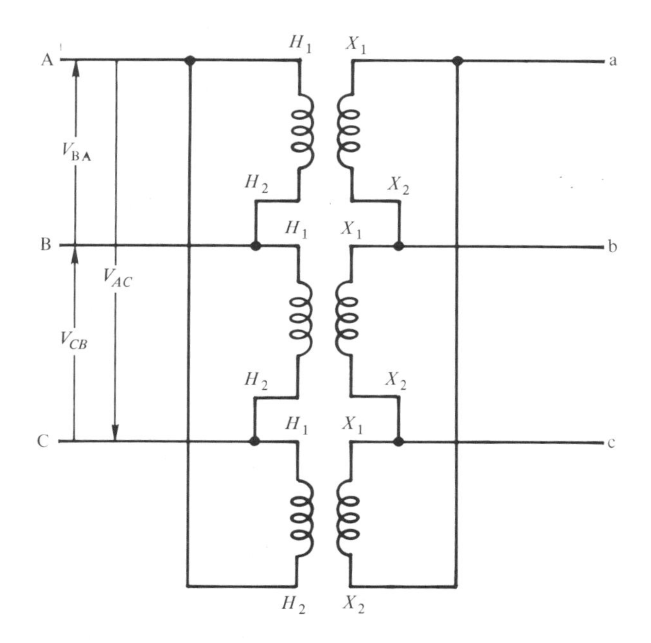

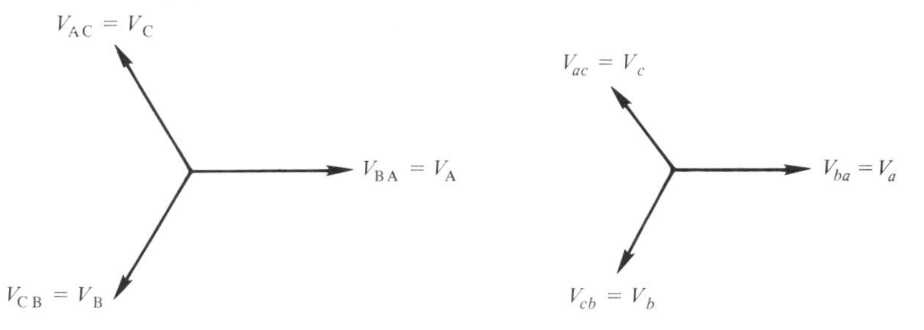

FIGURE 4-22. Delta-delta connection with subtractive polarities.

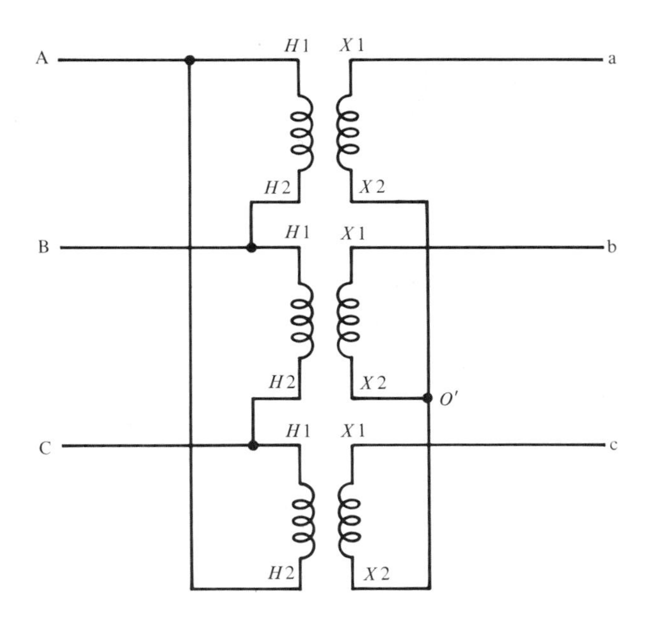

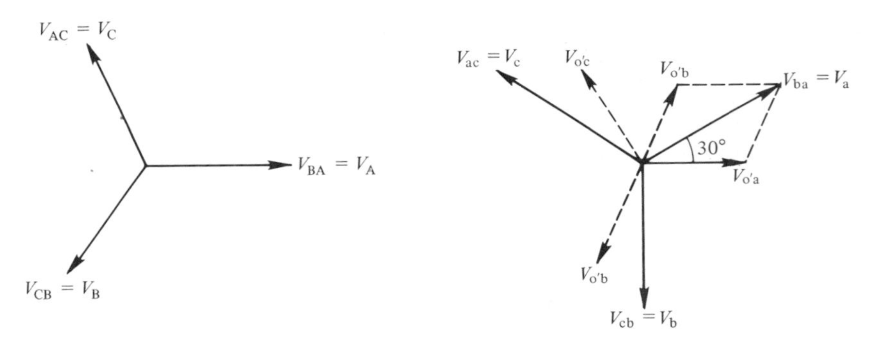

FIGURE 4-23. Delta-wye connection.

one transformer can be out of service without an interruption in system operation except that the capacity with two transformers is proportionately reduced. The secondary voltages also tend to be slightly unbalanced with the loss of one transformer. The delta/delta connection provides a path for the flow of the third-harmonic magnetizing current, the current in each coil being in phase with that of the other. Hence third-harmonic currents and voltages do not appear on the line.

Delta/Wye and Wye/Delta Connections

Delta/wye and wye/delta connections of three-phase transformers are shown in Figs. 4-23 and 4-24, respectively. The polarities in both cases are subtractive. Notice the 30° phase shift between the line and phase voltages in the two cases.

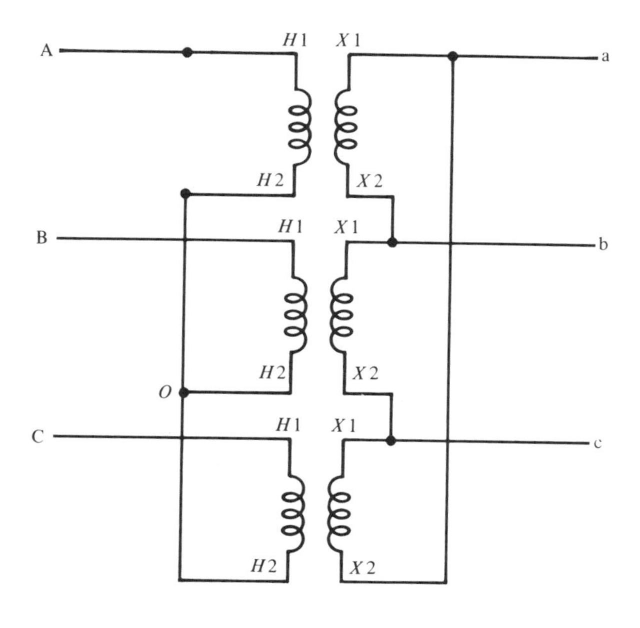

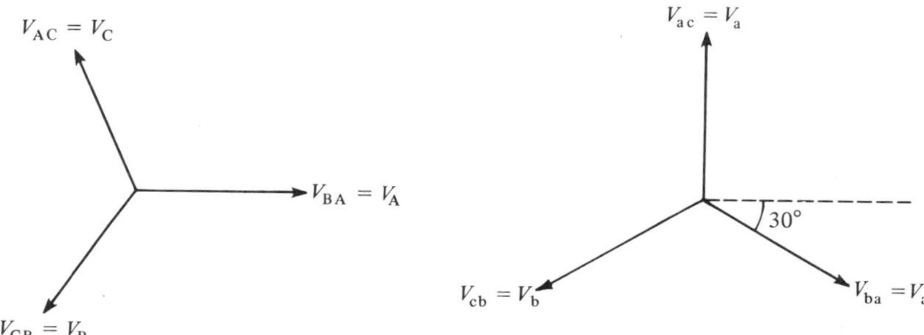

FIGURE 4-24. Wye-delta connection.

This shift in the delta-wye connection is opposite to that in the wye/delta connection. These connections are particularly suited for high-voltage systems. The delta/wye connection is used for stepping up and the wye/delta for stepping down the voltage. The wye connection on the high-voltage side permits the grounding of the neutral. The delta connection offers a path for the flow of the third-harmonic currents.

4-9 SPECIAL TRANSFORMER CONNECTIONS

In Section 4-8 we have studied three-phase transformer connections, where the primaries and/or the secondaries are connected in wye and/or delta. Such con-

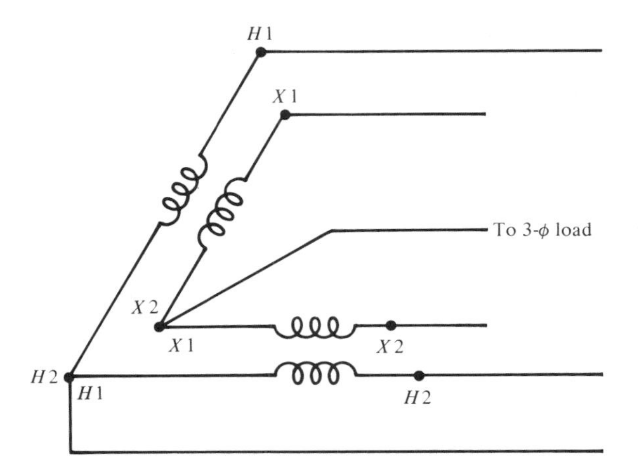

FIGURE 4-25. Open-delta connection.

nections involved three transformers. There are certain three-phase transformer connections that require only two transformers. We will discuss three such connections here.

Open-Delta Connection

Recall from Section 4-8 that in the delta/delta connection, if one of the transformers is removed, the remaining two transformers can provide a three-phase system. In practice, such a connection is known as the *open-delta* or *V connection,* and is shown in Fig. 4-25. Obviously, the rating of an open-delta transformer bank is less than that of a delta/delta bank (assuming that each transformer of the two banks has the same rating). If each transformer of a delta/delta system is rated at V volts and I amperes, the line current for the system will be $\sqrt{3}\, I$. But for an open-delta system the line current cannot exceed I. Consequently, the combined rating of two transformers connected in open-delta form amounts to $(I/\sqrt{3}I)$ or 58 percent of the rating of the original three transformers.

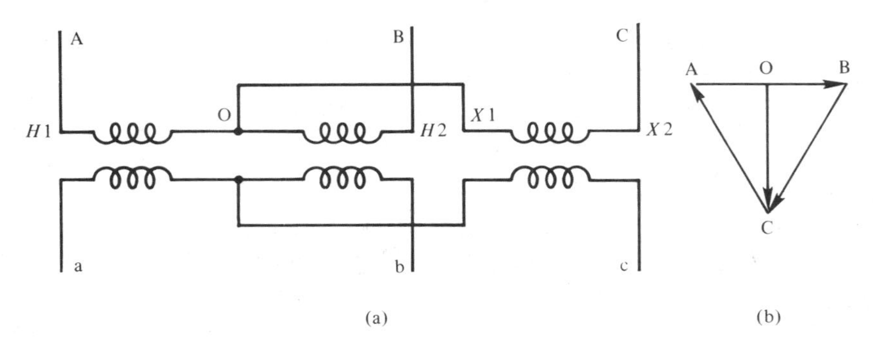

FIGURE 4-26. (a) T-connection; (b) phasor diagram.

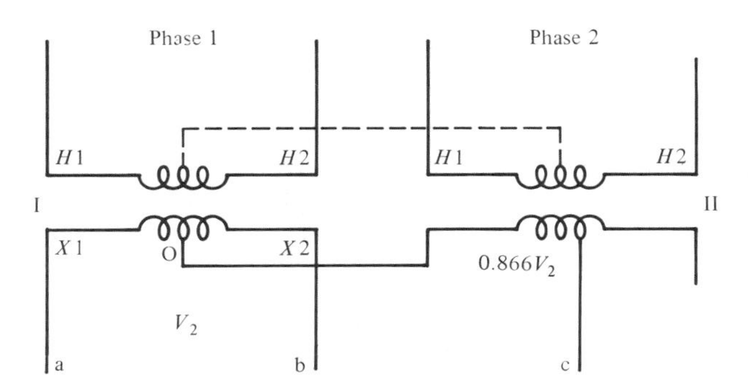

FIGURE 4-27. Scott connection.

T Connection

The T connection also offers a method of using only two transformers in a three-phase system. This connection is shown in Fig. 4-26(a). One transformer, having a center tap at O, is connected to terminals A and B of the three-phase system. This transformer is called the *main transformer*. The second transformer, called the *teaser transformer,* has one terminal connected to point O of the main transformer and the other to C of the three-phase source, as indicated in Fig. 4-26(a). From the phasor diagram in Fig. 4-26(b), it may be seen that the voltage $\mathbf{V}_{OC}$ across the teaser transformer $= (\sqrt{3}/2)\mathbf{V}_{AB} = 0.866\mathbf{V}_{AB}$. The ratio of the transformation of the two transformers being the same, we obtain a three-phase balanced system at the secondary. It is important that relative polarities be properly maintained, as given in Fig. 4-26(a). With the correct polarities, the currents in each of the halves of the main transformer flow in opposite directions.

Scott Connection

The Scott connection is a special connection used to obtain a two-phase system from a three-phase system, and vice versa. The connection is shown in Fig. 4-27. Transformer I has a midpoint tap and has a voltage rating V_2. Transformer II has a tap at a point corresponding to 86.6 percent of V_2.

There are numerous other transformer connections for various applications, but these are not considered here.

4-10

PARALLEL OPERATION OF THREE-PHASE TRANSFORMERS

In operating three-phase transformers in parallel, it is most important that there be no phase displacement between the voltages at the secondary leads, which are being connected in parallel. To verify that this condition is fulfilled, we may refer to the phasor diagrams of three-phase transformer connections. For instance, the phasor diagrams of Figs. 4-20 and 4-22 indicate that a wye/wye-connected transformer bank will operate in parallel with a delta/delta-connected bank if the line-to-line voltages of the two banks are the same. Notice from the respective phasor diagrams that the two phasor diagrams are identical. On the other hand, a wye/wye, or a delta/delta bank will not satisfactorily operate in parallel with either a wye/delta or a delta/wye bank. In such connections, when the two sets of primary leads are connected in parallel and to the same source, there will always be a phase displacement between the two sets of secondary leads.

4-11

AUTOTRANSFORMERS

In contrast to the two-winding transformers considered so far, the autotransformer is a single-winding transformer having a tap brought out at an intermediate point. Thus, as shown in Fig. 4-28, *ac* is the single winding (wound on a laminated core) and *b* is the intermediate point where the tap is brought out. Like a two-winding transformer, the autotransformer may be used as either a step-up or a step-down operation. Considering a step-down arrangement, let the primary applied (terminal) voltage be V_1, resulting in a magnetizing current and a core flux, φ_m. Let the secondary be open-circuited. Then the primary and secondary voltages obey the same rules as in a two-winding transformer, and we have

$$\frac{V_1}{V_2} = \frac{E_1}{E_2} = \frac{N_1}{N_2} = a$$

with $a > 1$ for step-down.

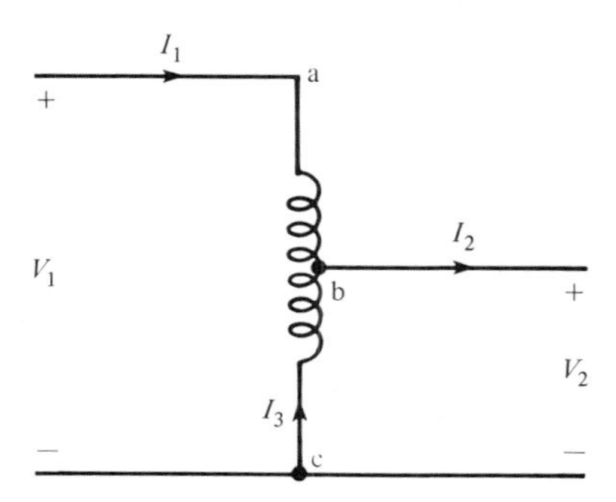

FIGURE 4-28. A step-down autotransformer.

Furthermore, ideally,

$$V_1 I_1 = V_2 I_2$$

and

$$\frac{V_1}{V_2} = \frac{I_2}{I_1} = a$$

Neglecting the magnetizing current, we must have the mmf balance equation as

$$N_2 I_3 = (N_1 - N_2) I_1$$

or

$$I_3 = \frac{N_1 - N_2}{N_2} I_1 = (a - 1) I_1 = I_2 - I_1$$

which agrees with the current-flow directions shown in Fig. 4-28.

The apparent power delivered to the load may be written as

$$P = V_2 I_2 = V_2 I_1 + V_2 (I_2 - I_1) \tag{4-39}$$

In (4-39) the power is considered to consist of two parts:

$$V_2 I_1 \equiv P_c \equiv \text{conductively transferred power through } bc$$

$$V_2 (I_2 - I_1) \equiv P_i \equiv \text{inductively transferred power through } ab$$

These powers are related to the total power by

$$\frac{P_i}{P} = \frac{I_2 - I_1}{I_2} = \frac{a - 1}{a} \tag{4-40}$$

and

$$\frac{P_c}{P} = \frac{I_1}{I_2} = \frac{1}{a} \tag{4-41}$$

where $a > 1$.

It may be shown that for a step-up transformer the power ratios are obtained as follows:

$$P = V_1 I_1 = V_1 I_2 + V_1(I_1 - I_2)$$

implying that the total apparent power consists of two parts:

$$P_c = V_1 I_2 = \text{conductively transferred power}$$

$$P_i = V_1(I_1 - I_2) = \text{inductively transferred power}$$

Hence

$$\frac{P_i}{P} = \frac{I_1 - I_2}{I_1} = 1 - a \tag{4-42}$$

and

$$\frac{P_c}{P} = \frac{I_2}{I_1} = a \tag{4-43}$$

where $a < 1$.

EXAMPLE 4-14

A two-winding 10-kVA 440/110-V transformer is reconnected as a step-down 550/440-V autotransformer. Compare the voltampere rating of the autotransformer with that of the original two-winding transformer, and calculate P_i and P_c.

Refer to Fig. 4-28. The rated current in the 110-V winding (or in *ab*) is

$$I_1 = \frac{10{,}000}{110} = 90.91 \text{ A}$$

Current in the 440-V winding (or in *bc*) is

$$I_3 = I_2 - I_1 = \frac{10{,}000}{440} = 22.73 \text{ A}$$

which is the rated current of the winding *bc*. The load current is

$$I_2 = I_1 + I_3 = 90.91 + 22.73 = 113.64 \text{ A}$$

Check: For the autotransformer

$$a = \frac{550}{440} = 1.25$$

and

$$I_2 = aI_1 = 1.25 \times \frac{10{,}000}{110} = 113.64 \text{ A}$$

which agrees with I_2 calculated above. Hence the rating of the autotransformer is

$$P_{\text{auto}} = V_1 I_1 = 550 \times \frac{10{,}000}{110} = 50 \text{ kVA}$$

Inductively supplied apparent power is

$$P_i = V_2(I_2 - I_1) = \frac{a-1}{a} P = \frac{1.25 - 1}{1.25} \times 50 = 10 \text{ kVA}$$

which is the voltampere rating of the two-winding transformer. The conductively supplied power is

$$P_c = \frac{P}{a} = \frac{50}{1.25} = 40 \text{ kVA}$$

■

EXAMPLE 4-15

Repeat Example 4-14 for a 440/550-V step-up connection.

The step-up connection is shown in Fig. 4-29. The rating of winding *ab* is 110 V and the load current I_2 flows through *ab*. Hence

$$I_2 = \frac{10{,}000}{110} = 90.91 \text{ A}$$

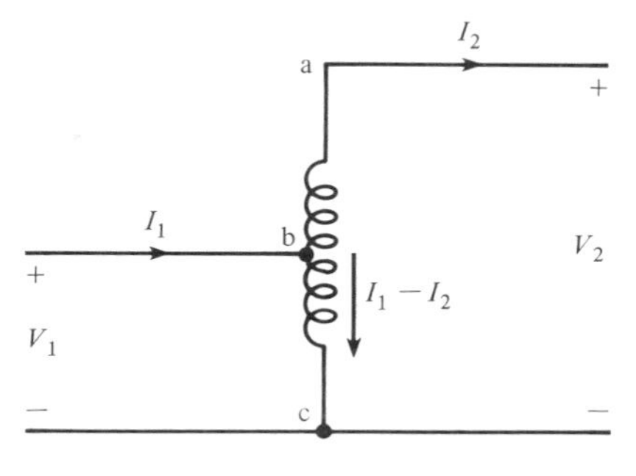

FIGURE 4-29. A step-up autotransformer.

The output voltage is

$$V_2 = 550 \text{ V}$$

Thus the voltampere rating of the autotransformer is

$$V_2 I_2 = 550 \times \frac{10{,}000}{110} = 50 \text{ kVA}$$

which is the same as in Example 4-14.

$$\text{power transferred conductively} = V_1 I_2 = 440 \times 90.91 = 40 \text{ kVA}$$

$$\text{power transferred inductively} = 50 - 40 = 10 \text{ kVA}$$

■

Consequently, a two-winding transformer connected as an autotransformer will have a voltampere rating $a/(a - 1)$ times its rating as a two-winding transformer.

4-12 THREE-WINDING TRANSFORMERS

Generally, large power transformers have three windings. The third winding is known as a *tertiary winding*, which may be used for the following purposes:

1. To supply a load at a voltage different from the secondary voltage.
2. To provide a low impedance for the flow of certain abnormal currents, such as third harmonic currents.
3. To provide for the excitation of a regulating transformer.

Wherever more than one secondary voltage is required, it can be secured more economically from a third winding than from an additional transformer.

A three-winding transformer is represented schematically in Fig. 4-30. When

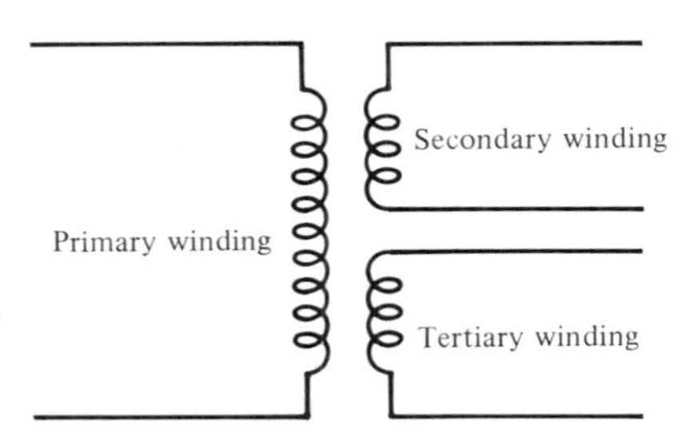

FIGURE 4-30. Three-winding autotransformer.

a three-winding transformer operates on load, each winding interacts inductively with each of the others. Thus there are three mutual inductances and three self- (or leakage) inductances to be considered. The analysis, therefore, becomes rather cumbersome and is not considered here.

4-13

INSTRUMENT TRANSFORMERS

Instrument transformers are of two kinds: current transformers (CTs) and potential transformers (PTs). These are used to supply power to ammeters, voltmeters, wattmeters, relays, and so on. Instrument transformers are used for (1) reducing the measured quantity to a low value which can be indicated by standard instruments (a standard voltmeter may be rated at 120 V and an ammeter at 5 A); and (2) isolating the instruments from high-voltage sources for safety. A connection diagram of a CT and a PT with an ammeter, a voltmeter, and a wattmeter is shown in Fig. 4-31. The load on the instrument transformer is called the *burden*. Depending on the burden, instrument transformers are rated from 25 to 500 VA. However, a PT or a CT is much (two to six times) bigger than a power transformer of the same rating.

An ideal instrument transformer has no phase difference between the primary and secondary voltages (or currents), which are independent of the burden. Like the ideal power transformer discussed earlier, the voltage ratio of an ideal PT is exactly equal to its turns ratio. The current ratio of an ideal CT is exactly equal to the inverse of the turns ratio. In practice, however, load-dependent ratio and phase-angle errors are present in instrument transformers.

The principle of operation of an instrument transformer is no different from that of an ordinary power transformer. Thus they have similar equivalent circuits and phasor diagrams, as shown in Fig. 4-32. It is clear from this diagram that the secondary impedance drop causes a phase displacement α, and the primary

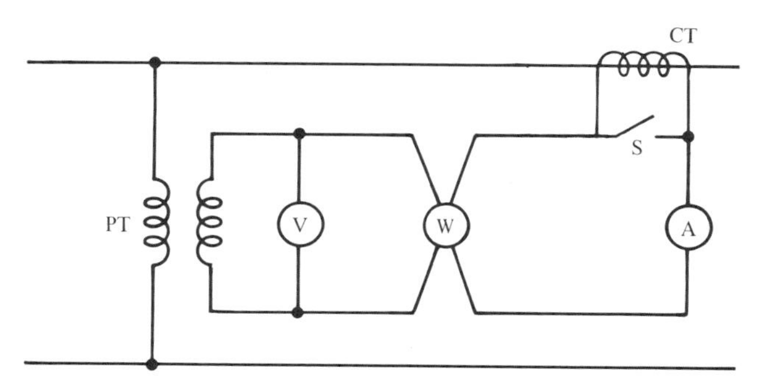

FIGURE 4-31. Instrument–transformer connections.

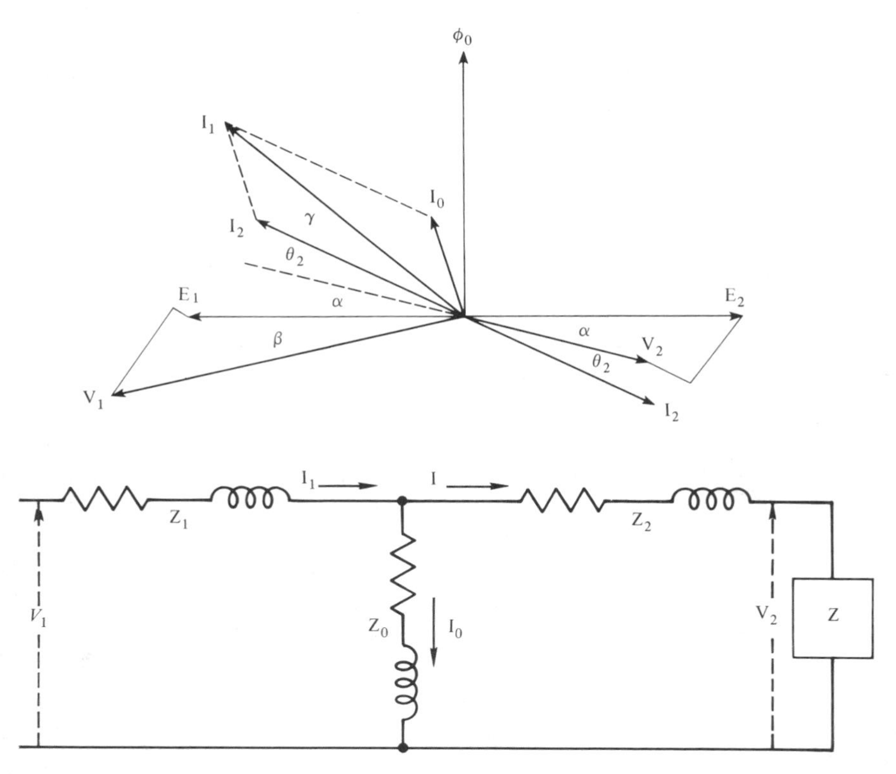

FIGURE 4-32. Phasor diagram and equivalent circuit of an instrument transformer.

impedance drop a phase displacement β; the exciting current I_0 causes a further phase displacement γ, so that the angle between the primary voltage and current is $(\theta_2 + \alpha + \beta + \gamma)$, compared with an angle θ_2 between the secondary voltage and current. Thus the transformer introduces a phase-angle error $(\alpha + \beta + \gamma)$. Moreover, V_1 and V_2 will be only approximately in the ratio of the number of turns. The significance of these errors is indicated in Fig. 4-33. Calibration curves of this type are furnished by the manufacturer. In order to nullify or reduce the errors, instrument transformers are designed with (1) small leakage reactances and low resistances which reduce angles α and β; (2) low flux densities and good transformer iron, which reduces the exciting current I_0 and therefore angle γ; and (3) less than a nominal turn ratio, which compensates for the ratio error. Compensating impedances may also be provided, so that the burden can be kept constant as instruments are put in or out of the circuit. For a constant burden, the instruments may be calibrated, or corrected, against the load.

Provision must be made to short-circuit the secondary of a current transformer before removing any instruments, for otherwise, dangerously high voltages may

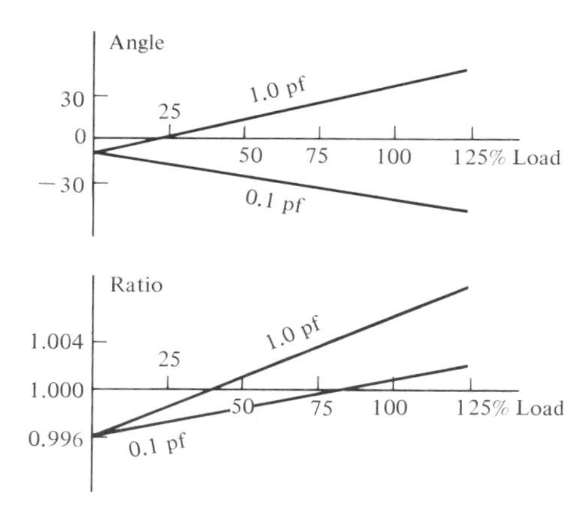

FIGURE 4-33. Phase-angle and ratio errors of an instrument transformer.

occur. It is clear from the equivalent circuit of Fig. 4-32 that if the secondary is open-circuited, the primary current, of fixed magnitude as determined by the load, must flow through the exciting impedance and act entirely as a magnetizing current to the transformer. This results in high flux density and correspondingly high voltage. Oscillograms of such a voltage show it to be peaked (see the following section) by a dominating third harmonic, as would be expected.

4-14

THIRD HARMONICS IN TRANSFORMERS

In earlier sections we mentioned the presence of third harmonic currents in transformers. First, we show a sine wave (or the fundamental) and its third harmonic in Fig. 4-34. The resultant, which is the sum of the fundamental and the third harmonic, is also shown in the figure. Next, let us assume that the core flux of a transformer is sinusoidal. The corresponding core flux density, $B(t)$, is shown in Fig. 4-35, which also shows the core saturation $B(H)$ characteristic. From these $B(t)$ and $B(H)$ curves we obtain, graphically, the waveform of the corresponding magnetizing current. The construction procedure and the current waveform are given in the figure. Now, compare the resultant waveform of Fig. 4-34 and the magnetizing current waveform of Fig. 4-35. The similarity between the two indicates that the magnetizing current does, indeed, contain third harmonics if the transformer core operates under a saturated condition.

Having established the presence of the third harmonic in a transformer, let us now consider their mutual relationships in a three-phase transformer. In a three-phase system, the third harmonic of the phase displaced from the reference phase by 120° is displaced by $3 \times 120° = 360°$. Also, the third harmonic in

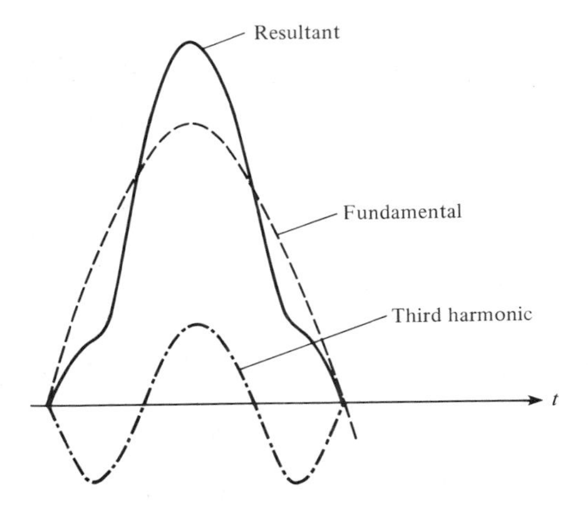

FIGURE 4-34. Resultant waveform of a fundamental and third harmonic.

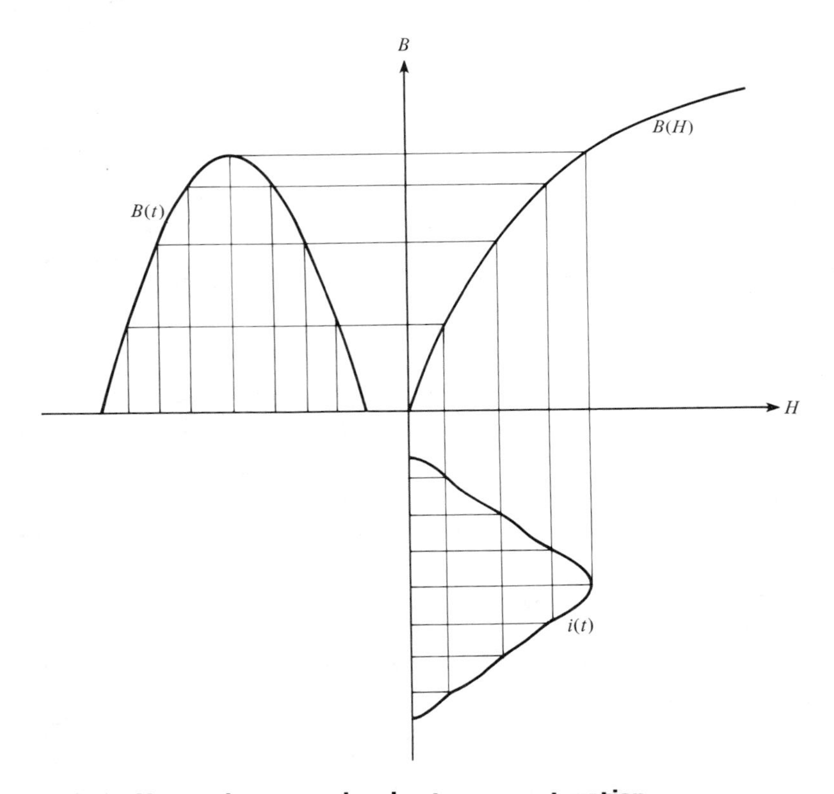

FIGURE 4-35. Harmonic generation due to core saturation.

the phase displaced from the reference phase by 240° is displaced by $3 \times 240° = 720°$. Consequently, in a three-phase transformer, the third harmonics in the three transformers are in phase. In a delta connection, third harmonics of a current add to each other and form circulating currents around the delta. This connection is, therefore, called a *trap* for third harmonics. In an ungrounded wye connection, the third harmonics of voltage add to each other and tend to push the neutral off the geometric neutral (see Fig. 4-21).

PROBLEMS

4-1 The core of a magnetic circuit is made in the form of a circular ring having a mean radius of 5 cm. A 200-turn coil is wound uniformly on this core. If the coil carries a 4-A current, determine (a) the mmf and (b) the magnetic field intensity in the core.

4-2 The relative permeability of the material of the ring of Problem 4-1 is 100, and the mean area of cross section of the ring is 16 cm^2. Calculate (a) the core flux density, (b) the core flux, (c) the reluctance, and (d) the permanence of the core.

4-3 For the ring of Problems 4-1 and 4-2, the following data for a $I\varphi$ relationship are obtained experimentally:

I(A)	0	2	4	6	8	10	12
φ(mWb)	0	0.256	0.512	0.720	0.980	1.20	1.380

where I is the coil current and φ is the core flux. Plot the BH curve for the core from the data above.

4-4 From the data of Problem 4-3, calculate the relative permeability of the core when (a) the coil current is 2 A and (b) the core flux is 1.0 mWb. (c) Determine the incremental permeability when the coil current changes from 8 A to 10 A.

4-5 For the magnetic circuit shown in Fig. 4P-5, assume that the iron portion is infinitely permeable. Coil N_1 carries 8 A of current. Calculate the flux densities in the gaps g_1, g_2, and g_3. Given: $g_1 = g_2 = 2$ mm, $g_3 = 4$ mm. $A_3 = 2A_1 = 2A_2 = 10$ cm^2, $N_1 = 60$ turns, $N_2 = 80$ turns, and $N_3 = 120$ turns. The coils N_2 and N_3 are unexcited.

4-6 Refer to the magnetic circuit of Fig. 4P-5. If all three coils are excited simul-

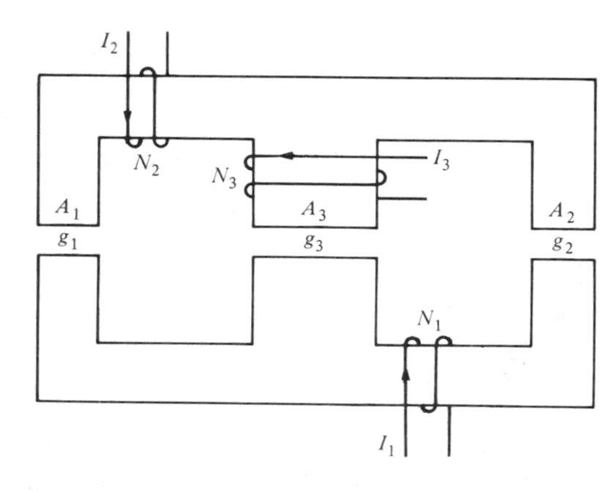

FIGURE 4P-5. PROBLEM 4-5.

taneously such that $I_1 = 8$ A, $I_2 = 6$ A, and $I_3 = 4$ A, with the directions of currents as shown, what is the flux density in gap g_3?

4-7 In Fig. 4P-5 gap g_3 is closed and only coil N_1 is excited with 8 A of current. Determine the flux densities in gaps g_1 and g_2. Next, if g_2 is also closed (other data remaining unchanged), what is the flux density in gap g_1?

4-8 Reconsider Fig. 4P-5. Using the data of Problem 4-5, evaluate the self- and mutual inductances of the three coils.

4-9 The eddy-current loss of a given solid core is 80 W at 50 Hz. What would be the eddy-current loss at 60 Hz if the core flux density remained the same at the two frequencies?

4-10 A 220/110-V 60-Hz ideal transformer is rated at 5 kVA. Calculate (a) the turns ratio and (b) the primary and secondary currents at (1) full load and (2) a 3-kW load at 0.8 lagging power factor.

4-11 A 10-Ω resistance is connected across the secondary of a 220/110-V 60-Hz ideal transformer. The secondary voltage is 110 V. Calculate (a) the primary current and (b) the equivalent resistance referred to the primary. (c) Determine the power dissipated at this equivalent resistance if the primary current calculated in part (a) flows through this resistance.

4-12 The secondary winding of a 60-Hz transformer has 50 turns. If the induced voltage in this winding is 220 V, determine the maximum value of the core flux.

4-13 On no-load a 60-Hz transformer having a 300-turn primary takes 60 W of power and 0.8 A of current at an input voltage of 110 V. The resistance of the winding is 0.2 Ω. Calculate (a) the core loss; (b) the magnetizing reactance, X_m; and (c) the core-loss equivalent resistance, R_c.

4-14 Repeat the calculations of Example 4-7 using the circuit referred to the secondary.

4-15 Refer to the data of the transformer of Example 4-7. Using the equivalent circuit referred to the secondary, calculate the efficiency of the transformer at full-load and 0.8 leading power factor.

4-16 Verify that the circuit shown in Fig. 4-12 is a valid equivalent circuit of a transformer referred to its secondary side.

4-17 Two 1150/115-V transformers are given. Show the interconnections of these transformers for (a) 2300/115-V operation and (b) 1150/115-V operation. Indicate the polarity markings.

4-18 A 25-kVA 440/220-V 60-Hz transformer has a constant core loss of 710 W. The equivalent resistance and reactance referred to the primary are 0.16 Ω and 0.66 Ω, respectively. Plot the following curves: (a) core loss versus primary current; (b) I^2R loss versus primary current; and (c) efficiency versus primary current. The primary current ranges from 0 to 100 A. What is the I^2R loss when the efficiency of the transformer is maximum? (*Note:* This is a special case of the general rule that the efficiency of a transformer is maximum when the I^2R losses in the two windings equal the core losses.)

4-19 A 75-kVA transformer has an iron loss of 1 kW and a full-load copper loss of 1 kW. Calculate the transformer efficiency at unity power factor at (a) full-load and (b) 50 percent of full-load.

4-20 The transformer of Problem 4-10 operates on full-load at unity power factor for 8 h, on no-load for 8 h, and on one-half load at unity power factor for 8 h during one day. Determine the all-day efficiency.

4-21 The maximum efficiency of a 100-kVA transformer is 98.4 percent and occurs at 90 percent full-load. Calculate the efficiency of the transformer at unity power factor at (a) full-load and (b) 50 percent of full-load.

4-22 The results of open-circuit and short-circuit tests on a 10-kVA 440/220-V 60-Hz transformer are as follows:

Open circuit test: high-voltage side open: $V_o = 220$ V, $I_o = 1.2$ A, and $W_o = 150$ W.

Short-circuit test: low-voltage side short-circuited: $V_s = 20.5$ V, $I_s = 42$ A, and $W_s = 140$ W.

Determine the parameters of the approximate equivalent circuit from these data.

4-23 Calculate the primary voltage of the transformer of Problem 4-22 when the secondary is supplying full-load at 0.8 lagging power factor. What is the efficiency of the transformer at this load?

4-24 A 1000-kVA transformer has a 94 percent efficiency at full-load and at 50 percent full-load. The power factor is unity in both cases. (a) Segregate the losses. (b) For unity power factor and 75 percent full-load, determine the efficiency of the transformer.

4-25 The primary and secondary voltages of an autotransformer are 440 V and 360 V, respectively. If the secondary current is 80 A, determine the primary current.

4-26 A 5-kVA 220/220V 60-Hz two-winding transformer has a full-load efficiency of 96 percent at unity power factor. The iron loss at 60 Hz is 60 W. This transformer is next connected as an autotransformer and delivers a 5-kVA unity power factor load at 220 V. The primary of the autotransformer is connected to a 440-V source. Calculate (a) the primary current and (b) the efficiency of the autotransformer.

4-27 A 100-kVA 42,000/2400-V 60-Hz transformer is operated in parallel with a 75-kVA 42,000/2400-V 60-Hz transformer. The respective impedances referred to the secondary are $(0.5 + j4)$ ohms and $(0.8 + j6)$ ohms. The total load on the transformer is 120 kVA at 0.8 lagging power factor. Calculate the load on each transformer.

4-28 Calculate the primary and secondary currents for the two transformers of Problem 4-27.

4-29 Three single-phase transformers are connected to make a three-phase delta/wye bank. The line voltages of the bank are 220 V and 400 V on the primary and secondary, respectively. The load on the transformer is 15 kW, 0.8 power factor lagging, wye connected. Determine the primary and secondary line and phase currents.

4-30 A 100-kVA 2200/220-V wye/wye three-phase 60-Hz transformer bank has an iron loss of 1500 W and a copper loss of 2700 W at full-load. Determine the efficiency of the transformer for 75 percent full-load at 0.8 power factor.

4-31 Two transformers, each rated at 50 kVA, are connected open-delta to form a three-phase system. What total kVA can be obtained from the open-delta bank without overloading the transformers?

4-32 Draw the voltage phasor diagrams of wye/wye- and delta/wye-connected transformers. Verify that the two sets of secondary voltages are 30° apart. Under these conditions, should these two banks be operated in parallel?

4-33 A three-phase 75-kVA 440-V 0.8 lagging power factor load is to be supplied by single-phase transformers having a turns ratio of 2. Calculate the currents in the windings of each transformer and the corresponding power factor if the transformers are connected in (a) open-delta and (b) delta/delta.

4-34 Two Scott-connected transformers are supplied from a 400-V three-phase source. The two-phase secondary supplies a 100-kVA load at 220 V. Determine the currents in the windings of the two transformers.

4-35 A current transformer (CT) and a potential transformer (PT) are connected to the current and voltage coils, respectively, of a wattmeter measuring power going into a load. The wattmeter reading is 240 W, and the turns ratios of the CT and the PT and 5 (primary): 100 (secondary) and 10:1, respectively. What is the actual power supplied to the load?

CHAPTER 5

DC Machines

5-1 INTRODUCTION

In Chapter 4 we studied the transformer. In a sense the transformer is an *energy transfer* device—energy is transferred from the primary to the secondary. During the energy transfer process, the form of energy remains unchanged; that is, the energy at both the input (primary) and output (secondary) terminals is electrical. In contrast, in a rotating electric machine, electrical energy is converted into mechanical form, and vice versa, depending on the mode of operation of the machine. An *electric motor* converts electrical energy into mechanical energy, whereas an *electric generator* is used to convert mechanical energy into electrical energy. For this reason, electric machines are also called *electromechanical energy converters*. In essence, the process of electromechanical energy conversion can be expressed as

$$\text{electrical energy} \quad \underset{\text{generator}}{\overset{\text{motor}}{\rightleftharpoons}} \quad \text{mechanical energy}$$

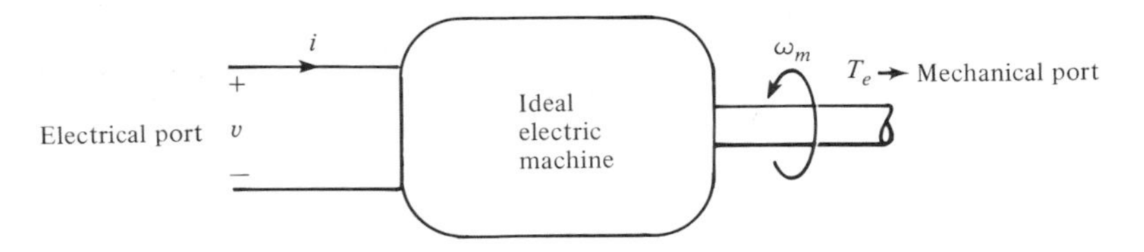

FIGURE 5-1. A general representation of an electric machine.

Schematically, an ideal electric machine may be represented by Fig. 5-1, for which we have, over a certain time interval Δt,

$$(vi)\,\Delta t = (T_e \omega_m)\,\Delta t$$

or (5-1)

$$vi = T_e \omega_m$$

where v and i are, respectively, the voltage and current at the electrical port, and T_e and ω_m are, respectively, the torque (in newton-meters, N-m) and angular rotational velocity (in radians per second, rad/s) at the mechanical port. We wish to reiterate that (5-1) is valid for an ideal machine in that the machine is lossless. (We will have more to say later about losses in machines.)

The remainder of the book will be concerned with electric machines, beginning with dc machines. The dc machine is the first machine devised for electromechanical energy conversion. The simplest form of the dc machine is the Faraday disk, which is discussed in the next section.

5-2 THE FARADAY DISK AND FARADAY'S LAW

Based on the principle of electromagnetic induction, in 1832 Faraday demonstrated that if a copper disk is rotated in an axially directed magnetic field (produced by a permanent magnet), with sliding contacts, or brushes, mounted at the rim and at the center of the disk, a voltage will be available at the brushes. Such a machine is shown in Fig. 5-2 and is commonly known as the Faraday disk or *homopolar* generator. Another name for the homopolar machine is the *acyclic* machine. The device shown in Fig. 5-2 is also capable of operating as a motor. It is a homopolar machine because the conductors (which may be imagined in form of spokes of the disk) are under the influence of the magnetic field of one polarity at all times. In contrast, in *heteropolar* machines the moving conductors are under the influence of magnetic fields of opposite polarities in an alternating fashion, as we shall soon see.

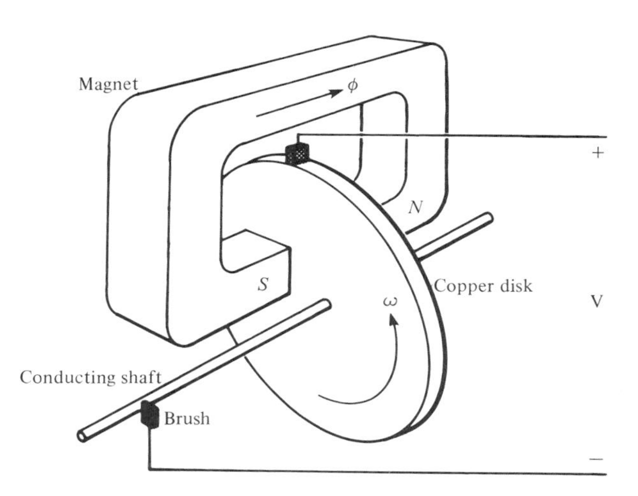

FIGURE 5-2. Faraday disk or homopolar machine.

As mentioned earlier, the operation of the homopolar generator is based on Faraday's law of electromagnetic induction. The law states that an emf is induced in a circuit placed in a magnetic field if either (1) the magnetic flux linking the circuit is time varying, or (2) there is a relative motion between the circuit and the magnetic field such that the conductors comprising the circuit cut across the magnetic flux lines. The first form of the law, stated as (1), is the basis of operation of transformers (see Chapter 4). The second form, stated as (2), is the basic principle of operation of electric generators, including the homopolar generator.

Consider a conductor of length l located at right angles to a uniform magnetic field of flux density B as shown in Fig. 5-3. Let the conductor be connected with fixed external connections to form a closed circuit. These external connections are shown in the form of conducting rails in Fig. 5-3. The conductor can slide on the rails, and thereby the flux linking the circuit changes. Hence, according to Faraday's law, an emf will be induced in the circuit, and a voltage, v, will be measured by the voltmeter. If the conductor moves with a velocity u (m/s) in a direction at right angles to B and l both, the area swept by the conductor in 1 second is lu. The flux in this area is Blu, which is also the flux linkage (since in effect, we have a single-turn coil formed by the conductor, the rails, and the voltmeter). In other words, flux linkage per unit time is Blu, which is, then, the induced emf, e. We write this in the equation form as

$$e = Blu \tag{5-2}$$

This form of Faraday's law is also known as the *flux-cutting rule*. Stated in words, an emf, e, as given by (5-2), is induced in a conductor of length l if it "cuts" magnetic flux lines of density B by moving at right angles to B at a

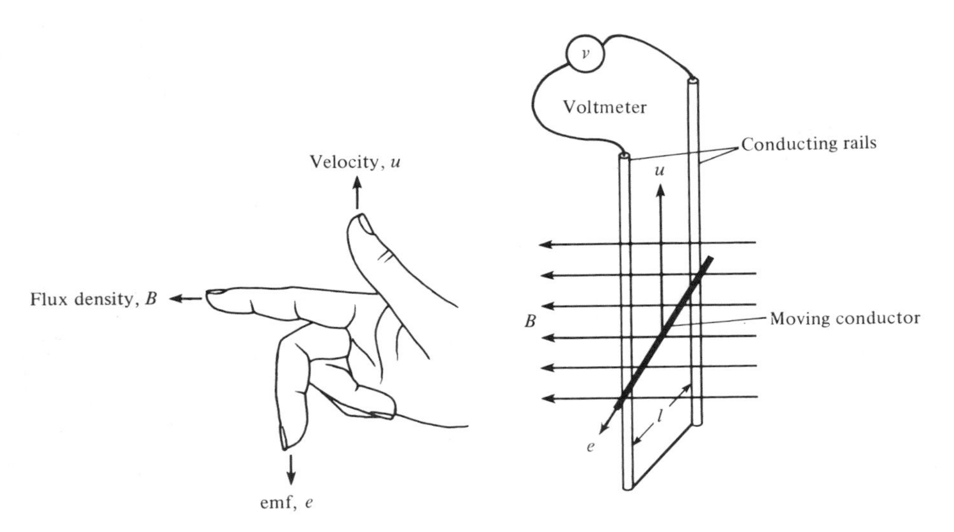

FIGURE 5-3. The right-hand rule and generator action.

velocity u (at right angles to l). The mutual relationships among e, B, l, and u are given by the *right-hand rule*, as shown in Fig. 5-3. Compare with the left-hand rule of Fig. 3-2.

EXAMPLE 5-1

A conducting disk of 0.5 m radius rotates at 1200 r/min in a uniform magnetic field of 0.4 T. Calculate the voltage available between the rim and the center of the disk.

SOLUTION

Referring to (5-2), in this problem we have $B = 0.4$ T, $l = 0.5$ m, and $u = r\omega$, where $\omega = 2\pi n/60$ rad/s is the angular velocity of the disk. Substituting $n = 1200$ r/min and $r = 0.5$ m, we obtain the linear velocity of the rim as

$$u_{\text{rim}} = 0.5 \times 2\pi \times \frac{1200}{60} = 20\pi \text{ m/s}$$

But the linear velocity of the center of the disk is zero. Hence the average velocity of the disk (or a radial conductor) is given by

$$u = \tfrac{1}{2}(20\pi + 0) = 10\pi \text{ m/s}$$

and, from (5-2), we have

$$e = 0.4 \times 0.5 \times 10\pi = 6.28 \text{ V}$$

■

Let us now summarize the salient points of this section: (1) The Faraday disk operates on the principle of electromagnetic induction, enunciated by Faraday. (2) There are two forms of expression of Faraday's law: time rate of change of flux linkage (with a circuit), or flux cut (by a conductor) results in an induced emf, but the latter is contained in the former. (3) Electric generators operate on the flux cutting principle. (4) Example 5-1 shows that rather a small voltage (12.76 V) is induced in a disk of large diameter (1 m) rotating at a reasonably high speed (1200 r/min). Thus the Faraday disk in its primitive form is not a practical type of dc generator. Indeed, present-day dc generators have little resemblance to the Faraday disk. However, the primitive Faraday disk generator has been developed into present-day homopolar, or acyclic, generators, which find applications requiring very large currents at very low voltages. A typical example is that of the chlorine-cell line in the chemical industry. Such electrochemical loads require a few thousand amperes of current per volt. Another example of application of the homopolar generator is in an aluminum-pot line requiring a current of over 150,000 A at 400 V.

5-3 HETEROPOLAR OR CONVENTIONAL DC MACHINE

Consider an N-turn coil, rotating at a constant angular velocity ω in a uniform magnetic field of flux density B, as shown in Fig. 5-4(a). Let l be the axial length of the coil and r be its radius. The emf induced in the coil can be found by application of (5-2). However, we must be careful in determining u in (5-2). Recall from the preceding section that u is the velocity at right angles to B. In the system under consideration, we have a rotating coil. Thus u in (5-2) corresponds to that component of velocity which is at right angles to B. We illustrate the components of velocities in Fig. 5-4(b), where the tangential velocity $u_t = r\omega$ has been resolved into a component u_1, along the direction of the flux, and another component u_2, across (or perpendicular to) the flux. The latter component "cuts" the magnetic flux, and is the component responsible for the emf induced in the coil. The u in (5-2) should then be replaced by u_2. The next term in (5-2) that needs careful consideration is l_1, the effective length of the conductor. For the N-turn coil of Fig. 5-4(a), we effectively have $2N$ conductors in series (since each coil side has N conductors and there are two coil sides). If l_1 is the length of each conductor, the total effective length of the N-turn coil is $2Nl_1$, which should be substituted for l in (5-2). The form of (5-2) for the N-turn coil then becomes

$$e = B(2Nl_1)u_2 \tag{5-3}$$

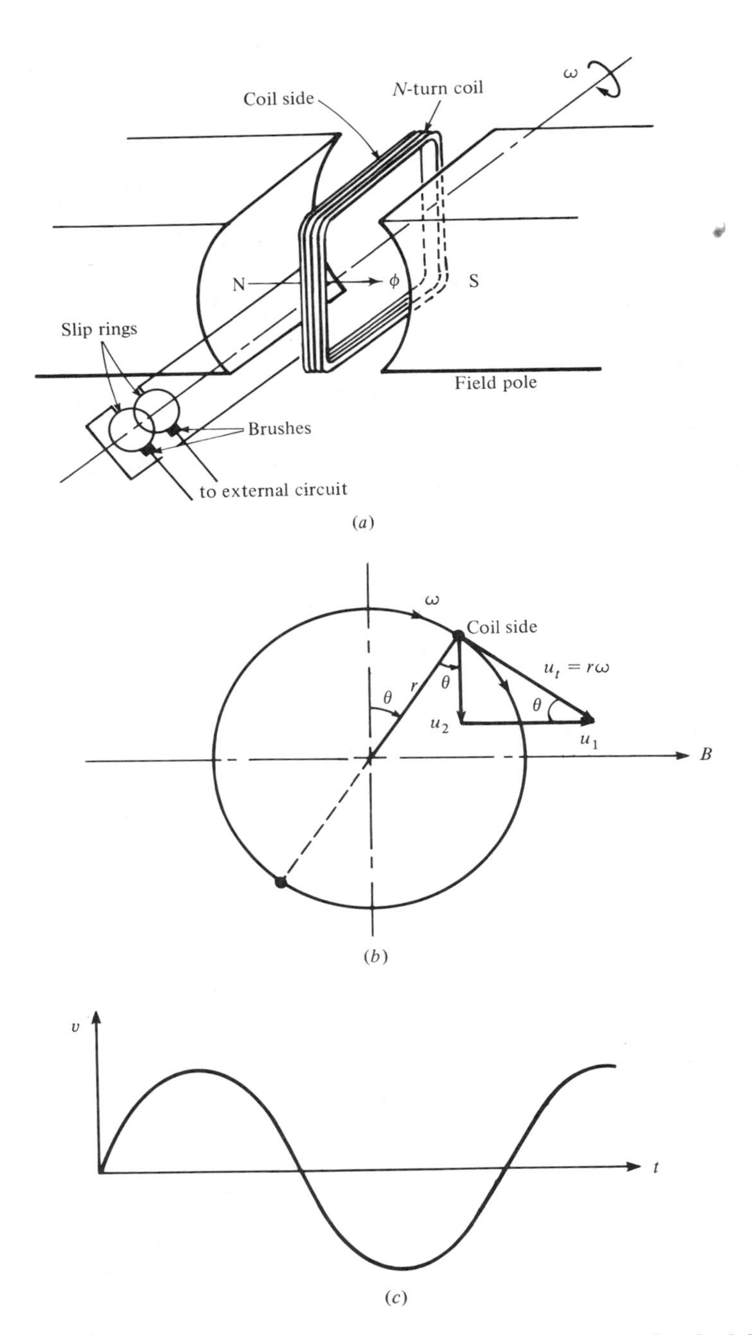

FIGURE 5-4. (a) An elementary generator; (b) resolution of velocities into "parallel" and "perpendicular" components; (c) voltage wave form at the brushes.

From Fig. 5-4(b), we have

$$u_2 = u_t \sin \theta = r\omega \sin \theta \tag{5-4}$$

If the coil rotates at a constant angular velocity ω, then

$$\theta = \omega t \tag{5-5}$$

Consequently, (5-3) to (5-5) yield

$$e = 2BNl_1 r\omega \sin \omega t = E_m \sin \omega t \tag{5-6}$$

where $E_m = 2BNl_1 r\omega$. A plot of (5-6) is shown in Fig. 5-4(c). The conclusion is that a sinusoidally varying voltage will be available at the slip rings, or brushes, of the rotating coil shown in Fig. 5-4(a). The brushes reverse polarities periodically.

In order to obtain a unidirectional voltage at the coil terminals, we replace the slip rings of Fig. 5-4(a) by the *commutator segments* shown in Fig. 5-5(a). It can be readily verified by applying the right-hand rule that in the arrangement shown in Fig. 5-5(a) the brushes will maintain their polarities regardless of the position of the coil. In other words, brush *a* will always be positive, and brush *b* will always be negative for the given relative polarities of the flux and the direction of rotation. Thus the arrangement shown in Fig. 5-5 forms an elementary *heteropolar* dc machine. The main characteristic of a heteropolar dc machine is that the emf induced in a conductor, or a current flowing through it, has its direction reversed as it passes from a north-pole to a south-pole region. This reversal process is known as *commutation,* and is accomplished by the commutator-brush mechanism, which also serves as a connection to the external circuit. The voltage available at the brushes will be of the form shown in Fig. 5-5(b). In this respect, commutation is the process of rectification of the induced alternating emf in the coil into a dc voltage at the terminals.

The natural question to ask at this point is: What have we gained, compared to the Faraday disk, by introducing the complications of a commutator? The answer to this question lies in the fact that a full range of ratings can be realized from a heteropolar cylindrical configuration. Recall from Example 5-1 that we could get only 6.28 V at the terminals of a homopolar machine. Obtaining much higher voltages is no problem with a heteropolar machine, as we shall presently see. As a matter of convention, we will term the dc heteropolar commutator machine simply a dc machine (unless otherwise stated).

Before leaving the subject of the homopolar and the elementary heteropolar machines, let us recall (5-1), according to which these machines must also be able to operate (at least in principle) as motors. The operation of the homopolar

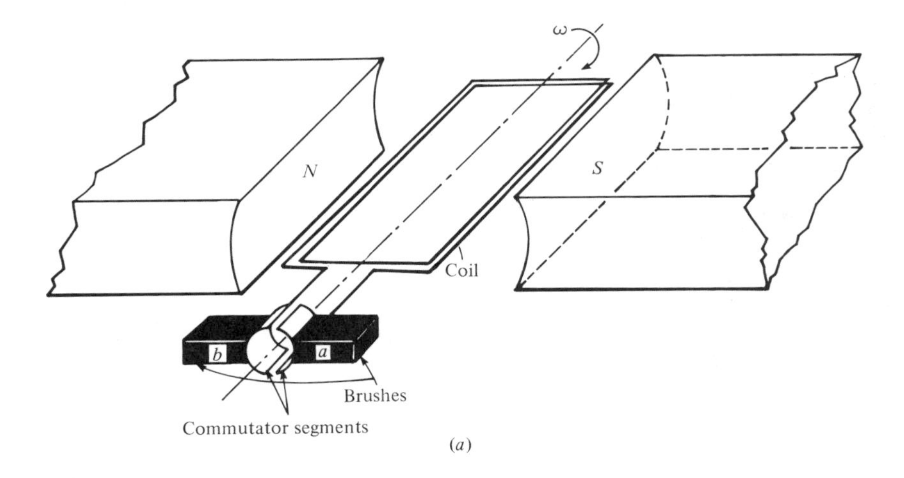

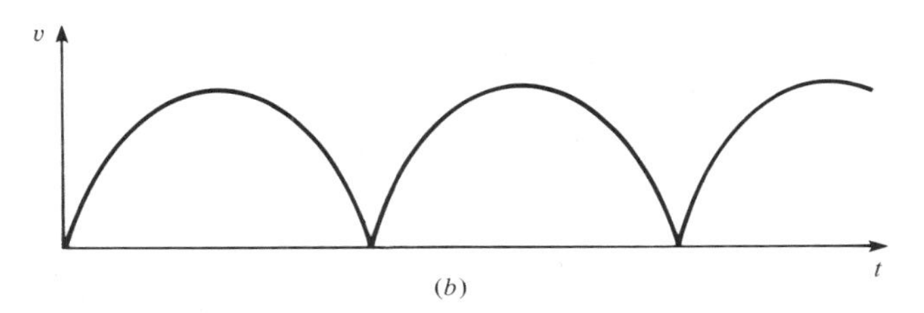

FIGURE 5-5. (a) An elementary dc generator; (b) output voltage at the brushes.

machine as a dc motor is left as an exercise for the reader. For the heteropolar machine, the production of a unidirectional torque, and hence operation as a dc motor, can be verified by referring to Fig. 5-6. Conductors a and b are the two sides of a coil. Thus a and b are connected in series. Notice from Fig. 5-6(a) that the current enters into conductor a through brush a and leaves conductor b

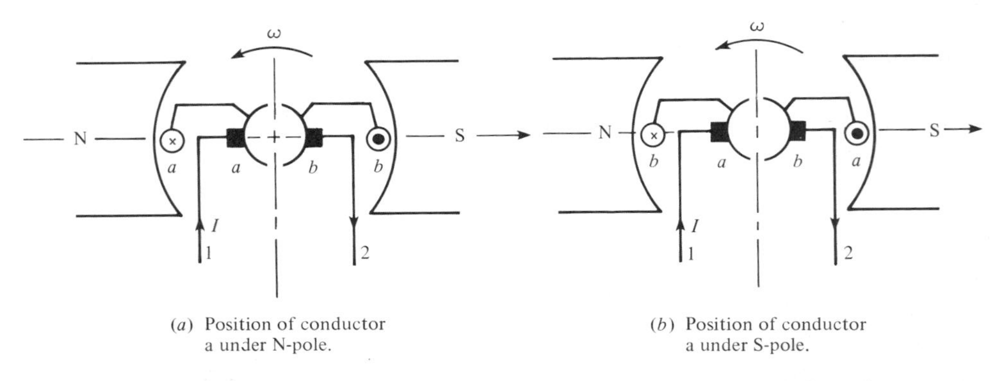

FIGURE 5-6. Production of a unidirectional torque and operation of an elementary dc motor.

through brush b. Applying the left-hand rule, the conductors will experience a force to produce a counterclockwise rotation. After half a rotation, the conductors interchange their respective position, as shown in Fig. 5-6(b). Now the current in conductor b is again in such a direction that the force on the coil will tend to rotate it in a counterclockwise direction. In other words, the torque is unidirectional and is independent of conductor position.

5-4 CONSTRUCTIONAL DETAILS

Before we consider some conventional dc machines, it is best that we study briefly the constructional features of their various parts and discuss the usefulness of these parts. We observe from section 5-3 that the basic elements of a dc machine are the rotating coil, a means for the production of flux, and the commutator-brush arrangement. In a practical dc machine the coil is replaced by the *armature winding* of the dc machine mounted on a cylindrical magnetic structure. The flux is provided by the *field winding* wound on field poles. Generally, the armature winding is placed on the rotating member—the *rotor*—and the field winding is on the stationary member—the *stator*. A common large dc machine is shown in Fig. 5-7, which, together with the schematic of Fig. 5-8, shows most of the important parts of a dc machine. The *field poles,* mounted on the

FIGURE 5-7. A dc machine. (Courtesy of General Electric Company)

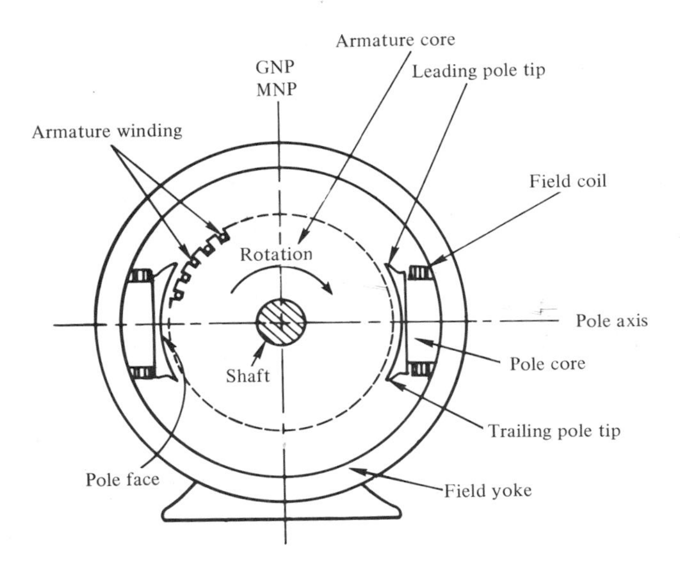

FIGURE 5-8. Parts of a dc machine.

stator, carry the field windings. Some machines carry more than one independent field winding on the same core. The cores of the poles are built of sheet-steel laminations. Because the field windings carry direct current, it is not necessary to have the pole cores laminated. It is, however, necessary for the pole faces to be laminated because of their proximity to the armature windings. (Use of laminations for the cores as well as for the pole faces facilitates assembly.) The rotor or the armature core, which carries the rotor or armature windings, is generally made of sheet-steel laminations. These laminations are stacked together to form a cylindrical structure. On its outer periphery the armature (or rotor) has *slots* in which the armature coils that make up the armature winding are located. For mechanical support, protection from abrasion, and for greater electrical insulation, nonconducting slot liners are often wedged in between the coils and the slot walls. The areas of magnetic material between the slots are called *teeth*. A typical slot/tooth geometry for a large dc machine is shown in Fig. 5-9. The commutator is made of hard-drawn copper segments insulated from one another by mica. The details of the commutator assembly are given in Fig. 5-10. The armature windings are connected to the commutator segments over which the carbon brushes slide and serve as leads for electrical connection.

The armature winding may be a *lap winding* [Fig. 5-11(a)] or a wave winding [Fig. 5-11(b)], and the various coils forming the armature winding may be connected in a series–parallel combination. In practice, the armature winding is housed as two layers in the slots of the armature core. In large machines the coils are preformed in the shapes shown in Fig. 5-12, and are interconnected to form an armature winding. The coils span approximately a pole pitch, the distance between two consecutive poles.

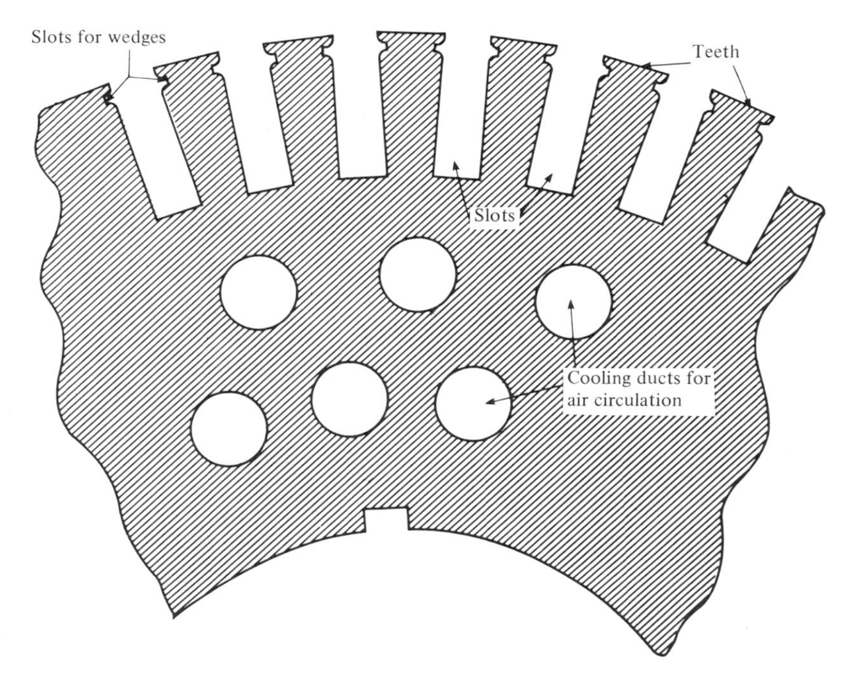

FIGURE 5-9. Portion of an armature lamination of a dc machine showing slots and teeth.

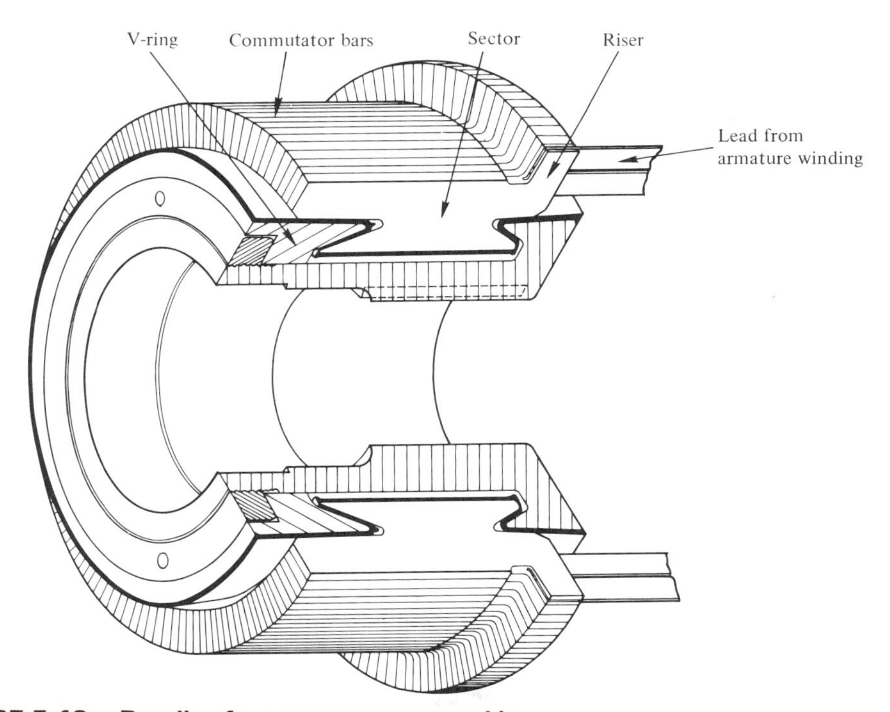

FIGURE 5-10. Details of a commutator assembly.

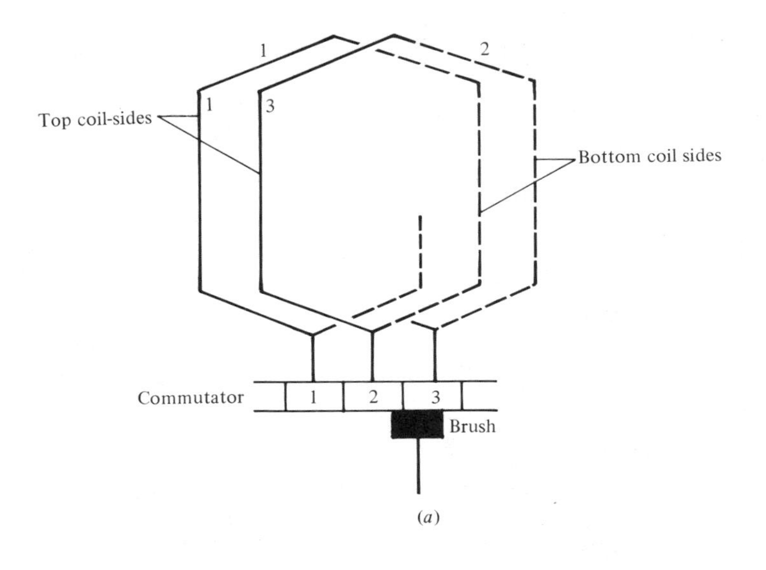

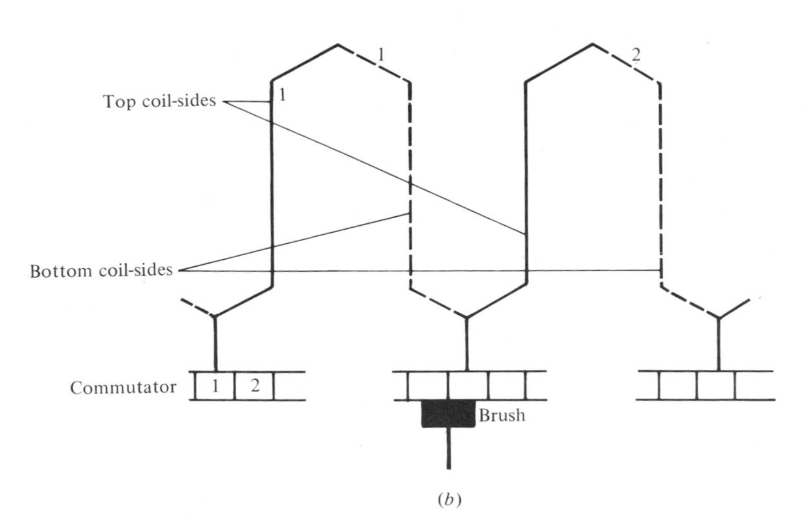

FIGURE 5-11. (a) Elements of lap winding; (b) elements of wave winding. Odd-numbered conductors are at the top and even-numbered conductors are at the bottom of the slots.

The winding layout of a double-layer lap-wound armature is shown in Fig. 5-13. By tracing the circuit, as given below the winding lay out, it is seen that the entire winding is divided into four parallel paths. The layout of a double-layer wave winding is shown in Fig. 5-14. Here we have shown a simplex winding. For duplex and multiplex windings, see Reference 7. In a simplex lap winding the number of paths in parallel a is equal to the number of poles p, whereas in a wave winding the number of parallel paths is always two. An

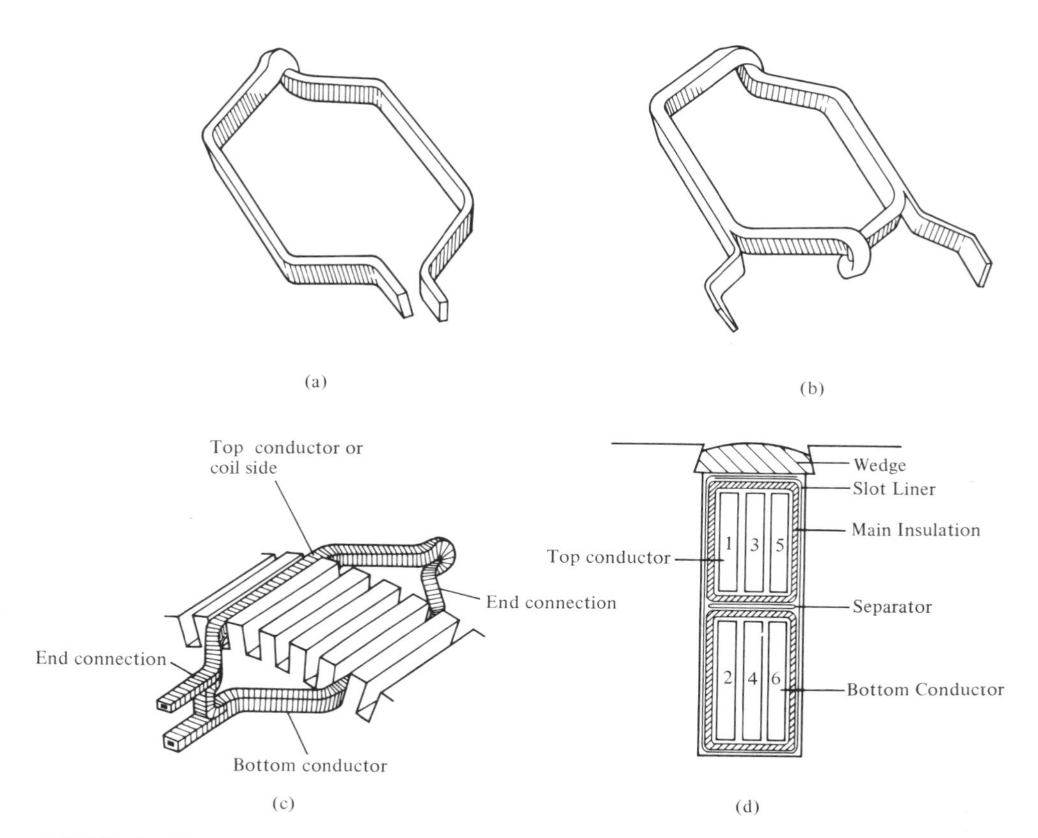

FIGURE 5-12. (a) Coil for a lap winding, made of a single bar; (b) multiturn coil for wave winding; (c) a coil in a slot; (d) slot details, showing several coils arranged in two layers.

assembled armature is shown in Fig. 5-15, and Fig. 5-16 shows an assembled field pole.

In addition to the armature and field windings, commutating poles and compensating windings are also found on large dc machines. These are shown in Fig. 5-17 and are used essentially to improve the performance of the machine, as we shall see later.

5-5 CLASSIFICATION ACCORDING TO FORMS OF EXCITATION

Conventional dc machines that have a set of field windings and armature windings can be classified, on the basis of mutual electrical connections between the field and armature windings, as follows:

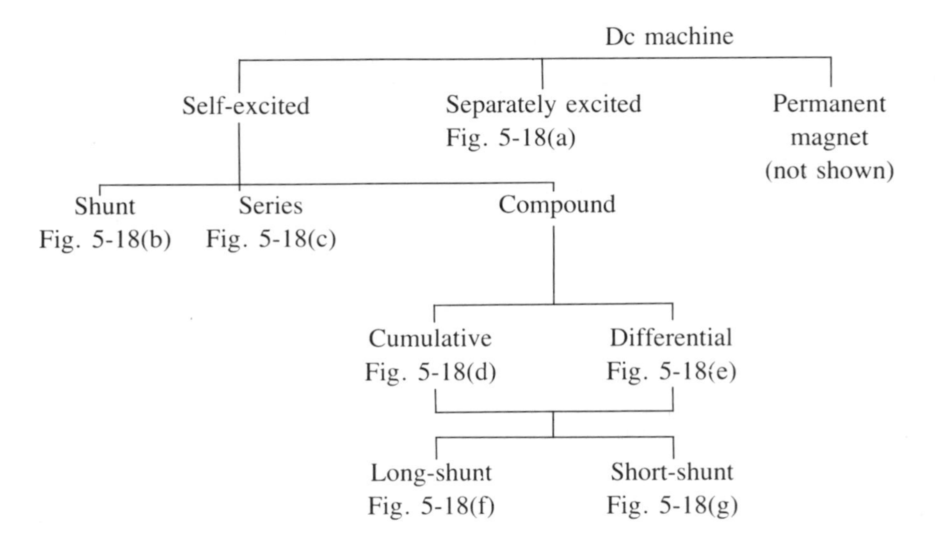

These interconnections of field and armature windings essentially determine the machine operating characteristics.

In a separately excited machine, shown in Fig. 5-18(a), there is no electrical interconnection between the field and the armature windings. On the other hand,

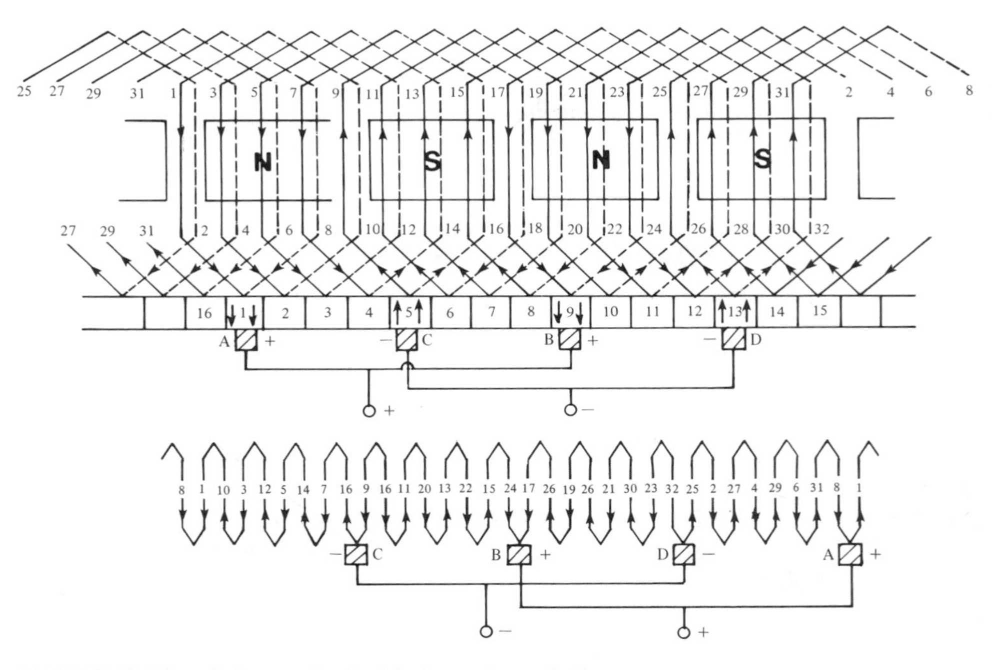

FIGURE 5-13. A four-pole double-layer lap winding.

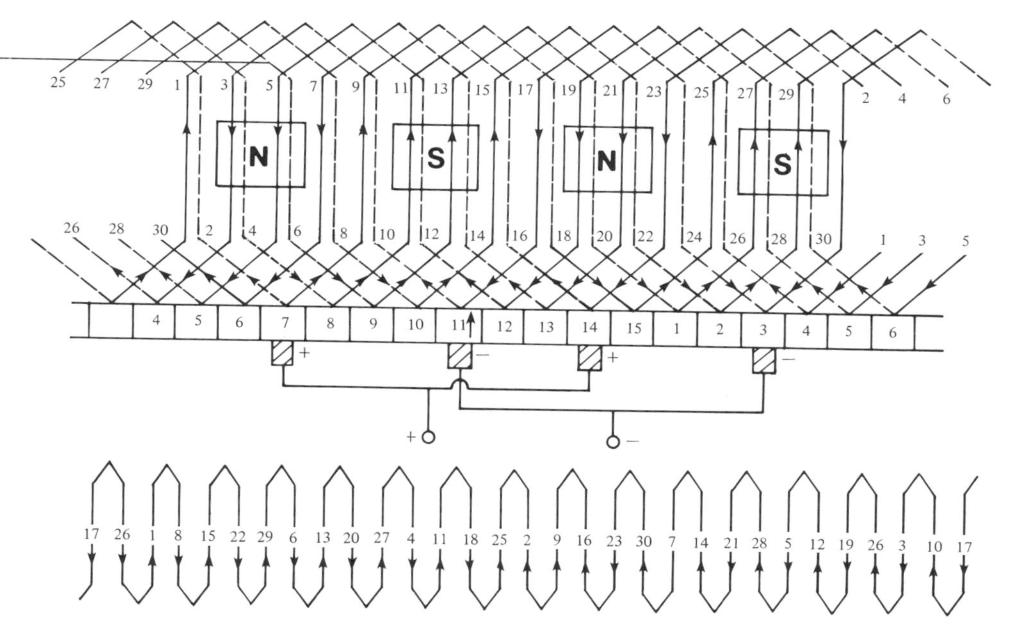

FIGURE 5-14. A four-pole double-layer wave winding.

FIGURE 5-15. Armature of a 2000-kW 450-rpm dc generator.

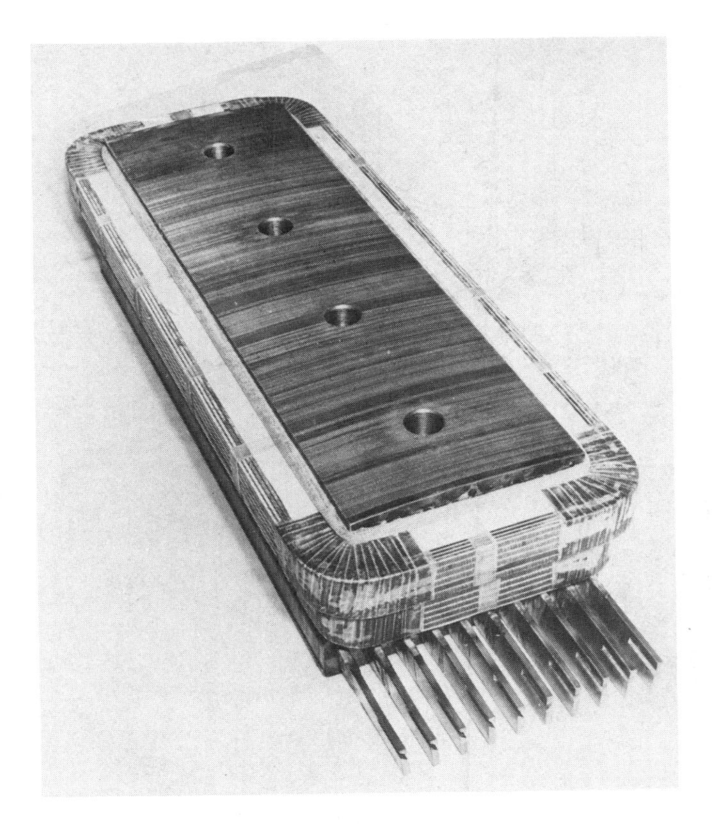

FIGURE 5-16. Field pole of a 2550-kW dc motor.

in a self-excited machine, the field winding is connected to the armature winding. A parallel connection between the field and armature windings results in the shunt machine of Fig. 5-18(b). The series machine has the field and the armature windings connected in series as in Fig. 5-18(c). A compound machine has both shunt and series field windings in addition to the armature winding. If the relative polarities of the shunt and series field windings are additive, as illustrated in Fig. 5-18(d), we obtain a cumulative compound machine. Notice from Fig. 5-18(d), that the two fields are shown to produce magnetic fluxes in the same direction. In a differential compound machine, the series field is in opposition to the shunt field, implying that the respective resulting fluxes are in opposition, as shown in Fig. 5-18(e). A differential or cumulative compound machine may have a long-shunt connection, in which case the shunt field is across the armature–series field combination, as given in Fig. 5-18(f). Figure 5-18(g) shows a short-shunt connection, where the shunt field is directly across the armature. Finally, in a permanent magnet machine, there is no field winding and the necessary magnetic flux is provided by the permanent magnet. Such dc machines are generally of fractional-horsepower rating.

FIGURE 5-17. Stator of a 1030-kW dc motor showing interpoles and compensating bars.

5-6 PERFORMANCE EQUATIONS

The three quantities of greatest interest in evaluating the performance of a dc machine are (1) the induced electromotive force (emf), (2) the electromagnetic torque developed by the machine, and (3) the speed corresponding to quantities 1 and/or 2. We will now derive the equations that enable us to determine these quantities.

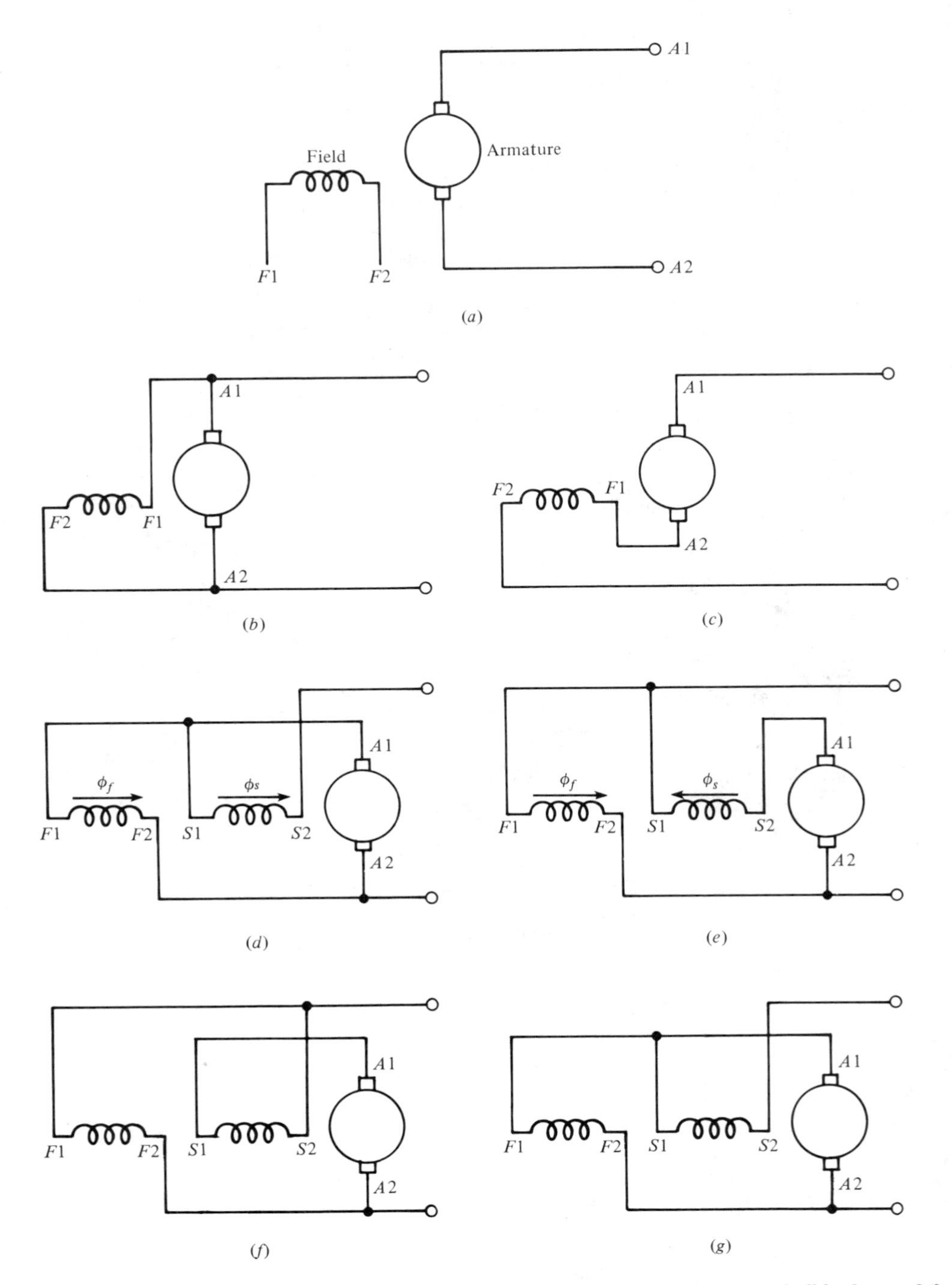

FIGURE 5-18. Classification of dc machines: (a) separately excited; (b) shunt; (c) series; (d) cumulative compound; (e) differential compound; (f) long-shunt; (g) short-shunt.

EMF Equation. The emf equation yields the emf induced in the armature of a dc machine. The derivation follows directly from Faraday's law of electromagnetic induction, according to which the emf induced in a moving conductor is the flux cut by the conductor per unit time.

Let us define the following symbols:

Z = number of active conductors on the armature

a = number of parallel paths in the armature winding

p = number of field poles

φ = flux per pole

n = speed of rotation of the armature, rpm

Then, with reference to Fig. 5-19,

$$\text{flux cut by one conductor in one rotation} = \varphi p$$

$$\text{flux cut by one conductor in } n \text{ rotations} = \varphi n p$$

$$\text{flux cut per second by one conductor} = \frac{\varphi n p}{60}$$

$$\text{number of conductors in series} = \frac{Z}{a}$$

$$\text{flux cut per second by } Z/a \text{ conductors} = \frac{\varphi n p}{60}\left(\frac{Z}{a}\right)$$

Hence

$$\text{emf induced in the armature winding}$$

$$= E = \left(\frac{\varphi n Z}{60}\right) \times \frac{p}{a} \quad \text{V} \qquad (5.7)$$

Equation (5-7) is known as the *emf equation* of a dc machine. In (5-7) we have separated p/a from the other terms because p and a are related to each other for the two types of windings.

EXAMPLE 5-2

Determine the voltage induced in the armature of a dc machine running at 1750 r/min and having four poles. The flux per pole is 25 mWb, and the armature is lap-wound with 728 conductors.

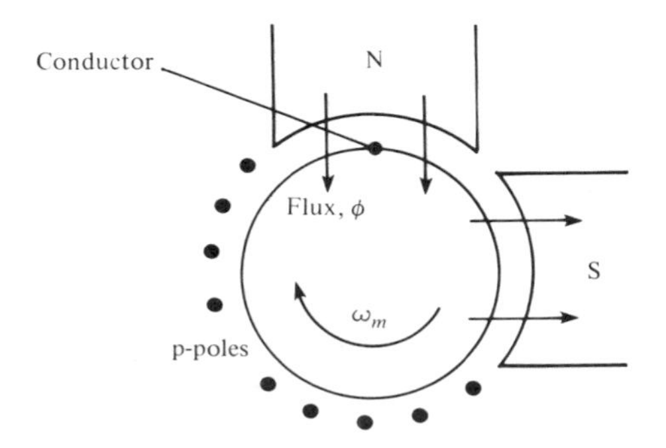

FIGURE 5-19. A conductor rotating at a speed of ω_m in the field of *p*-poles.

SOLUTION

Since the armature is lap-wound, $p = a$, and (5-7) becomes

$$E = \frac{\varphi n Z}{60} = \frac{25 \times 10^{-3} \times 1750 \times 728}{60} = 530.8 \text{ V}$$ ■

Torque Equation. The mechanism of torque production in a dc machine has been considered earlier (see Fig. 5-6), and the electromagnetic torque developed by the armature can be evaluated either from the *Bll* rule (see Section 5.1) or from (5-1). Here we choose the latter approach. Regardless of the approach, it is clear that for torque production we must have a current through the armature, as this current interacts with the flux produced by the field winding. Let I_a be the armature current and E the voltage induced in the armature. Thus the power at the armature electrical port (see Fig. 5-1) is EI_a. Assuming that this entire electrical power is transformed into mechanical form, we rewrite (5-1) as

$$EI_a = T_e \omega_m \tag{5-8}$$

where T_e is the electromagnetic torque developed by the armature and ω_m is its angular velocity in rad/s. The speed n (in r/min) and ω_m (in rad/s) are related by

$$\omega_m = \frac{2\pi n}{60} \tag{5-9}$$

Hence from (5-7)–(5-9) we obtain

$$\frac{\omega_m}{2\pi} \varphi Z \frac{p}{a} I_a = T_e \omega_m$$

which simplifies to

$$T_e = \frac{Zp}{2\pi a} \varphi I_a \tag{5-10}$$

which is known as the *torque equation*. An application of (5-10) is illustrated by the next example.

EXAMPLE 5-3

A lap-wound armature has 576 conductors and carries an armature current of 123.5 A. If the flux per pole is 20 mWb, calculate the electromagnetic torque developed by the armature.

SOLUTION

For lap winding, we have $p = a$. Substituting this and other given numerical values in (5-10) yields

$$T_e = \frac{576}{2\pi} \times 0.02 \times 123.5 = 226.4 \text{ N-m}$$

■

EXAMPLE 5-4

If the armature of Example 5-3 rotates at an angular velocity of 150 rad/s, what is the induced emf in the armature?

SOLUTION

To solve this problem we use (5-8) rather than (5-7). From Example 5-3, we have $I_a = 123.5$ A and $T_e = 226.4$ N-m. Hence (5-8) gives

$$E \times 123.5 = 226.4 \times 150$$

or

$$E = 275 \text{ V}$$

■

Speed Equation and Back EMF. The emf and torque equations discussed above indicate that the armature of a dc machine, whether operating as a generator or as a motor, will have an emf induced while rotating in a magnetic field, and will develop a torque if the armature carries a current. In a generator, the induced emf is the internal voltage available from the generator. When the generator supplies a load, the armature carries a current and develops a torque. This torque opposes the torque of the prime mover (such as a diesel engine).

In motor operation, the developed torque of the armature supplies the load connected to the shaft of the motor and the emf induced in the armature is termed the *back emf*. This emf opposes the terminal voltage of the motor.

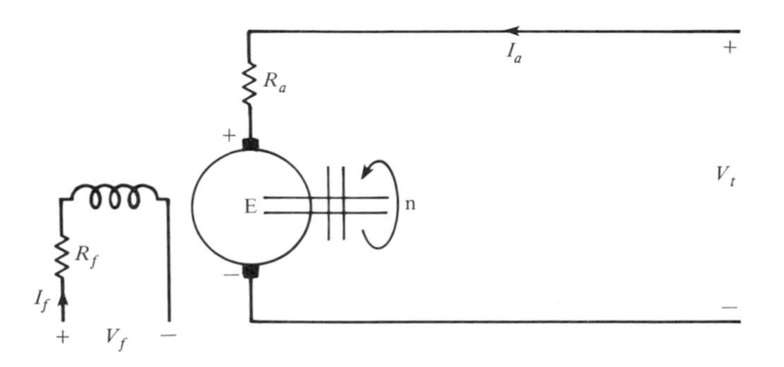

FIGURE 5-20. Equivalent circuit of a separately-excited motor.

Referring to Fig. 5-20, which shows the equivalent circuit of a separately excited dc motor running at speed n while taking an armature current I_a at a voltage V_t. From this circuit we have

$$V_t = E + I_a R_a \tag{5-11}$$

Substituting (5-7) in (5-11), putting $k_1 = Zp/60a$, and solving for n, we get

$$n = \frac{V_t - I_a R_a}{k_1 \varphi} \tag{5-12}$$

This equation is known as the *speed equation,* as it contains all the factors that affect the speed of a motor. In a later section we shall consider the influence of these factors on the speed of dc motors. For the present, let us focus our attention on k_1 and φ. The term k_1 replacing $Zp/60a$ is a design constant in the sense that once the machine has been built, Z, p, and a cannot be altered. The magnetic flux φ is controlled primarily by the field current I_f (Fig. 5-20). If the magnetic circuit is unsaturated, φ is directly proportional to I_f. Thus we may write

$$\varphi = k_f I_f \tag{5-13}$$

where k_f is a constant. We may combine (5-12) and (5-13) to obtain

$$n = \frac{V_t - I_a R_a}{k I_f} \tag{5.14}$$

where $k = k_1 k_f =$ a constant. This form of the speed equation is more meaningful because all the quantities in (5-14) can be conveniently measured. [In contrast, in (5-12), it is very difficult to measure φ.]

EXAMPLE 5-5

A 250-V shunt motor has an armature resistance of 0.25 Ω and a field resistance of 125 Ω. At no-load the motor takes a line current of 5.0 A while running at 1200 rpm. If the line current at full-load is 52.0 A, what is the full-load speed?

SOLUTION

The motor equivalent circuit is shown in Fig. 5-21. The field current, $I_f = 250/125 = 2.0$ A.

At no-load:

$$\text{armature current, } I_a = 5.0 - 2.0 = 3.0 \text{ A}$$

$$\text{back emf, } E_1 = V_t - I_a R_a = 250 - 3 \times 0.25 = 249.25 \text{ V}$$

$$\text{speed, } n_1 = 1200 \text{ r/min} \quad \text{(given)}$$

At full-load:

$$\text{armature current} = 52.0 - 2.0 = 50.0 \text{ A}$$

$$\text{back emf, } E_2 = 250 - 50 \times 0.25 = 237.5 \text{ V}$$

$$\text{speed, } n_2 = \text{unknown}$$

Now,

$$\frac{n_2}{n_1} = \frac{E_2}{E_1} = \frac{237.5}{249.25}$$

Hence

$$n_2 = \frac{237.5}{249.25} \times 1200 = 1143 \text{ r/min}$$

■

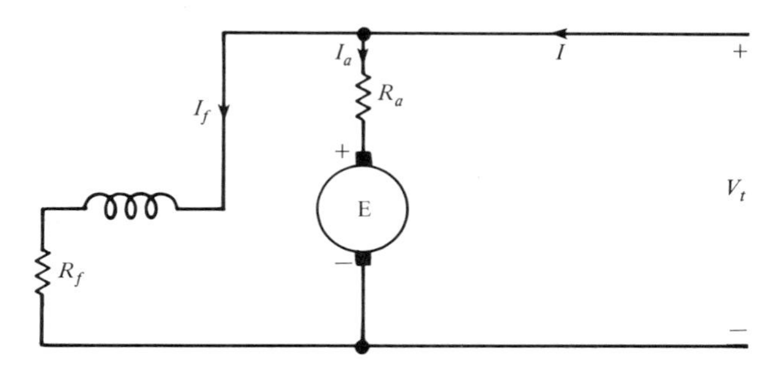

FIGURE 5-21. Equivalent circuit of a shunt motor.

ARMATURE REACTION

In the discussions so far we have assumed no interaction between the fluxes produced by the field windings and by the current-carrying armature windings. In reality, however, the situation is quite different. Consider the two-pole machine shown in Fig. 5-22(a). If the armature does not carry any current (that is, when the machine is on no-load), the air-gap field takes the form shown in Fig. 5-22(b). The geometric neutral plane and magnetic neutral plane (GNP and

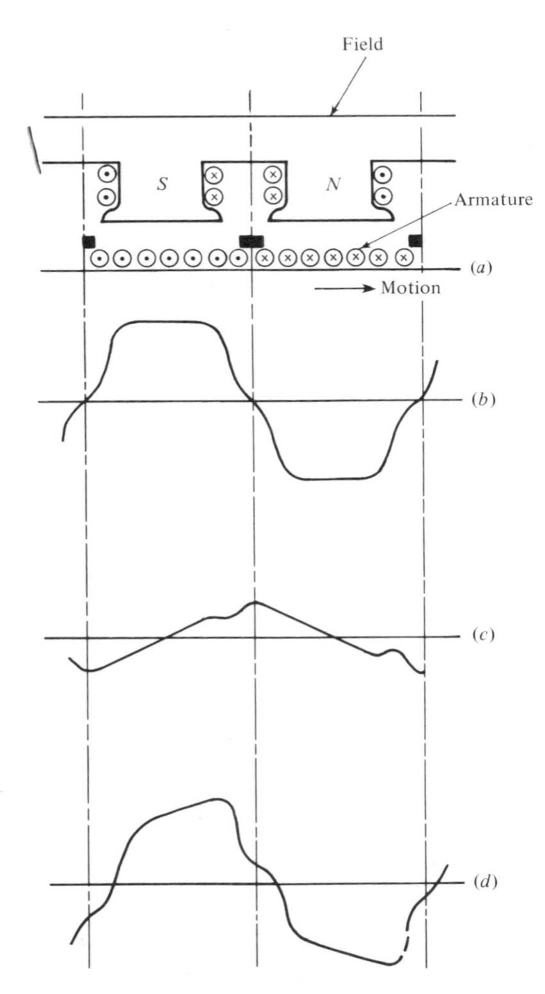

FIGURE 5-22. Airgap fields in a dc machine: (a) two-pole machine, showing armature and field mmf's; (b) flux-density distribution due to field mmf; (c) flux-density distribution due to armature mmf; (d) resultant flux-density distribution [curve (b) + curve (c)].

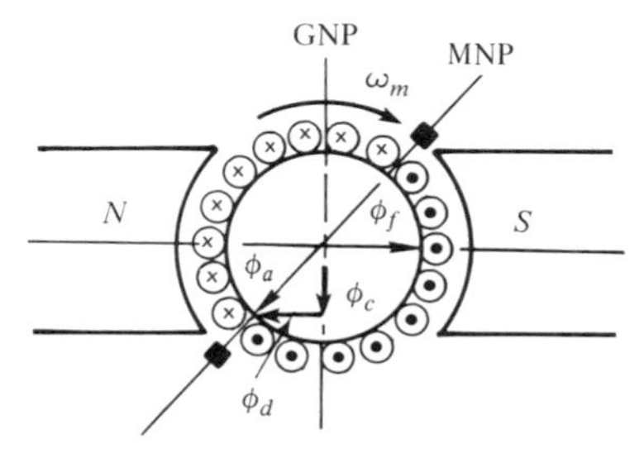

ϕ_a = Flux due to armature MMF
ϕ_c = Flux due to cross-magnetization
ϕ_d = Flux due to demagnetization
ϕ_f = Flux due to field MMF

(*a*)

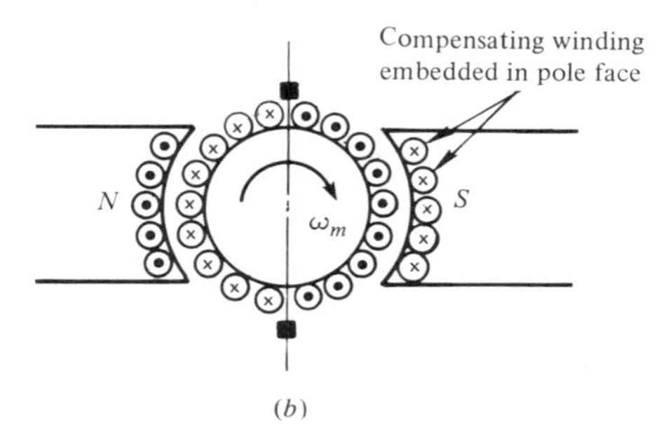

(*b*)

FIGURE 5-23. (a) Armature reaction resolved into cross and demagnetizing components; (b) neutralization of cross-magnetizing component by compensating winding.

MNP) respectively, are coincident. (*Note:* Magnetic flux lines enter the MNP at right angles.) Noting the polarities of the induced voltages in the conductors, we see that the brushes are located at the MNP for maximum voltage at the brushes. We now assume that the machine is on "load" and that the armature carries current. The direction of flow of current in the armature conductors depends on the location of the brushes. For the situation shown in Fig. 5-22(b), the direction of the current flow is the same as the direction of the induced voltages. In any event, the current-carrying armature conductors produce their own magnetic fields, as shown in Fig. 5-22(c), and the air-gap field is now the resultant of the fields due to the field and armature windings. The resultant air-gap field is thus distorted and takes the form shown in Fig. 5-22(d). The interaction of the fields due to the armature and field windings is known as *armature reaction*. As a consequence of armature reaction the air-gap field is distorted and the MNP is no longer coincident with the GNP. For maximum voltage at the terminals, the brushes have to be located at the MNP. Thus one undesirable effect of armature reaction is that the brushes must be shifted constantly, since the shift of the MNP from the GNP depends on the load (which is presumably always changing). The effect of armature reaction can be analyzed in terms of cross-magnetization and demagnetization, as shown in Fig. 5-23(a). We just mentioned the effect of cross-magnetization resulting in the distortion of the air-gap field and requiring the shifting of brushes according to the load on the machine. The effect of demagnetization is to weaken the air-gap field. All in all, therefore, armature reaction is not a desirable phenomenon in a machine.

The effect of cross-magnetization can be neutralized by means of compensating windings, as shown in Fig. 5-23. These are conductors embedded in pole faces, connected in series with the armature windings and carrying currents in an opposite direction to that flowing in the armature conductors under the pole face (Fig. 5-23). Once cross-magnetization has been neutralized, the MNP does not shift with load and remains coincident with the GNP at all loads. The effect of demagnetization can be compensated for by increasing the mmf on the main field poles. By neutralizing the net effect of armature reaction, we imply that there is no "coupling" between the armature and field windings.

5-8 REACTANCE VOLTAGE AND COMMUTATION

In discussing the action of the commutator in Section 5-3, we indicated that the direction of flow of current in a coil undergoing commutation reverses by the time the brushes move from one commutator segment to the other. This is represented schematically in Fig. 5-24. The flow of current in coil *a* for three different instants is shown. We have assumed that the current fed by a segment is proportional to the area of contact between the brush and the commutator segment. Thus, for satisfactory commutation, the direction of flow of current in coil *a* must completely reverse [Fig. 5-24(a) and (c)] by the time the brush moves from segment 2 to segment 3. The ideal situation is represented by the straight line in Fig. 5-25 and may be termed straight-line commutation. Because coil *a* has some inductance L, the change of current ΔI in a time Δt induces a voltage $L\,(\Delta I/\Delta t)$ in the coil. According to Lenz's law, the direction of this voltage, called *reactance voltage*, is opposite to the change (ΔI) that is causing it. As a result, the current in the coil does not completely reverse by the time the brush moves from one segment to the other. The balance of the "unreversed" current jumps over as a spark from the commutator to the brush, and thereby the commutator wears out because of pitting. This departure from ideal commutation is also shown in Fig. 5-25.

The directions of the (speed-) induced voltage, current flow, and reactance voltage are shown in Fig. 5-26(a). Note that the direction of the induced voltage depends on the direction of rotation of the armature conductors and on the direction of the air-gap flux. It is determined from the right-hand rule. Next,

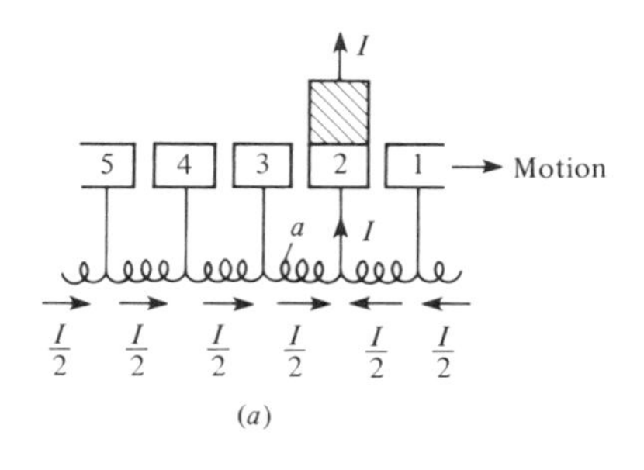

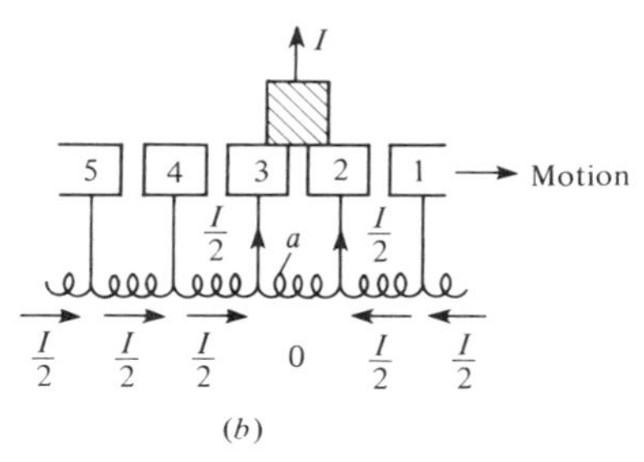

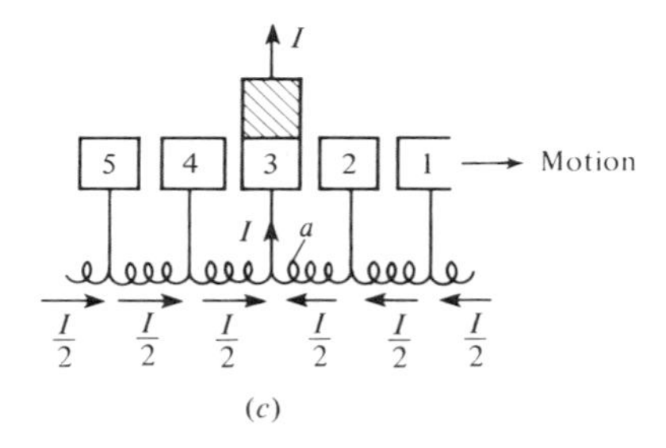

FIGURE 5-24. Coil *a* undergoing commutation.

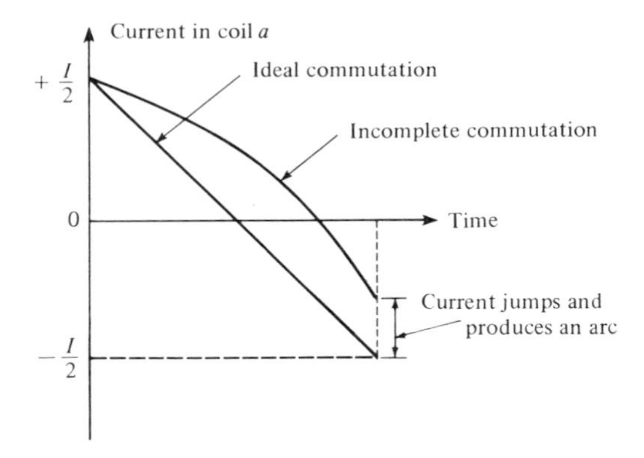

FIGURE 5-25. Commutation in coil *a*.

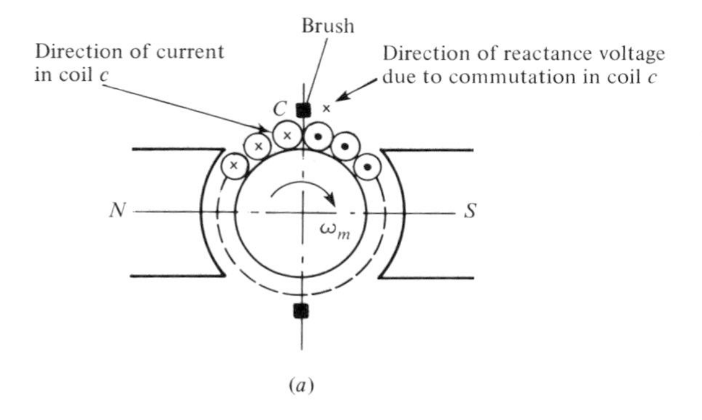

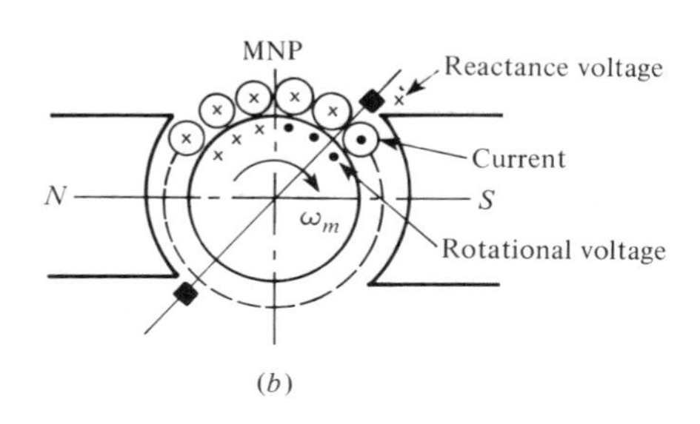

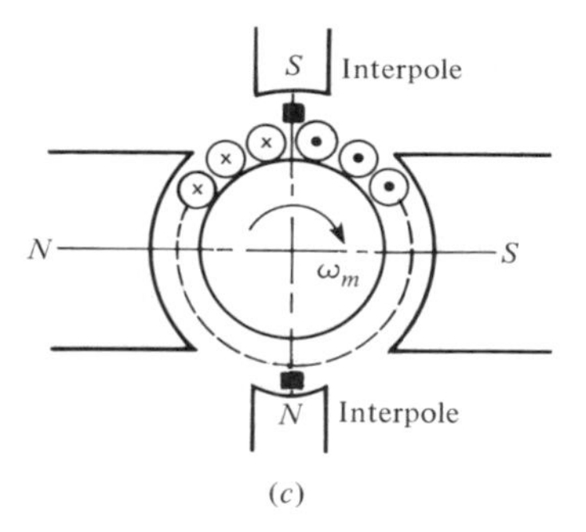

FIGURE 5-26. Reactance voltage and its neutralization: (a) reactance voltage and current in coil *c*, rotational voltage $\simeq$ 0; (b) reactance voltage, rotational voltage, and current in coil *c*; (c) interpoles.

the direction of the current flow depends on the location of the brushes (or tapping points). Finally, the direction of the reactance voltage depends on the change in the direction of the current flow and is determined from Lenz's law. For the brush position shown in Fig. 5-26(a), observe that the reactance voltage retards the current reversal. If the brushes are advanced in the direction of rotation (for generator operation), we may notice, from Fig. 5-26(b), that the reactance voltage is in the same direction as the (speed-) induced voltage, and therefore the current reversal is not opposed. We may further observe that the coil undergoing commutation, being near the tip of the south pole, is under the influence of the field of a weak south pole. From this argument, we may conclude that commutation improves if we advance the brushes. But this is not a very practical solution. The same—perhaps better—results can be achieved if we keep the brushes at the GNP, or MNP, as in Fig. 5-26(a), but produce the "field of a weak south pole" by appropriately winding and connecting an auxiliary field winding, as shown in Fig. 5-26(c). The poles producing the desired field for better commutation are known as *commutating poles* or *interpoles*.

5-9 VOLTAGE BUILDUP IN A SHUNT GENERATOR

Saturation plays a very important role in governing the behavior of dc machines. It is extremely difficult to take into account the effects of saturation in the dynamical equations of motion. For the time being, let us consider qualitatively the consequences of saturation on the operation of a self-excited shunt generator.

A self-excited shunt machine is shown in Fig. 5-21. We write the steady-state equations for the operation of the machine as a generator. From the circuit shown in Fig. 5-21 we have

$$V_t = R_f I_f$$

and

$$E = V_t + I_a R_a = I_f R_f + I_a R_a$$

These equations are represented by the straight lines shown in Fig. 5-27(a). Notice that the voltages V_t and E will keep building up and no equilibrium point can be reached. On the other hand, if we include the effect of saturation, as in Fig. 5-27(b), point P defines an equilibrium, because at this point the field-resistance line intersects the saturation curve. A deviation from P to P' or P'' would immediately show that at P' the voltage drop across the field is greater than the induced voltage, which is not possible; and at P'' the induced voltage is greater than the field-circuit voltage drop, which is not possible either.

The small voltage 0 V shown in Fig. 5-27(b) results from the residual magnetism of the field poles. Evidently, without this remanent flux the shunt generator will not build up any voltage. Also shown in Fig. 5-27(b) is the critical resistance. A field resistance greater than the critical resistance (for a given speed) would not let the shunt generator build up any appreciable voltage. Finally, we should ascertain that the polarity of the field winding is such that a

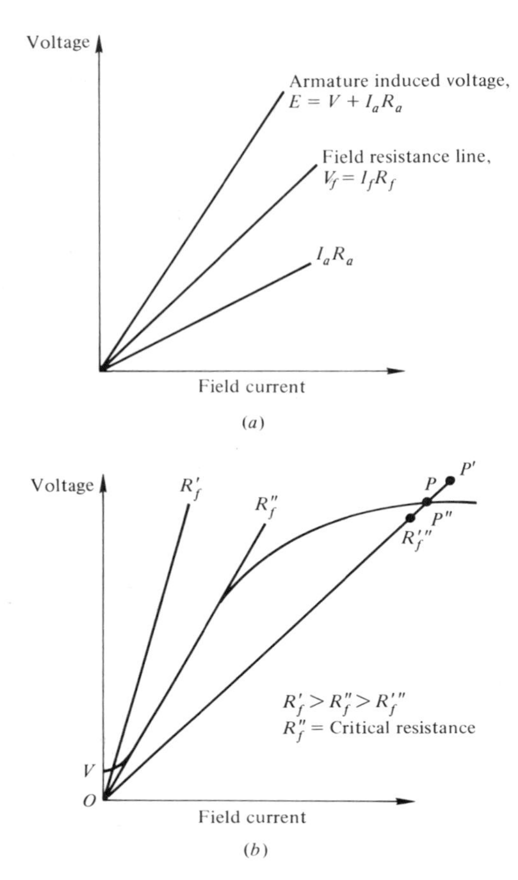

FIGURE 5-27. No-load characteristic of a shunt generator: (a) no stable operating point for the shunt generator; (b) stable no-load voltage of a shunt generator.

current through it produces a flux that aids the residual flux. If it does not, the two fluxes tend to neutralize each other and the machine voltage will not build up. To summarize, the conditions for voltage buildup in a shunt generator are the presence of residual flux, field-circuit resistance less than the critical resistance, and appropriate polarity of the field winding.

5-10 GENERATOR CHARACTERISTICS

No-load and load characteristics of dc generators are usually of interest in determining their potential applications. Of the two, load characteristics are of greater importance. As the names imply, no-load and load characteristics correspond, respectively, to the behavior of the machine when it is supplying no power (open-circuited, in the case of a generator) and when it is supplying power to an external circuit.

The only no-load (or open-circuit) characteristics that are meaningful are those of the shunt and separately excited generators. We have discussed the no-load characteristic of a shunt generator as a voltage buildup process in Section 5-9. For the separately excited generator, the no-load characteristic corresponds to the magnetization, or saturation, characteristic—variation of E (or V_t) under open-circuit conditions as a function of the field current, I_f. This characteristic is illustrated in Fig. 5-28, where the symbols are as shown in Fig. 5-20.

Turning now to the load characteristics, we define them as a variation of the terminal voltage as a function of load current supplied by the generator. These characteristics are fairly straightforward to obtain if we can identify the causes of voltage drops in dc generators. The main causes of the voltage drop in generators are:

1. *Armature resistance drop:* This is an I_aR_a drop due to the resistance of the armature.
2. *Brush contact drop:* The mechanical contact between the brushes and the commutator offers an electrical resistance. Consequently, when a current flows through the brush, a voltage drop occurs. Usually, this voltage drop is taken as a constant (of 2 V).
3. *Armature reaction voltage drop:* From Section 5-7 we recall that armature reaction has a demagnetizing component, which opposes the main field mmf, resulting in a reduction of flux. The reduced flux will, in turn, reduce the armature induced emf and hence the terminal voltage.

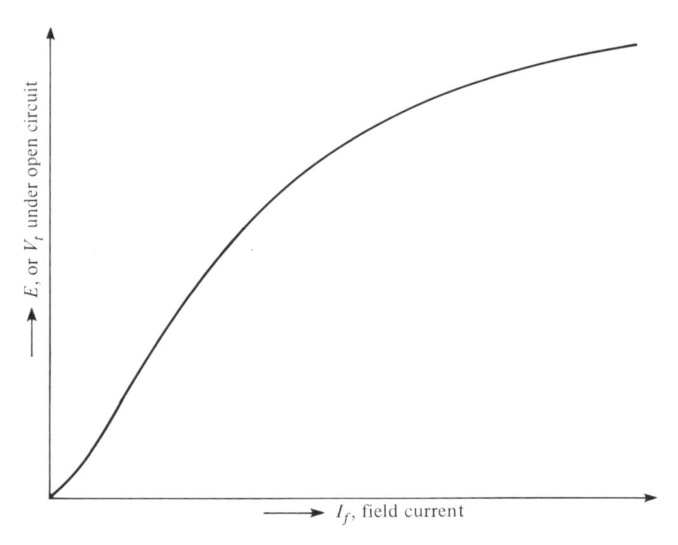

FIGURE 5-28. No-load characteristic of a separately excited generator.

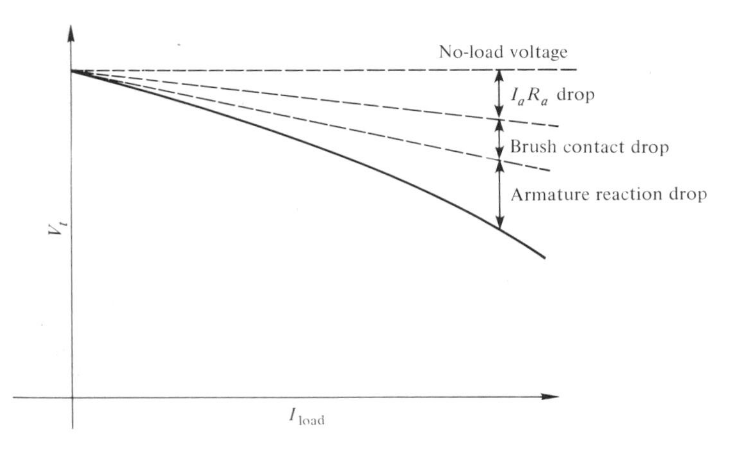

FIGURE 5-29. Load characteristic of a separately-excited generator.

4. *The cumulative effects of factors 1 to 3.* In self-excited (shunt, series, and compound) generators, these effects further lower the terminal voltage.

In view of the above, let us refer to Fig. 5-29, which shows the load characteristic of a separately excited dc generator. Notice that the terminal voltage on load differs from the no-load voltage by the three voltage drops mentioned in (1) to (3). The load characteristics of self-excited generators are shown in Fig. 5-30. The shunt generator has a characteristic similar to that of a separately excited generator, except for the cumulative effect mentioned in (4). If the shunt generator is loaded beyond a certain point, it breaks down in that the terminal voltage collapses. In a series generator, the load current flows through the field winding. This implies that the field flux, and hence the induced emf, increases with the load until the core begins to saturate magnetically. A load beyond a

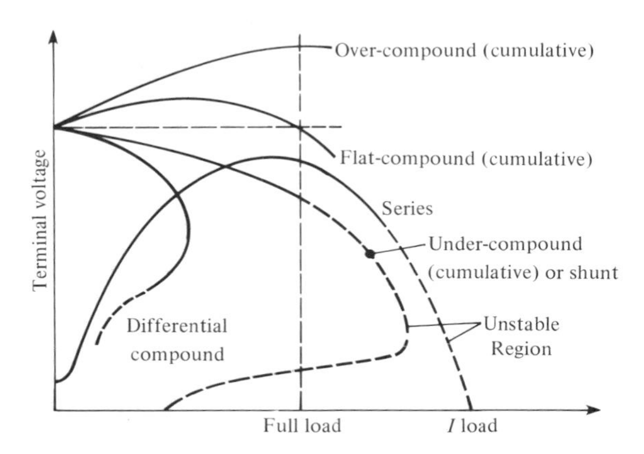

FIGURE 5-30. Load characteristics of dc generators.

certain point would also result in a collapse of the terminal voltage of the series generator. Compound generators have combined characteristics of shunt and series generators. In a differential compound generator, shunt and series fields are in opposition. Hence the terminal voltage drops very rapidly with the load. On the other hand, cumulative compound generators have shunt and series fields aiding each other. The two field mmf's could be adjusted such that the terminal voltage on full-load is less than the no-load voltage, as in an under-compound generator; or the full-load voltage may be equal to the no-load voltage, as in a flat-compound generator. Finally, the terminal voltage on full-load may be greater than the no-load voltage, as in an over-compound generator.

EXAMPLE 5-6

A 50-kW 250-V short-shunt compound generator has the following data: $R_a = 0.06\ \Omega$, $R_{se} = 0.04\ \Omega$, and $R_f = 125\ \Omega$. Calculate the induced armature emf at rated load and terminal voltage. Take 2 V as the total brush-contact drop.

SOLUTION

The equivalent circuit of the generator is shown in Fig. 5-31, from which

$$I_L = \frac{50 \times 10^3}{250} = 200 \text{ A}$$

$$I_L R_{se} = (200)(0.04) = 8 \text{ V}$$

$$V_f = 250 + 8 = 258 \text{ V}$$

$$I_f = \frac{258}{125} = 2.06 \text{ A}$$

$$I_a = 200 + 2.06 = 202.06 \text{ A}$$

$$I_a R_a = (202.06)(0.06) = 12.12 \text{ V}$$

$$E = 250 + 12.12 + 8 + 2 = 272.12 \text{ V}$$

■

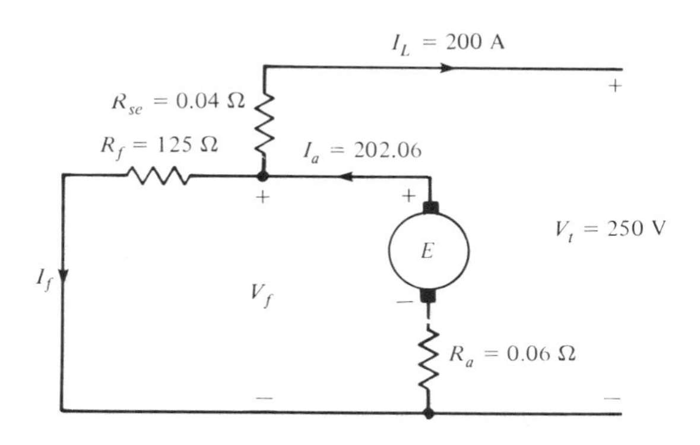

FIGURE 5-31. Example 3-6.

5-11 MOTOR CHARACTERISTICS

Among the various characteristics of dc motors, their torque–speed characteristics are most important from a practical standpoint. The torque and speed equations derived earlier govern the motor characteristics. From these equations (and after accounting for magnetic saturation) it follows that the shunt, series, and cumulative compound motors have the torque–speed characteristics of the forms shown in Fig. 5-32. The governing equations also yield the motor speed–current characteristics of Fig. 5-33.

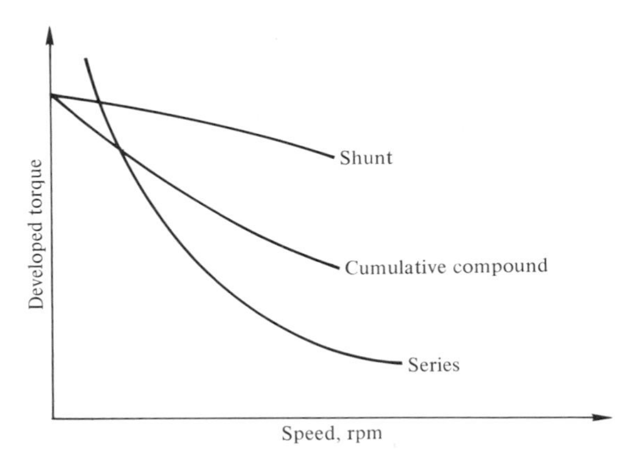

FIGURE 5-32. Torque–speed characteristics of dc motors.

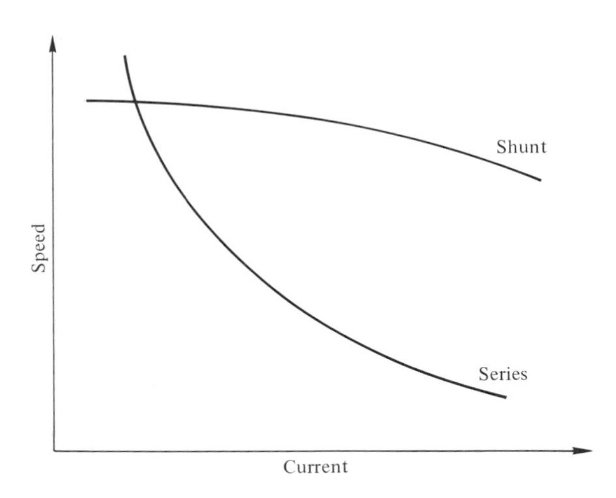

FIGURE 5-33. Speed–current characteristics.

Among the various dc motors, the series motor has been primarily used for traction applications for many years, and is still used in electric vehicles, electric trains, overhead cranes and automatic starter motors. Compared to other types of motors, the series motor yields the highest torque per unit input current. Because the field winding is connected in series with the armature winding (of a series motor), the torque equation may be written as $T_e = kI_a^2$ where k is a constant. This relationship yields the characteristics of the series motor shown in Figs. 5-32 and 5-33.

In Fig. 5-32 we have also shown the developed power versus speed for shunt and series motors. As we recall from Section 5-1, developed power is simply the product of developed torque (in N-m) and the angular speed (in rad/s). From Fig. 5-32 and 5-33 it is clear that a series motor on no-load will run at a dangerously high speed. Similarly, it follows from the speed equation that loss of field excitation in a shunt motor will result in overspeeding of the motor.

5-12

STARTING AND CONTROL OF MOTORS

In addition to certain operational conveniences, the basic requirements for satisfactory starting of a dc motor are (1) sufficient starting torque, and (2) armature current, within safe limits, for successful commutation and for preventing the armature from overheating. The second requirement is obvious from the speed equation, according to which the armature current, I_a, is given by $I_a = V_t/R_a$ when the motor is at rest (n or $\omega_m = 0$). A typical 50-hp 230-V motor having an armature resistance of 0.05 Ω, if connected across 230 V, shall draw 4600 A of current. This current is evidently too large for the motor, which might be rated to take 180 A on full-load. Commonly, double the full-load current is allowed to flow through the armature at the time of starting. For the motor under consideration, therefore, an external resistance

$$R_{\text{ext}} = \left(\frac{230}{2 \times 180}\right) - 0.05 = 0.59\ \Omega$$

must be inserted in series with the armature to limit I_a within double the rated value.

In practice, the necessary starting resistance is provided by means of a starter. Typical three-point and four-point starters are shown in Figs. 5-34(a) and (b), respectively. Notice that at the time of starting, the resistance R_{ext} comes in series with the armature. As the motor speeds up this resistance is cut out in steps. When the entire resistance is cut out, the starter arm is held by the electromagnet M. When the supply is turned off, the starter arm is pulled back to the "off" position by the spring. Notice from Fig. 5-34(a) that if the field is

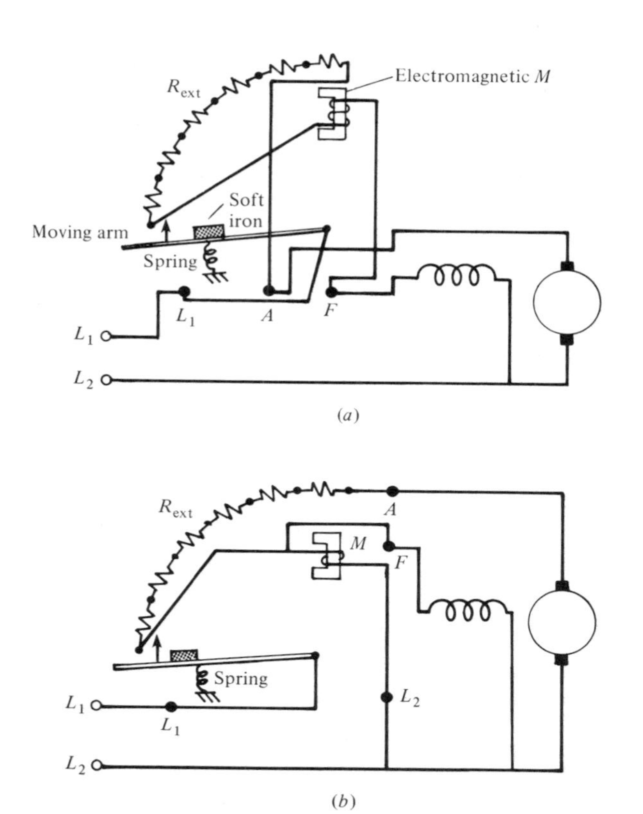

FIGURE 5-34. (a) A three-point starter; (b) a four-point starter.

opened, the electromagnet M is deenergized and the arm is pulled back to the "off" position. This provides field-failure protection. But if a large resistance is connected in series with the field circuit, the electromagnet may fail to hold the arm. On the other hand, there is no field-failure protection in a four-point starter, as is clear from Fig. 5-34(b). The electromagnet current is independent of the field current. In addition to the manual starter just described, there also exist pushbutton (or automatic) starters.

The three- and four-point starters just mentioned are old-fashioned starters. Modern starters are of the automatic pushbutton type. The switching of resistances is made automatically by contactors. Numerous versions of automatic starters for dc motors are commercially available. To illustrate the principle of operation of pushbutton starters, we consider only one type here, as shown in Fig. 5-35. Here, the contactors and the corresponding relays have the same numbers. For instance, a current in relay coil 1 will operate contactor 1. If this contactor is normally open, it will be closed when a predetermined current flows through the coil, and vice versa. Let us consider the sequence of operation. The start pushbutton is normally open and the stop pushbutton is normally closed.

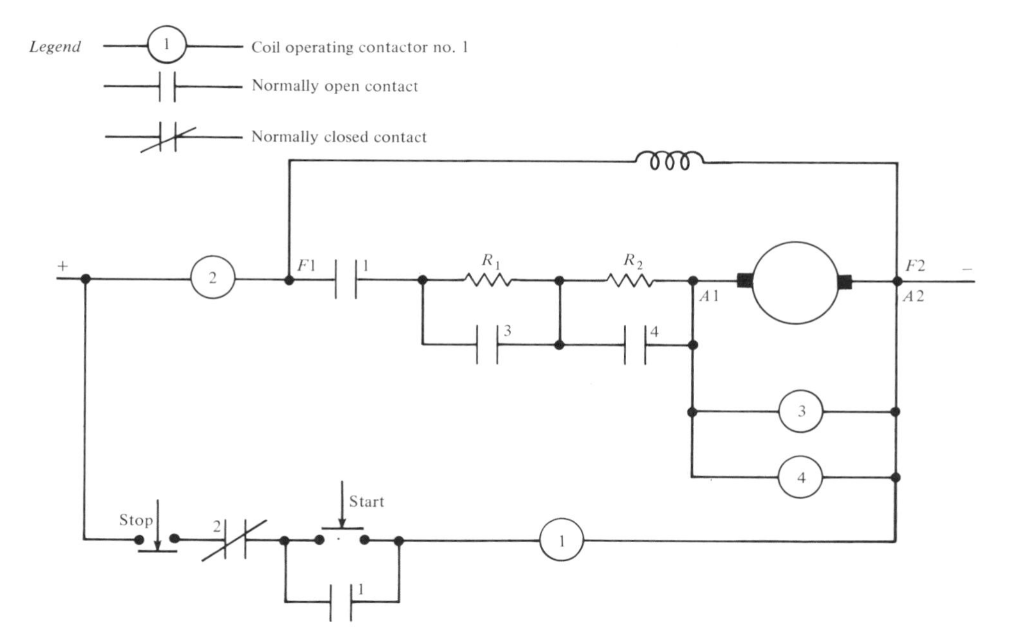

FIGURE 5-35. A push-button starter.

To start the motor, we push the start button, which makes a current flow to relay coil 1. This closes contactor 1 and current begins to flow through relay 2 (which is a current relay for over-current protection) and the armature. The armature current is limited by resistances R_1 and R_2 and the armature resistance. As the motor picks up speed, it develops a back emf and thereby a voltage is applied across relays 3 and 4. Relay 4 operates at a higher voltage than the operating voltage of relay 3. At a certain value of the back emf, relay 3 operates, contactor 3 closes, and R_1 is short-circuited. At a higher speed, relay 4 operates to close contactor 4 and short-circuit the resistance R_2. To stop the motor, we push the stop button to deenergize relay 1, thereby opening contactor 1 and open-circuiting the armature.

From the speed equation of a dc motor, it follows that the speed of the motor can be varied by varying (1) the field-circuit resistance to control I_f and hence the field flux, (2) the armature circuit resistance, and (3) the terminal voltage. Let us now consider the scope of each of these three methods. In method (1), an external variable resistance is connected in the field circuit. When this resistance is set at zero, the field current is limited only by the field resistance. Corresponding to this field current, we obtain a motor speed, n_1. Now, as the externally inserted resistance increases, the field current decreases and the motor speed increases (in accordance with the speed equation) to a corresponding speed n_2, where $n_2 > n_1$. In other words, by using an external resistance in series with the motor field, we can only increase the motor speed (from a minimum speed

n_1). By inserting a resistance in series with the armature, as in method (2), we can reduce the voltage across the armature and hence decrease the motor speed. Because the armature current is of a relatively large magnitude, compared to the field current, method (2) is a wasteful method (as shown by Example 5-8). Method (3) can be efficiently used either to increase or decrease the speed of the motor, but is feasible only if a variable voltage source is available. Method (3) in conjunction with method (1) constitutes the *Ward–Leonard system*, which is capable of providing a wide variation of speed in both forward and reverse directions. This method is presented in a simplified form in Example 5-9.

EXAMPLE 5-7

A 230-V shunt motor has an armature resistance of 0.05 Ω and a field resistance of 75 Ω. The motor draws 7 A of line current while running light at 1120 r/min. The line current at a certain load is 46 A. (a) What is the motor speed at this load? (b) At this load, if the field-circuit resistance is increased to 100 Ω, what is the new speed of the motor? Assume the line current to remain unchanged.

SOLUTION

(a) On no-load,

$$N_0 = 1120 \text{ r/min (given)}$$

$$I_f = \frac{230}{75} = 3.07 \text{ A}$$

$$I_a = 7 - 3.07 = 3.93 \text{ A}$$

The speed equation gives

$$1120 = \frac{230 - 3.93 \times 0.05}{3.07k}$$

or

$$k = 0.0668$$

On load (with $R_f = 75\ \Omega$):

$$I_f = 3.07 \text{ A}$$

$$I_a = 46 - 3.07 = 42.93 \text{ A}$$

$$n = \frac{230 - 42.93 \times 0.05}{3.07 \times 0.0668} = 1111 \text{ r/min}$$

(b) On load (with $R_f = 100\ \Omega$):

$$I_f = \frac{230}{100} = 2.3\ \text{A}$$

$$I_a = 46 - 2.3 = 43.7\ \text{A}$$

$$n = \frac{230 - 43.7 \times 0.05}{2.3 \times 0.0668} = 1483\ \text{rpm}$$

■

EXAMPLE 5-8

Refer to part (a) of Example 5-7. The no-load conditions remain unchanged. On load, the line current remains at 46 A, but a 0.1-Ω resistance is inserted in the armature. Determine the speed of the motor and the power dissipated in the 0.1-Ω resistance.

SOLUTION

In this case we have (from Example 5-7)

$$I_f = 3.07\ \text{A}$$

$$I_a = 42.93\ \text{A}$$

$$k = 0.0668$$

Thus, with $R_a = 0.05 + 0.1 = 0.15\ \Omega$, we have

$$I_a^2(0.1) = 42.93^2 \times 0.1 = 184.3\ \text{W}$$

$$n = \frac{230 - 42.93 \times 0.15}{3.07 \times 0.0688} = 1058\ \text{r/min}$$

■

EXAMPLE 5-9

The system shown in Fig. 5.36 is called the *Ward–Leonard system* for controlling the speed of a dc motor. Discuss the effects of varying R_{fg} and R_{fm} on the motor speed.

SOLUTION

Increasing R_{fg} decreases I_{fg} and hence E_g. Thus the motor speed will decrease. The opposite will be true if R_{fg} is decreased.

Increasing R_{fm} will increase the speed of the motor, as shown in Example 5-7. Decreasing R_{fm} will result in a decrease of the speed. ■

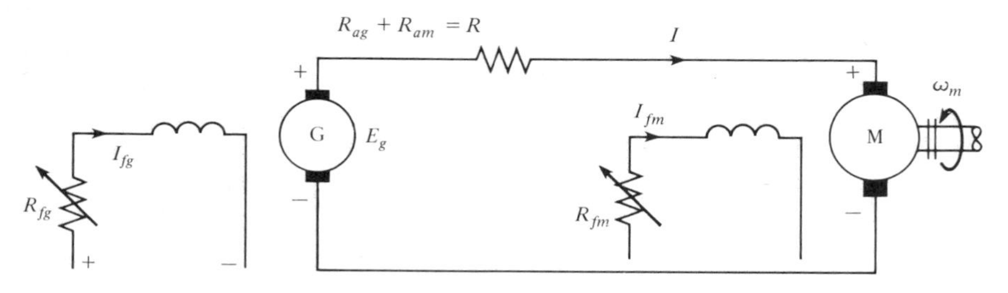

FIGURE 5-36. A Ward-Leonard system.

5-13

LOSSES AND EFFICIENCY

Besides the voltage–amperage and speed–torque characteristics, the performance of a dc machine is measured by its efficiency:

$$\text{efficiency} = \frac{\text{power output}}{\text{power input}} = \frac{\text{power output}}{\text{power output} + \text{losses}}$$

$$= \frac{\text{power input} - \text{losses}}{\text{power input}} \tag{5-15}$$

Efficiency may, therefore, be determined either from load tests or by determination of losses. The various losses are classified as follows:

1. *Electrical:* (a) Copper losses in various windings, such as the armature winding and different field windings; and (b) loss due to the contact resistance of the brush (with the commutator).
2. *Magnetic:* These are the iron losses and include the hysteresis and eddy-current losses in the various magnetic circuits, primarily the armature core and pole faces.
3. *Mechanical:* These include the bearing-friction, windage, and brush-friction losses.
4. *Stray-load:* These are other load losses not covered above. They are taken as 1 percent of the output (as a rule of thumb). Another recommendation often followed is to take the stray-load loss as 28 percent of the core loss at the rated output.

The power flow in a dc generator or motor is represented in Fig. 5-37, in which the symbols are as follows: V_t = terminal voltage, V; I_a = armature current, A; E = back or induced emf in the armature, V; V_f = voltage across the field winding, V; I_f = current through the field winding, A; T_e = electro-

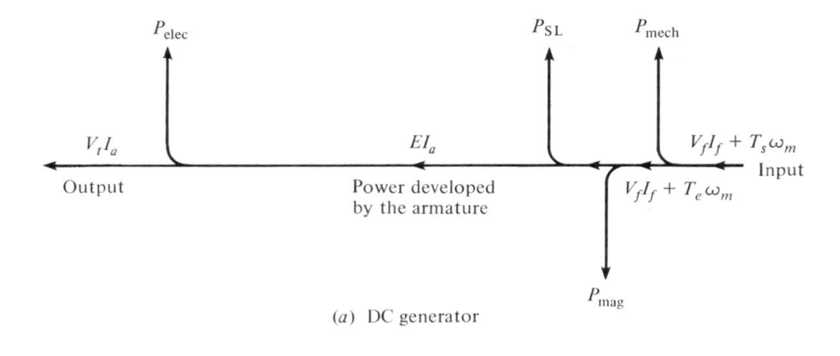

(*a*) DC generator

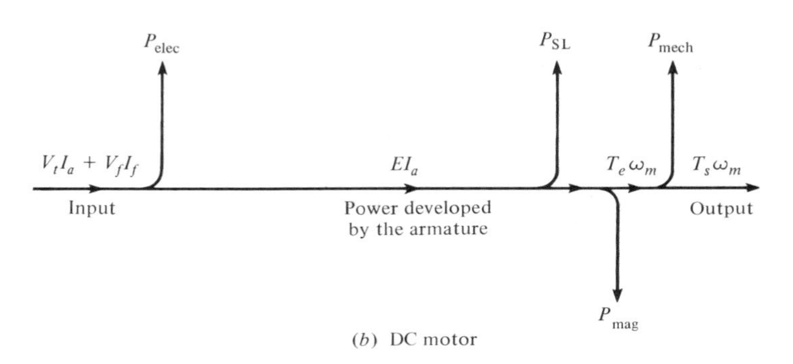

(*b*) DC motor

FIGURE 5-37. (a) Power flow in separately-excited dc generator; (b) power flow in separately-excited dc motor.

magnetic torque developed by the armature, N-m; T_s = torque available at the shaft, N-m; P_{elec} = electrical power, W; P_{SL} = stray-load loss, W; P_{mech} = rotational mechanical loss, W; P_{mag} = magnetic losses, W; and ω_m = mechanical angular velocity, rad/s.

EXAMPLE 5-10

A 10 hp 230-V shunt motor takes a full-load line current of 40 A. The armature and field resistances are 0.25 Ω and 230 Ω, respectively. The total brush-contact drop is 2 V and the core and friction losses are 380 W. Calculate the efficiency of the motor. Assume that stray-load loss is 1 percent of rated output.

SOLUTION

$$\text{input} = (40)(230) = 9200 \text{ W}$$

$$\text{field-resistance loss} = \left(\frac{230}{230}\right)^2 (230) = 230 \text{ W}$$

$$\text{armature-resistance loss} = (40 - 1)^2(0.25) = 380 \text{ W}$$

$$\text{core loss and friction loss} = 380 \text{ W}$$

$$\text{brush-contact loss} = (2)(39) = 78 \text{ W}$$

$$\text{stray-load loss} = \frac{10}{100} \times 746 = 75 \text{ W}$$

$$\text{total losses} = 1143 \text{ W}$$

$$\text{power output} = 9200 - 1143 = 8057 \text{ W}$$

$$\text{efficiency} = \frac{8057}{9200} = 87.6\%$$

■

5-14

TESTS ON DC MACHINES

Tests are performed on dc machines for the following purposes:

1. To obtain no-load and magnetization characteristics.
2. To obtain load characteristics, including the determination of losses and efficiency.
3. To evaluate temperature rise.
4. To assess commutation.

No-Load Tests. To determine the magnetization characteristic, the machine under test is driven as a generator at a constant speed. The field is excited separately and the armature is open-circuited. A plot of the open-circuit armature voltage as a function of the field current gives the magnetization characteristic. A typical magnetization characteristic is shown in Fig. 5-28.

On no-load, the input power is the sum of magnetic, mechanical, and electrical losses in the field and the armature. If we run the machine as a motor on no-load (at a constant speed by adjusting the armature voltage) and vary the field excitation, the input power will be the sum of a constant mechanical loss, a varying core loss, and electrical losses in the field and the armature. Subtracting the electrical losses from the input power gives the constant mechanical loss and the variable core loss. The core loss is approximately proportional to the square of the armature voltage. If we plot the sum of the core loss and the mechanical loss versus (armature voltage)^2, we will obtain a straight line. The intercept of this straight line (by extrapolation) with the vertical (Fig. 5-38) gives the mechanical losses.

Load Tests. Small dc machines may be tested on load by loading them directly by a mechanical brake or a dynamometer. Larger machines are evaluated for their characteristics by segregation of losses (see Example 5-10). Direct testing of large machines is uneconomical and often not feasible. In such cases,

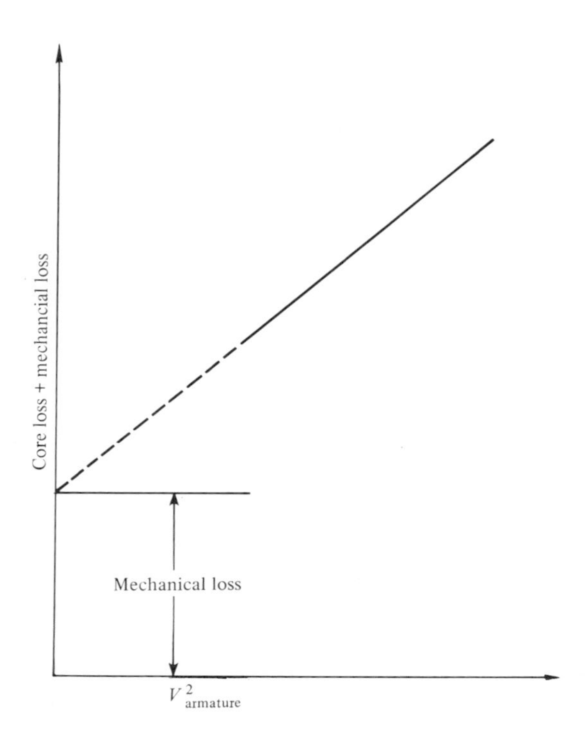

FIGURE 5-38. Separation of mechanical loss.

back-to-back tests may be performed if two identical machines are available. This test is rather specialized and we will not consider it here.

Temperature Rise. Losses in electric machines are dissipated as heat and result in a temperature rise in the machine. The temperature rise depends on the rate of heat generation, heat transfer, and thermal capacity. The rating of a machine is determined by the temperature rise. Heat-run tests on small machines can be performed by loading them as generators. Temperatures within the machine can be monitored by strategically located thermocouples. Large machines are tested for temperature rise by the back-to-back test.

Commutation. There are a number of tests conducted to assess the performance of the commutator of a dc machine. These include a check of the brush location, contact drop, armature faults, and black-band test. A simple test to check the brush location is to find by means of probes the two adjacent commutator segments between which no instantaneous voltage is observable when the field is suddenly turned on and turned off.

The contact drop is illustrated by Fig. 5-39, and checks the uniformity of current density under the brush. If the voltage drop between the commutator

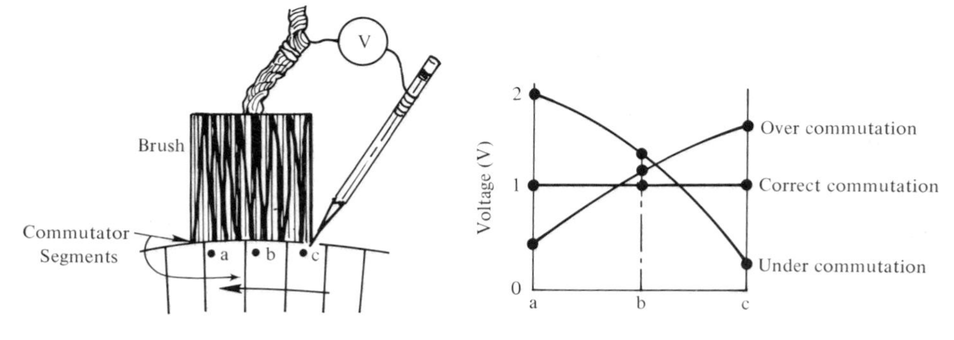

FIGURE 5-39. Contact-drop test.

and the brush at a few points along its width gives a straight line, we have a correct condition for commutation. Under- and over commutation are also shown in Fig. 5-39.

Faults in the armature winding can be detected by passing a current through the stationary armature and measuring the voltages between the segments around the commutator. If these voltages are equal, there is no fault. A low voltage indictes a short circuit and an open-circuited coil is indicated by zero voltage between all pairs of segments in the faulty path except the pair at the terminals of the faulty coil.

The black-band test is used to check the effectiveness of the commutating poles. In this test, for a range of fixed armature currents, the current in the commutating pole is reduced until sparking begins. The commutating pole current is then gradually increased until sparking occurs again. The results of this test are shown in Fig. 5-40. The range of sparkless commutation is called the *black band*. If the black band is asymmetrical at low armature currents, the brushes are incorrectly located; if tilted with increase of armature current, the machine is over- or under-commutated; and if very narrow at large armature currents, the commutating poles are saturated. The per unit current in Fig. 5-40 is the ratio of the actual current to the full-load armature current.

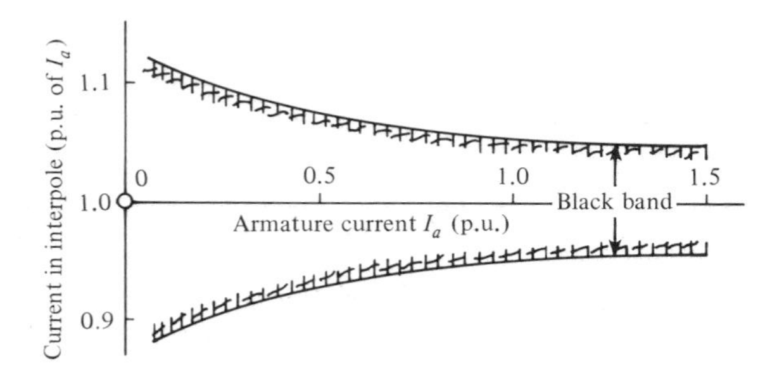

FIGURE 5-40. Black band.

5-15

CERTAIN APPLICATIONS

Dc machines find applications in the following: electric traction and diesel-electric locomotives; large rolling mills; electrochemical plants and metal refining plants; elevators; earth-moving equipment; battery charging; ship, train, and aircraft auxiliaries; isolated experimental stations; excitors for synchronous machines (see Chapter 7); automatic control systems; wind-generating systems; and so on.

The choice of a dc motor for a specific application depends on the nature of the load. For instance, permanent-magnet dc motors are preferred for actuators requiring high peak and steady power and fast response. They are basic drives in aircraft control systems. For traction loads, dc series motors are ideally suited. Series motors are used to drive cranes, hoists, and high-inertia loads. In contrast, the shunt motor is essentially a constant-speed motor. Its speed can be easily controlled by adjusting the field current or armature voltage. Hence it has numerous industrial applications, such as in driving pumps, compressors, punch presses, and so on.

Dc generators are used in the chemical industry for applications requiring electrolysis processing. Differential compound generators are used in welding.

5-16

PARALLEL OPERATION OF DC GENERATORS

Generators are operated in parallel for load sharing; that is when the demand exceeds the capacity of one generator, a second one is operated in parallel to provide an adequate supply for the load.

Shunt Generators in Parallel. The load (or external) characteristics of two shunt generators are shown in Fig. 5-41. The field excitation of the two generators is so adjusted that their open-circuit terminal voltages are equal. For a terminal voltage V_t, the load current I is divided into I_1, delivered by the first generator, and I_2, supplied by the second generator. Clearly, $I = I_1 + I_2$. The generator with the more drooping characteristic takes the smaller share of the load. Because of their drooping characteristics, shunt generators inherently work well in parallel. There is no tendency for an unstable operation.

If the open-circuit voltages of the two generators are not equal, as shown in Fig. 5-42, then at very light loads the machine with the smaller open-circuit voltage receives current (instead of supplying) and tends to act as a motor. Beyond point A in Fig. 5-42, both machines work as generators.

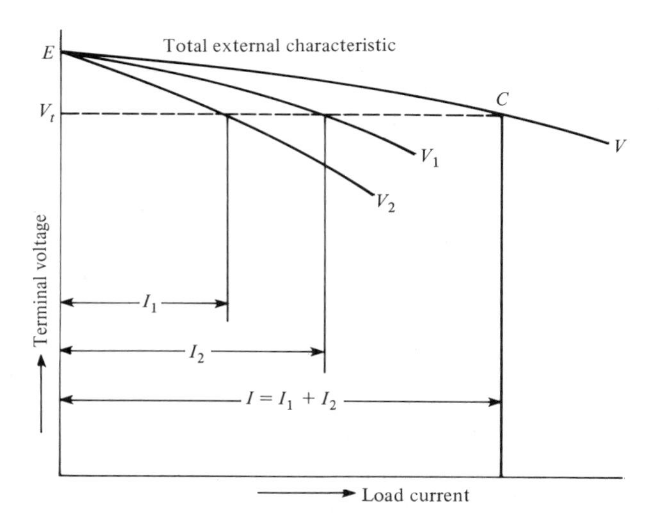

FIGURE 5-41. Load sharing between two shunt generators having the same no-load voltage.

Load sharing between shunt generators can be obtained graphically, as illustrated in Fig. 5-41 and 5-42. On the other hand, if it is assumed that the terminal voltage of a generator varies linearly with the current supplied by the generator, the following relations hold:

$$V_0 = \frac{E_1 R_{a2} + E_2 R_{a1}}{R_{a1} + R_{a2}}$$

$$I_0 = \frac{E_1 - E_2}{R_{a1} + R_{a2}}$$

$$I_1 = \frac{E_1 R_{a2} + (E_1 - E_2)R}{R_{a1} R_{a2} + (R_{a1} + R_{a2})R}$$

$$I_2 = \frac{E_2 R_{a1} + (E_2 - E_1)R}{R_{a1} R_{a2} + (R_{a1} + R_{a2})R}$$

$$I = \frac{E_1 R_{a2} + E_2 R_{a1}}{R_{a1} R_{a2} + (R_{a1} + R_{a2})R}$$

where V_0 is the no-load terminal voltage of the two generators in parallel; I_0 the no-load circulating current; E_1 and E_2 the no-load induced voltages of generators 1 and 2 respectively; I_1 and I_2 the currents supplied by the two generators; R_{a1}

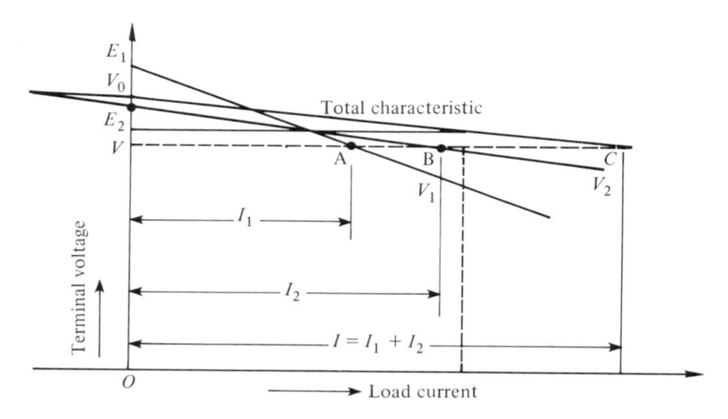

FIGURE 5-42. Load sharing between two shunt generators having unequal induced voltage.

and R_{a2} the respective armature circuit resistances; I the load current; and R the load resistance. The terminal voltage, V, on load is then given by

$$V = \frac{E_1 R_{a2} + E_2 R_{a1}}{(R_{a1} R_{a2}/R) + (R_{a1} + R_{a2})}$$

Series and Cumulative Compound Generators in Parallel. Because series and cumulative compound generators do not have drooping characteristics, their parallel operation is inherently unstable. For their satisfactory operation in parallel either an equalizer connection (Fig. 5-43) must be used or the field windings must be cross-connected (Fig. 5-44). By the equalizer connection or the cross-connection, the two field windings are connected in parallel, and the flux in either machine is not directly proportional to its own armature current. If an equalizer connection is used, its resistance must be small compared to the resistances of the series-field windings.

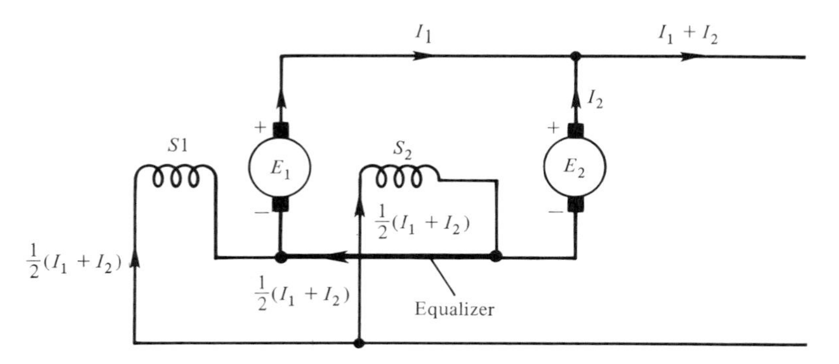

FIGURE 5-43. Series generators in parallel, showing equalizer connections.

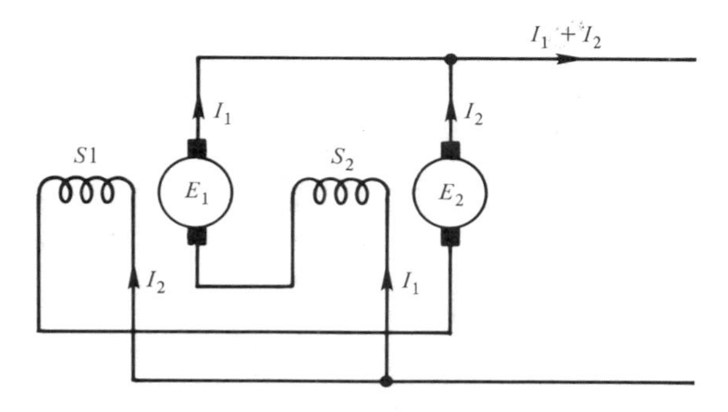

FIGURE 5-44. Series generators in parallel, showing cross-connected series fields.

REFERENCES

1. **R. G. Kloeffler, R. M. Kerchner,** and **J. L. Brenneman,** *Direct Current Machinery*. New York: Macmillan, 1948.
2. **M. G. Say** and **E. O. Taylor,** *Direct Current Machines*. New York: Halsted Press, 1980.

PROBLEMS

5-1 The flux per pole of a four-pole generator is 75 mWb. The generator runs at 900 r/min. Determine the induced voltage if (a) the armature has 32 conductors connected as lap winding, and (b) the armature has 30 conductors connected as wave winding.

5-2 Determine the flux per pole of a six-pole generator required to generate 240 V at 500 rpm. The armature has 120 slots, with eight conductors per slot, and is lap connected.

5-3 If the armature current in the generator of Problem 5-4 is 25 A, what electromagnetic torque is developed?

5-4 ✓ The armature and field resistances of a 240-V shunt generator are 0.2 Ω and 200 Ω, respectively. The generator supplies a 9600-W load. Determine the induced voltage, taking 2 V as the total brush-contact voltage drop.

5-5 The armature, shunt field, and the series field resistances of a compound generator are 0.02 Ω, 80 Ω, and 0.03 Ω, respectively. The generator induced voltage is 510 V and the terminal voltage is 500 V. Calculate the power supplied to the load (at 500 V) if the generator has (a) a long-shunt connection and (b) a short-shunt connection.

5-6 A four-pole shunt-connected generator has a lap-connected armature with 728 conductors. The flux per pole is 25 mWb. If the generator supplies two hundred 110-V 75-W bulbs, determine the speed of the generator. The field and armature resistances are 110 Ω and 0.075 Ω, respectively.

5-7 A shunt generator delivers 50 A of current to a load at 110 V, at an efficiency of 85 percent. The total constant losses are 480 W (excluding the field-circuit losses), and the shunt-field resistance is 65 Ω. Calculate the armature resistance.

5-8 For the generator of Problem 5-7, plot a curve for efficiency versus armature current. At what armature current is the efficiency maximum?

5-9 A separately excited six-pole generator has a 30-mWb flux per pole. The armature is lap wound and has 534 conductors. This supplies a certain load at 250 V while running at 1000 r/min. At this load, the armature copper loss is 640 W. Calculate the load supplied by the generator. Take 2 V as the total brush-contact drop.

5-10 In a dc machine, the hysteresis and eddy-current losses at 1000 r/min are 10,000 W at a field current of 7.8 A. At a speed of 750 r/min and 7.8-A of field current, the total iron losses become 6000 W. Assuming that the hysteresis loss is directly proportional to the speed and that the eddy-current loss is proportional to the square of the speed, determine the hysteresis and eddy-current losses at 500 r/min at 7.8 A of field current.

5-11 The saturation characteristic of a dc shunt generator is as follows:

Field Current (A)	1	2	3	4	5	6	7
Open-Circuit Voltage (V)	53	106	150	192	227	252	270

The generator speed is 900 r/min. At this speed, what is the maximum field-circuit resistance such that the self-excited shunt generator would not fail to build up?

5-12 To what value will the no-load voltage of the generator of Problem 5-11 build up, at 900 rpm, for a field-circuit resistance of 42 Ω?

5-13 A 240-V separately excited dc machine has an armature resistance of 0.25 Ω. The armature current is 56 A. Calculate the induced voltage for (a) generator operation and (b) motor operation.

√ **5-14** The field and armature winding resistances of a 400-V dc shunt machine are 120 Ω and 0.12 Ω, respectively. Calculate the power developed by the armature if (a) the machine takes 50 kW while running as a motor, and (b) the machine delivers 50 kW while running as a generator.

5-15 The field and armature resistances of a 220-V series motor are 0.2 Ω and 0.1 Ω, respectively. The motor takes 30 A of current while running at 700 r/min. If the total iron and friction losses are 350 W, determine the motor efficiency.

5-16 A 400-V shunt motor delivers 23 kW of power at the shaft at 1200 r/min while drawing a line current of 62 A. The field and armature resistances are 200 Ω and 0.05 Ω, respectively. Assuming 1 V of contact drop per brush, calculate (a) the torque developed by the motor and (b) the motor efficiency.

√ **5-17** A 400-V series motor, having an armature circuit resistance of 0.5 Ω, takes 44 A of current while running at 650 r/min. What is the motor speed for a line current of 36 A?

5-18 A 220-V shunt motor having an armature resistance of 0.2 Ω and a field resistance of 110 Ω takes 4 A of line current while running on no-load. When loaded, the motor runs at 1000 r/min while taking 42 A of current. Calculate the no-load speed.

5-19 The machine of Problem 5-18 is driven as a shunt generator at 1000 r/min to deliver a 44-kW load at 220V. If the machine takes 44 kW while running as in Problem 5-18, what is its speed when operating as a generator?

√ **5-20** A 220-V shunt motor having an armature resistance of 0.2 Ω and a field resistance of 110 Ω takes 4 A of line current while running at 1200 r/min on no-load. On load, the input to the motor is 15 kW. Calculate (a) the speed, (b) the developed torque, and (c) the efficiency at this load.

5-21 A 400-V series motor has a field resistance of 0.2 Ω and an armature resistance of 0.1 Ω. The motor takes 30 A of current at 1000 r/min while developing a torque T. Determine the motor speed if the developed torque is 0.6 T.

5-22 A shunt machine, while running as a generator at no-load, has an induced voltage of 260 V at 1200 r/min. Its armature and field resistances are 0.2 Ω and 110 Ω, respectively. If the machine is run as a shunt motor, it takes 4 A at 220 V on no-load. At a certain load the motor takes 30 A at 220 V. However, on load, armature reaction weakens the field by 3 percent. Calculate the motor speed and the efficiency at the load specified.

5-23 The machine of Problem 5-22 is run as a motor. It takes 25 A of current at 800 r/min. What resistance must be inserted in the field circuit to increase the motor speed to 1000 r/min? The torque on the motor for the two speeds remains unchanged.

5-24 The motor of Problem 5-22 runs at 600 r/min while taking 40 A at a certain load. If a 0.8-Ω resistance is inserted in the armature circuit, determine the motor speed provided that the torque on the motor remains constant.

5-25 A 220-V shunt motor delivers 40 hp on full-load at 950 r/min, and has an efficiency of 88 percent. The armature and field resistances are 0.157 Ω and 110 Ω, respectively. Determine (a) the starting resistance such that the starting line current does not exceed 1.6 times the full-load current and (b) the starting torque.

5-26 A 220-V series motor runs at 750 r/min while taking 15 A of current. What is the motor speed if it takes 10 A of current? The torque on the motor is such that it increases as the square of the speed.

5-27 A 500 V, 450 r/min, 750 kW (name plate data), separately excited dc generator operates at rated conditions with a rotational loss of 12,180 W. Armature reaction is negligible, $R_a = 0.007\ \Omega$, and $R_f = 35\ \Omega$. For rated conditions find (a) the back *emf*; (b) power converted from mechanical to electrical form; (c) input shaft torque; (d) efficiency if the field resistance draws 14 A current. If speed were changed to half rated value (225 r/min) without adjustment of field current, find (e) the maximum electrical power output possible without overheating armature windings.

CHAPTER 6

Synchronous Machines

6-1 INTRODUCTION

The bulk of electric power for everyday use is produced by polyphase synchronous generators, which are the largest single-unit electric machines in production. For instance, synchronous generators with power ratings of several hundred megavoltamperes (MVA) are fairly common, and it is expected that machines of several thousand megavoltamperes will be in use in the near future. These are called synchronous machines because they operate at constant speeds and constant frequencies under steady-state conditions. Like most rotating machines, synchronous machines are capable of operating both as a motor and as a generator. They are used as motors in constant-speed drives, and where a variable-speed drive is required, a synchronous motor is used with an appropriate frequency changer such as an inverter or cycloconverter. As generators, several synchronous machines often operate in parallel, as in a power station. While operating in parallel, the generators share the load with each other; at a given time one of the generators may not carry any load. In such a case, instead of shutting down the generator, it is allowed to ''float'' on the line as a synchronous motor on no-load.

The operation of a synchronous generator is based on Faraday's law of electromagnetic induction, and an ac synchronous generator works very much like a dc generator, in which the generation of emf is by the relative motion of conductors and magnetic flux. Clearly, however, a synchronous generator does not have a commutator as does a dc generator. A synchronous generator in its elementary form was shown in Fig. 5-4. The two basic parts of a synchronous machine are the magnetic field structure, carrying a dc-excited winding, and the armature. The armature often has a three-phase winding in which the ac emf is generated. Almost all modern synchronous machines have stationary armatures and rotating field structures. The dc winding on the rotating field structure is connected to an external source through slip rings and brushes. Some field structures do not have brushes but, instead, have brushless excitation by rotating diodes. In some respects the stator carrying the armature windings is similar to the stator of a polyphase induction motor (discussed in Chapter 7).

6-2 SOME CONSTRUCTION DETAILS

Some of the factors that dictate the form of construction of a synchronous machine are discussed next.

Form of Excitation. Notice from the preceding remarks that the field structure is usually the rotating member of a synchronous machine and is supplied with a dc-excited winding to produce the magnetic flux. This dc excitation may be provided by a self-excited dc generator mounted on the same shaft as the rotor of the synchronous machine. Such a generator is known as an *exciter*. The direct current thus generated is fed to the synchronous machine field winding. In slow-speed machines with large ratings, such as hydroelectric generators, the exciter may not be self-excited. Instead, a pilot exciter, which may be self-excited or may have a permanent magnet, activates the exciter. A hydroelectric generator and its rotor and exciters are shown in Fig. 3-2. The maintenance problems of direct-coupled dc generators impose a limit on this form of excitation at about a 100-MW rating.

An alternative form of excitation is provided by silicon diodes and thyristors, which do not present excitation problems for large synchronous machines. The two types of solid-state excitation systems are:

1. Static systems that have stationary diodes or thyristors, in which the current is fed to the rotor through slip rings.
2. Brushless systems that have shaft-mounted rectifiers that rotate with the rotor, thus avoiding the need for brushes and slip rings. Figure 6-1 shows a brushless excitation system.

FIGURE 6-1. Rotor of a 3360 kVA 6 kV brushless synchronous generator with rotating diodes. (Courtesy Brown Boveri Company)

Field Structure and Speed of Machine. We have already mentioned that the synchronous machine is a constant-speed machine. This speed is known as synchronous speed. For instance a 60-Hz two-pole synchronous machine must run at 3600 r/min, whereas the synchronous speed of a 12-pole 60-Hz machine is only 600 r/min. The rotor field structure consequently depends on the speed rating of the machine. Therefore, turbogenerators, which are high-speed machines, have *round* or *cylindrical rotors* [see Figs. 6-2 and 3-1(b)]. Hydroelectric and diesel-electric generators are low-speed machines and have *salient pole rotors,* as depicted in Figs. 3-2, 6-1, and 6-3. Such rotors are less expensive than round rotors to fabricate. They are not suitable for large high-speed machines, however, because of the excessive centrifugal forces and mechanical stresses that develop at speeds around 3600 r/min.

Stator. The stator of a synchronous machine carries the armature or load winding. We recall from Chapter 5 that the armature of a dc machine has a winding that is distributed around the periphery of the armature. Thus slot-embedded conductors, covering the entire surface of the armature and interconnected in a predetermined manner, constitute the armature winding of a dc machine. Similarly, in a synchronous machine, the armature winding is formed

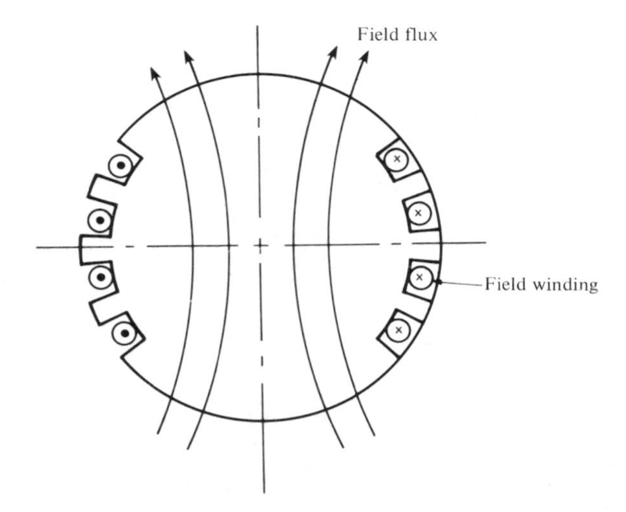

FIGURE 6-2. Field winding on a round or cylindrical rotor.

by interconnecting the various conductors in the slots spread over the periphery of the stator of the machine. Often, more than one independent winding is on the stator. An arrangement of a three-phase stator winding is shown in Fig. 6-4. Notice that the three phases are displaced from each other in space, as the windings are distributed in the slots over the entire periphery of the stator. Each slot contains two coil sides. For instance, slot 1 has coil sides of phases A and B, whereas slot 2 contains two layers (or two coil sides) of phase A only. Such

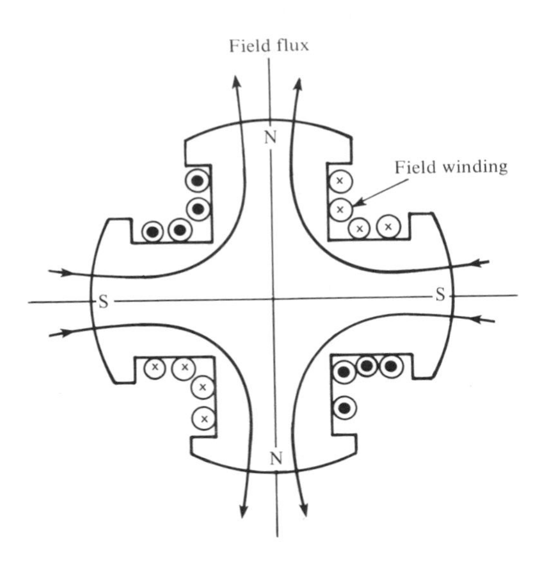

FIGURE 6-3. Field winding on a salient rotor.

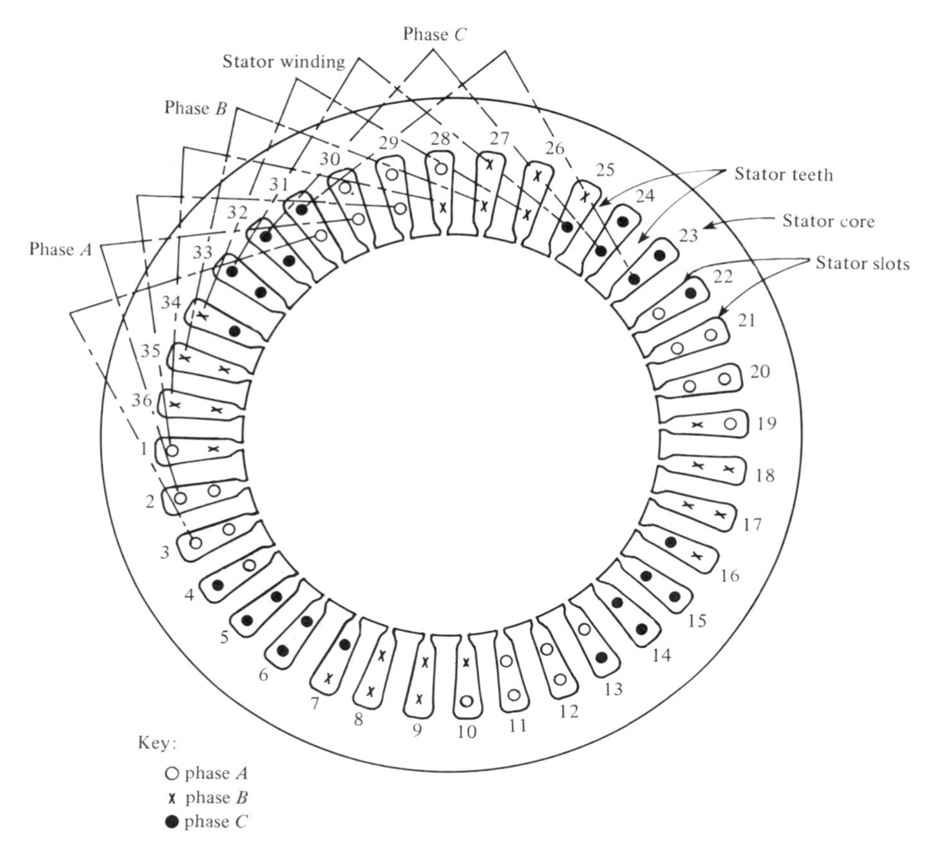

FIGURE 6-4. Stator windings.

a winding is known as a *double-layer* winding. Furthermore, it is a four-pole winding laid in 36 slots and we thus have three slots per pole per phase.

In order to produce the four-pole flux, each coil should have a span (or *pitch*) of one-fourth of the periphery. In practice, the pitch is made a little less and, as shown in Fig. 6-4, each coil embraces eight teeth. The coil pitch is about 89 percent of the pole pitch, and the winding is, therefore, a *fractional-pitch* (or *chorded*) *winding*. There is essentially no difference between the stator of a round-rotor machine and that of a salient-rotor machine. The stators of waterwheel or hydroelectric generators, however, usually have a large-diameter armature compared to other types of generators. The stator core consists of punchings of high-quality laminations having slot-embedded windings, as shown in Fig. 6-4. Mounting of stator conductors in slots of one stator half to make the armature winding of a large synchronous machine is shown in Fig. 6-5.

Cooling. Because synchronous machines are often built in extremely large sizes, they are designed to carry very large currents. A typical armature current

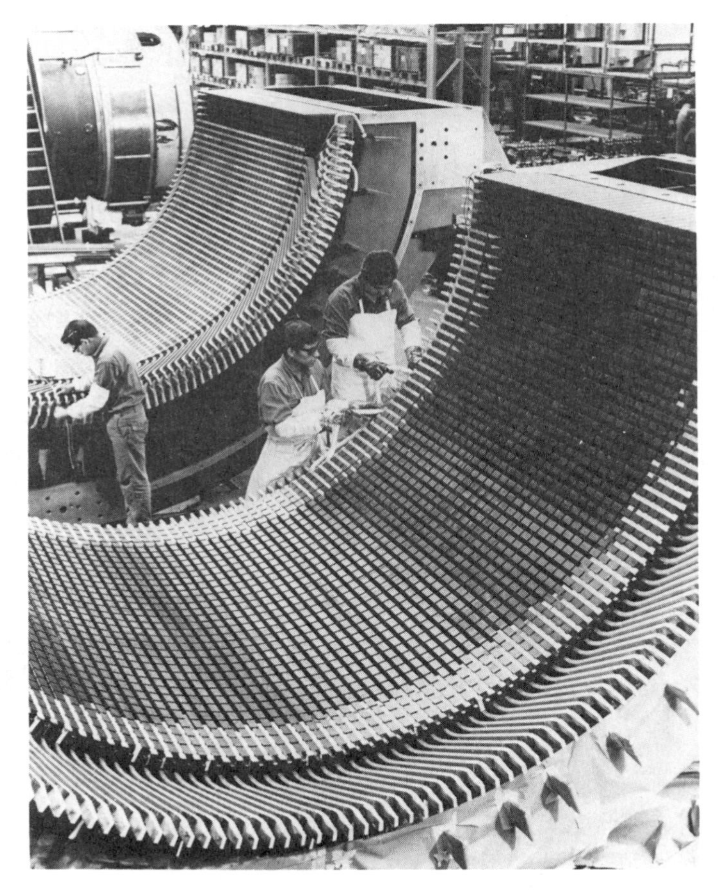

FIGURE 6-5. Mounting stator conductors in slots on one stator half of a synchronous motor.

density may be of the order of 10 A/mm^2 in a well-designed machine. Also, the magnetic loading of the core is such that it reaches saturation in many regions. The severe electric and magnetic loadings in a synchronous machine produce heat that must be appropriately dissipated. Thus the manner in which the active parts of a machine are cooled determines its overall physical structures. In addition to air, some of the coolants used in synchronous machines are water, hydrogen, and helium.

Damper Bars. So far we have mentioned only two electrical windings of a synchronous machine: the three-phase armature winding and the field winding. We also pointed out that, under steady state, the machine runs at a constant speed, that is, at synchronous speed. However, like other electric machines, a synchronous machine undergoes transients during starting and abnormal con-

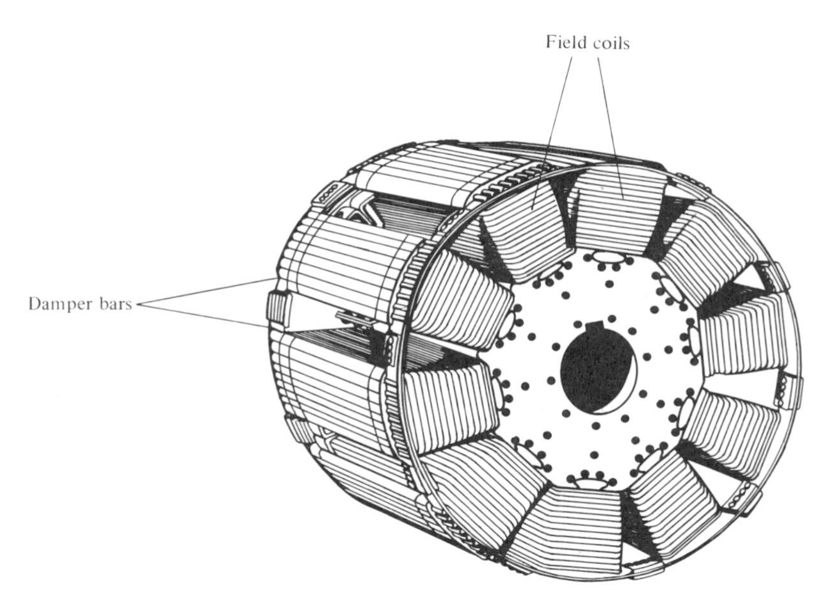

FIGURE 6-6. A salient rotor showing the field windings and damper bars (shaft not shown).

ditions. During transients, the rotor may undergo mechanical oscillations and its speed deviates from the synchronous speed, which is an undesirable phenomenon. To overcome this, an additional set of windings, resembling the cage of an induction motor, is mounted on the rotor. When the rotor speed is different from the synchronous speed, currents are induced in the damper windings. The damper winding acts like the cage rotor of an induction motor, producing a torque to restore the synchronous speed. Also, the damper bars provide a means of starting the machine, which is otherwise not self-starting. Figures 3-1(b) and 6-6 show the damper bars on round and salient rotors, respectively.

6-3 MAGNETOMOTIVE FORCES AND FLUXES DUE TO ARMATURE AND FIELD WINDINGS

In general, we may say that the behavior of an electric machine depends on the interaction between the magnetic fields (or fluxes) produced by various mmf's acting on the magnetic circuit of the machines. For instance, in a dc motor the torque is produced by the interaction of the flux produced by the field winding and the flux produced by the current-carrying armature conductors. In a dc generator also, we must consider the effect of the interaction between the field and armature mmf's (as discussed in Section 5-7).

As mentioned in the preceding section, the main sources of fluxes in a synchronous machine are the armature and the field mmf's. In contrast with a dc machine, in which the flux due to the armature mmf is stationary in space, the fluxes due to each phase of the armature mmf pulsate in time and the resultant flux rotates in space, as will be demonstrated later. For the present we shall consider the mmf's produced by a single full-pitch coil having N turns, as shown in Fig. 6-7, where the slot opening is negligible. Clearly, the machine has two poles [Fig. 6-7(a)]. The mmf has a constant value of Ni between the coil sides, as shown in Fig. 6-7(b). Traditionally, the magnetic effects of a winding in an electric machine are considered on a per pole basis. Thus if i is the current in the coil, the mmf per pole is $Ni/2$, which is plotted in Fig. 6-7(c). The reason for such a representation is that Fig. 6-7(c) also represents a flux-density distribution, but to a different scale. Obviously, the flux density over one pole (say the north pole) must be opposite to that over the other (south) pole, thus keeping

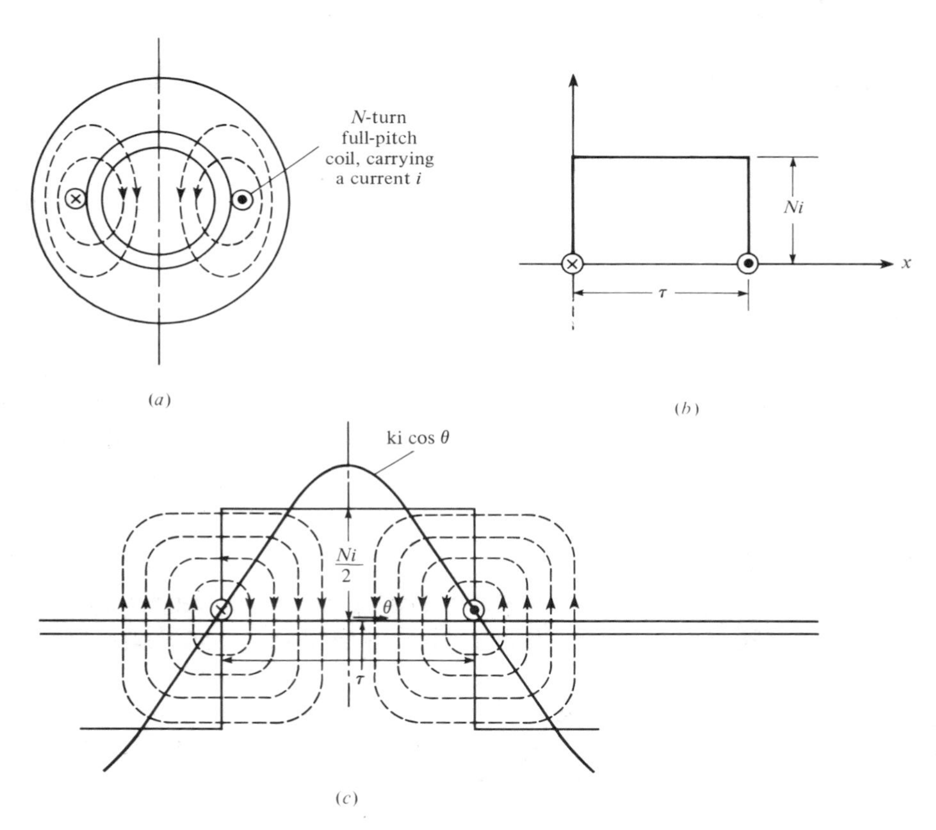

FIGURE 6-7. Flux and mmf produced by a concentrated winding. (a) Flux lines produced by an *N*-turn coil; (b) mmf produced by the *N*-turn coil; (c) mmf per pole.

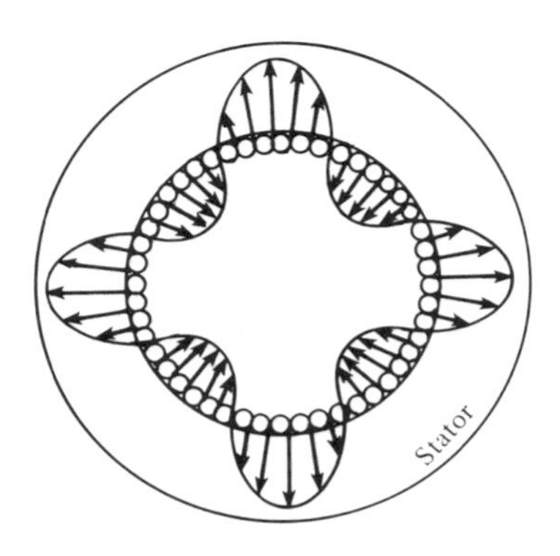

FIGURE 6-8. Four cycles of mmf correspond to eight poles or 1440 electrical degrees.

the flux entering the rotor equal to that leaving the rotor surface. Comparing parts (b) and (c), we notice that the representation of the mmf curve with positive and negative areas [part (c)] has the advantage that it gives the flux-density distribution, which must contain positive and negative areas. The mmf distribution shown in part (c) may be resolved into its harmonic components. The mmf of one phase of the winding shown in Fig. 6-4 can be obtained by appropriately adding the mmf's of individual coils (see Examples 6-2 and 6-3). In such properly designed armature windings, we may assume harmonics to be absent and the resultant mmf of each phase may be ideally taken as sinusoidal. Such an assumption considerably simplifies the mathematical analysis. A sinusoidally distributed mmf (or the fundamental component of the mmf $Ni/2$) is shown in Fig. 6-7(c). This mmf is expressed mathematically as

$$\mathcal{F} = ki \cos \theta \tag{6-1}$$

where k is a constant, i the current in the winding producing the mmf, and θ the angle measured with respect to the magnetic axis of the winding, as shown in Fig. 6-7(c). Notice from (6-1) that in one complete rotation in the air gap, corresponding to 360°, we obtain one complete cycle of the mmf. One cycle of mmf is said to correspond to 360 electrical degrees (in contrast to the 360 mechanical degrees in one rotation). It is possible to obtain more than one complete cycle of mmf in one complete rotation. For instance, in Fig. 6-8, the winding around the stator periphery is so arranged that we obtain four cycles of mmf. In other words, we obtain 1440 electrical degrees by going through 360 mechanical degrees. The ''cycles of mmf's'' are precisely designated by the number of poles, P. *One cycle of mmf corresponds to two poles and 360 electrical degrees,* and one complete rotation around the machine periphery corresponds to 360 mechanical degrees. Hence for a *P-pole* machine, the general relationship between the electrical degree θ and the mechanical degree θ_m is

$$\theta = \frac{P}{2}\,\theta_m \tag{6-2}$$

Obviously, $P/2$ is the number of *pole pairs* and is designated by p. Now we can generalize (6-1) for an mmf having p-pole pairs as

$$\mathcal{F} = ki \cos p\theta_m \tag{6-3}$$

Finally, if the winding current $i = I_m \sin \omega t$, then (6-3) becomes

$$\mathcal{F} = \mathcal{F}_m \sin \omega t \cos p\theta_m \tag{6-4}$$

where $\mathcal{F}_m = kI_m =$ amplitude of mmf wave.

Now let us consider the three-phase stator (or armature) winding of a synchronous machine to be excited by three-phase currents. As a result, the mmf's produced by the three phases are displaced from each other by 120° in time and space. (Recall from Fig. 6-4 that the armature windings of the three phases are

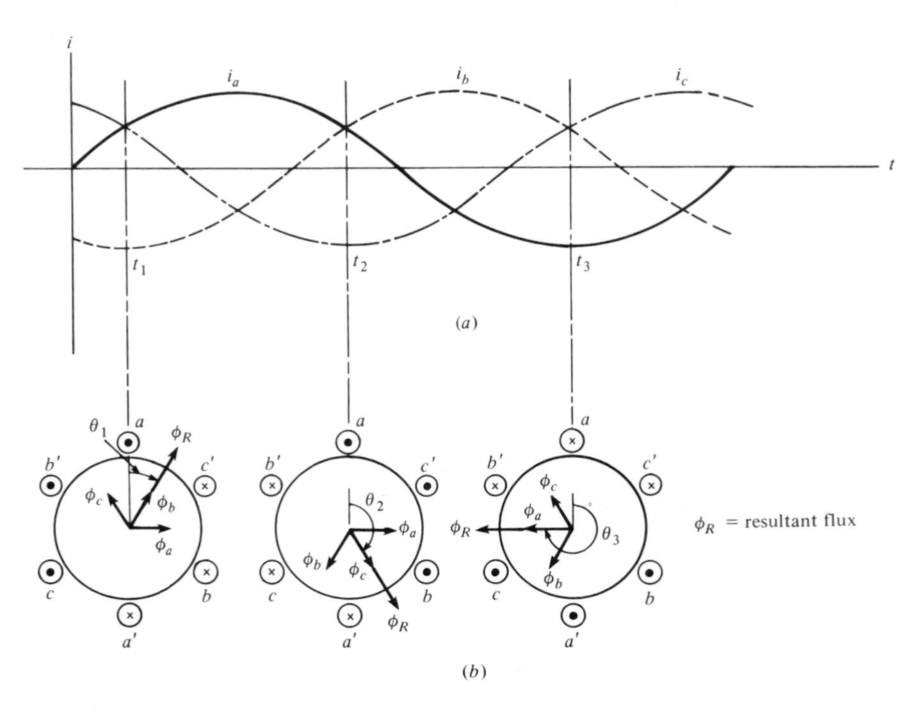

FIGURE 6-9. Production of a rotating magnetic field of a three-phase excitation. (a) Time diagram; (b) space diagram.

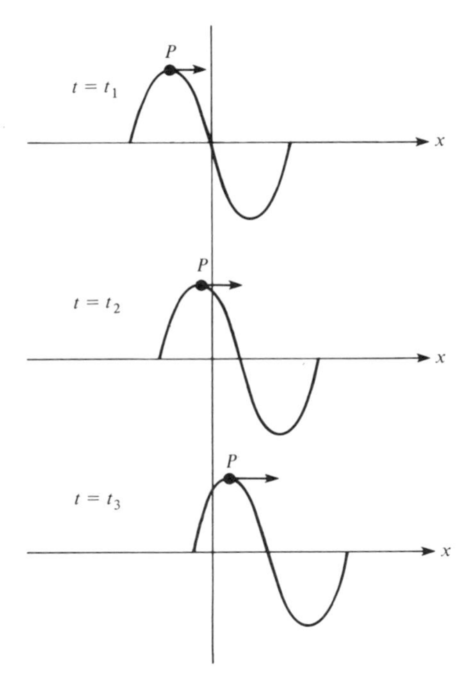

FIGURE 6-10. The function $\sin[\omega t - p\theta]$ at different time intervals $t_1 < t_2 < t_3$.

displaced from each other in space.) If we assume the mmf distribution in space to be sinusoidal, we may write for the three mmf's:

$$\mathcal{F}_a = \mathcal{F}_m \sin \omega t \cos p\theta_m$$

$$\mathcal{F}_b = \mathcal{F}_m \sin (\omega t - 120°) \cos (p\theta_m - 120°)$$

$$\mathcal{F}_c = \mathcal{F}_m \sin (\omega t + 120°) \cos (p\theta_m + 120°)$$

Observing that $\sin A \cos B = \frac{1}{2} \sin (A - B) + \frac{1}{2} \sin (A + B)$ and adding $\mathcal{F}_a$, $\mathcal{F}_b$, and $\mathcal{F}_c$, we obtain the resultant mmf as

$$\mathcal{F}(\theta, t) = 1.5\, \mathcal{F}_m \sin (\omega t - p\theta_m) \tag{6-5}$$

The magnetic field resulting from the mmf of (6-5) is a rotating magnetic field. Graphically, the production of the rotating field is illustrated in Fig. 6-9. Figure 6-10 shows the position of the resultant mmf at three different instants $t_1 < t_2 < t_3$. Notice that as time elapses, a fixed point P moves to the right, implying that the resultant mmf is a traveling wave of a constant amplitude. The magnetic field produced by this mmf in an electric machine is then known as a *rotating magnetic field*. We may arrive at the same conclusion by considering the resultant mmf at various instants, as shown in Fig. 6-9. From these diagrams it is clear that as we progress in time from t_1 to t_3, the resultant mmf rotates in space from θ_1 to θ_3. The existence of the rotating magnetic field is essential to the operation of a synchronous motor.

6-4 SYNCHRONOUS SPEED

To determine the velocity of the traveling field given by (6-5), imagine an observer traveling with the mmf wave from a certain point. To this observer, the magnitude of the mmf wave will remain constant (independent of time), implying that the right side of (6-1) would appear constant. Expressed mathematically, this would mean that

$$\sin(\omega t - p\theta) = \text{constant}$$

or

$$\omega t - p\theta = \text{constant}$$

Differentiating both sides with respect to t, we obtain

$$\omega - p\dot{\theta} = 0$$

or

$$\omega_m \equiv \dot{\theta} = \frac{\omega}{p} \tag{6-6}$$

This speed is known as the *synchronous speed,* which is the angular velocity of the mmf wave.

In (6-5) and (6-6), ω is the frequency of the stator mmf's. What is the significance of p? Notice that the given mmf's vary in space as $\cos p\theta$, indicating that for one complete travel around the stator periphery, the mmf undergoes p cyclic changes. Thus p may be considered as the order of harmonics, or the number of *pole pairs* in the mmf wave. If P is the *number of poles* then, obviously, $P = 2p$. Writing ω_m in terms of speed in r/min, n_s, and ω in terms of the frequency, f, we have

$$\omega_m = \frac{2\pi n_s}{60}$$

and

$$\omega = 2\pi f$$

Substituting for p, ω_m, and ω in (6-6) yields

$$n_s = \frac{120f}{P} \tag{6-7}$$

which is the *synchronous speed* in r/min.

An alternative form of (6-7) is

$$f = \frac{Pn_s}{120} \tag{6-8}$$

which implies that the frequency of the voltage induced in a synchronous generator having P poles and running at n_s r/min is f hertz. The same conclusion may be arrived at from Fig. 5-4: namely, in a two-pole machine, one cycle is generated in one rotation. Thus, in a P-pole machine, $P/2$ cycles are generated in one rotation; and in n_s rotations, $Pn_s/2$ cycles are generated. Since n_s rotations take 60 s, in 1 s $Pn_s/2 \times 60 = Pn_s/120$ cycles are generated, which is the frequency f.

EXAMPLE 6-1

For a 60-Hz generator, list four possible combinations of number of poles and speed.

SOLUTION

From (6-8), we must have $Pn_s = 7200 = 120 \times 60$. Hence we obtain the following table:

Number of Poles	Speed (r/min)
2	3600
4	1800
6	1200
8	900

■

EXAMPLE 6-2

An N-turn winding is made up of coils distributed in slots, such as the winding shown in Fig. 6-4. The voltages induced in these coils are displaced from one another in phase by the slot angle α. The resultant voltage at the terminals of the N-turn winding is then the phasor sum of the coil voltages. Find an expression for the *distribution factor*, k_d, where

$$k_d \equiv \frac{\text{magnitude of resultant voltage}}{\text{sum of magnitudes of individual coil voltages}} \tag{6-9}$$

SOLUTION

Let P be the number of poles, Q the number of slots, and m the number of phases. Then $Q = Pqm$, where q is the number of slots per pole per phase. The slot angle α is given (in electrical degrees) by

$$\alpha = \frac{(180°)P}{Q} = \frac{180°}{mq}$$

The phasor addition of voltages (for $q = 3$) is shown in Fig. 6-11, from the geometry of which we get

$$k_d = \frac{E_r}{qE_c} = \frac{2l \sin (q\alpha/2)}{q[2l \sin (\alpha/2)]} = \frac{\sin (q\alpha/2)}{q \sin (\alpha/2)} \tag{6-10}$$

which is the desired result. In (6-10), l is the radius of the arc of the circle, as shown in Fig. 6-11. ■

The distribution factors for a few three-phase windings are as follows:

Slot/Pole/Phase	2	3	4	5	6	8
k_d	0.966	0.960	0.958	0.957	0.957	0.956

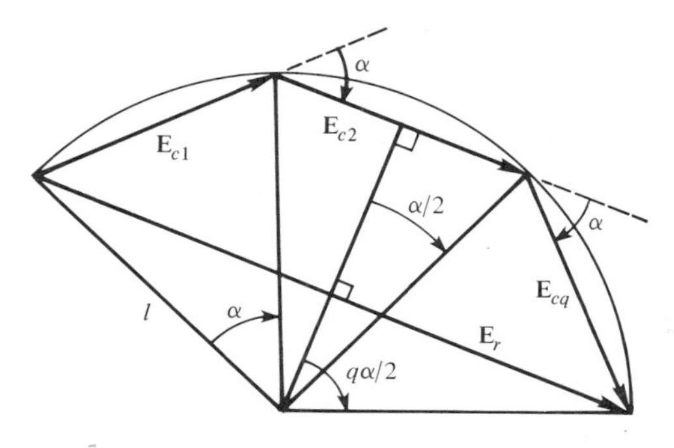

FIGURE 6-11. Phasor addition of individual-coil voltages to obtain the distribution factor.

EXAMPLE 6-3

The voltage induced in a fractional-pitch coil is reduced by a factor known as the *pitch factor*, k_p, as compared to the voltage induced in a full-pitch coil. Derive an expression for the pitch factor. Recall that the winding of Fig. 6-4 is a fractional-pitch winding.

SOLUTION

In a sinusoidally distributed flux density we show a full-pitch and a fractional-pitch coil in Fig. 6-12. The coil span of the full-pitch coil is equal to the pole pitch, τ. Let the coil span of the fractional-pitch coil be $\beta < \tau$, as shown. The flux linking the fractional-pitch coil will be proportional to the shaded area in the figure, whereas the flux linking the full-pitch coil is proportional to the entire area under the curve. The pitch factor is therefore the ratio of the shaded area to the total area:

$$k_p = \int_{(\tau-\beta)/2}^{(\tau+\beta)/2} \sin\frac{\pi x}{\tau}\,dx \Bigg/ \int_0^\tau \sin\frac{\pi x}{\tau}\,dx = \sin\frac{\pi\beta}{2\tau} \tag{6-11}$$

Notice that in (6-7), β and τ may be measured in any convenient unit and $\pi\beta/2\tau$ is in electrical radians. ■

EXAMPLE 6-4

Sometimes it is convenient to combine the distribution factor and the pitch factor as one factor, called the *winding factor*, k_w. Calculate the distribution factor, the pitch factor, and the winding factor, $k_w \equiv k_d k_p$, for the stator winding of Fig. 6-4.

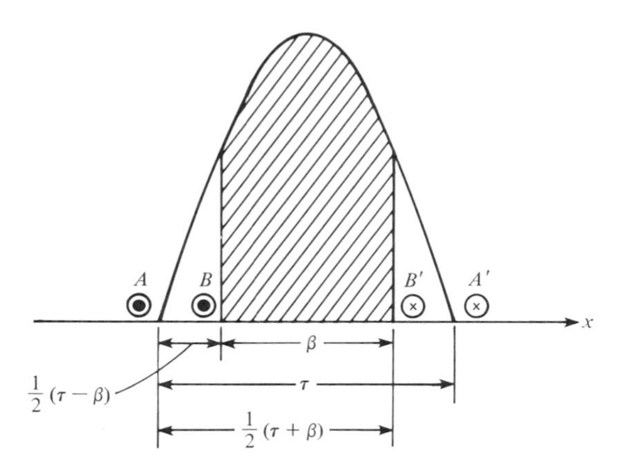

FIGURE 6-12. Determination of pitch factor.

SOLUTION

From Fig. 6-4, $m = 3$, $P = 4$, and $Q = 36$. Thus

$$q = \frac{36}{(4)(3)} = 3 \qquad \alpha = \frac{180^\circ}{(3)(3)} = 20^\circ$$

Substituting these in (6-10) yields

$$k_d = \frac{\sin 30^\circ}{3 \sin 10^\circ} = 0.96$$

Also, Fig. 6-4 shows that $\tau = 9$ slots and $\beta = 8$ slots. Hence, from (6-11),

$$k_p = \sin \frac{8\pi}{18} = \sin 80^\circ = 0.985$$

and

$$k_w = k_d k_p = (0.96)(0.985) = 0.945$$ ■

6-5 SYNCHRONOUS GENERATOR OPERATION

Like the dc generator, a synchronous generator functions on the basis of Faraday's law. If the flux linking the coil changes in time, a voltage is induced in a coil. Stated in another form, a voltage is induced in a conductor if it cuts magnetic flux lines (see Fig. 5-4). Considering the machine shown in Fig. 6-13 and assuming that the flux density in the air gap is uniform implies that sinusoidally varying voltages will be induced in the three coils aa', bb', and cc' if the rotor, carrying dc, rotates at a constant speed, n_s. Recalling from Chapter 3 that if φ is the flux per pole, ω is the angular frequency, and N is the number of turns in phase a (coil aa'), then the voltage induced in phase a is given by, according to (5-3),

$$\begin{aligned} e_a &= \omega N \varphi \sin \omega t \\ &= E_m \sin \omega t \end{aligned} \tag{6-12}$$

where $E_m = 2\pi f N \varphi$ and $f = \omega / 2\pi$ is the frequency of the induced voltage. Because phases b and c are displaced from phase a by $\pm$ 120° (Fig. 6-13), the corresponding voltages may be written as

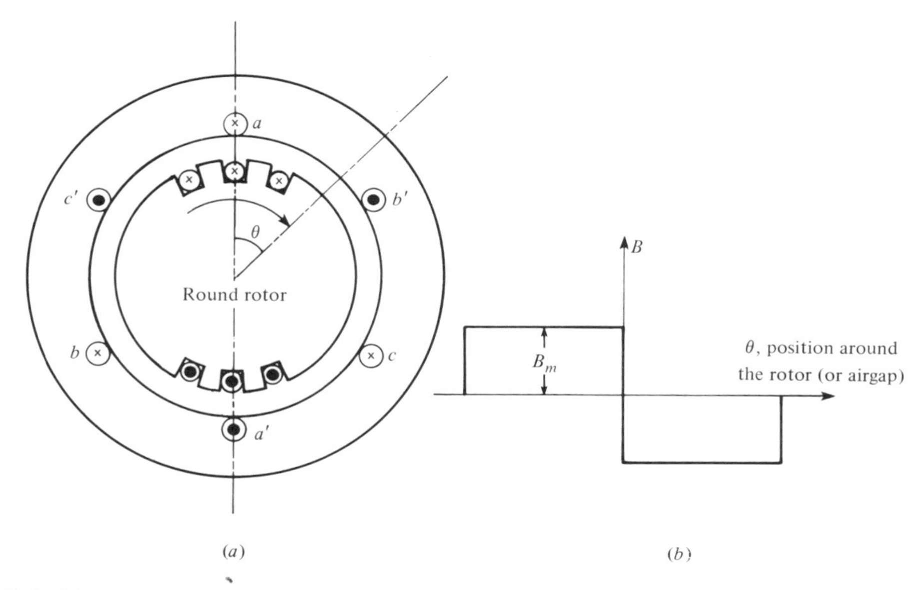

FIGURE 6-13. (a) A three-phase round rotor machine; (b) flux density distribution produced by the rotor excitation.

$$e_b = E_m \sin(\omega t - 120°)$$

$$e_c = E_m \sin(\omega t + 120°)$$

These voltages are sketched in Fig. 6-14 and correspond to the voltages from a three-phase generator.

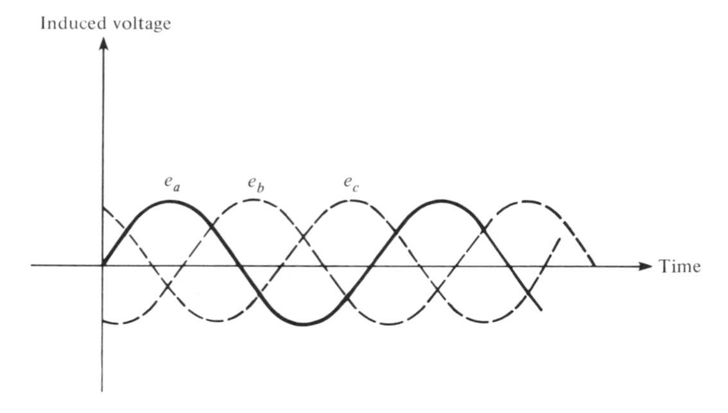

FIGURE 6-14. A three-phase voltage produced by a three-phase synchronous generator.

PERFORMANCE OF A ROUND-ROTOR SYNCHRONOUS GENERATOR

At the outset we wish to point out that we will study the machine on a per phase basis, implying a balanced operation. Thus let us consider a round-rotor machine operating as a generator on no-load. Variation of E_0 with I_f is shown in Fig. 6-15 and is known as the open-circuit characteristic of a synchronous generator. Let the open-circuit phase voltage be E_0 for a certain field current I_f. Here E_0 is the internal voltage of the generator. We assume that I_f is such that the machine is operating under unsaturated conditions. Next we short-circuit the armature at the terminals, keeping the field current unchanged (at I_f), and measure the armature phase current $\mathbf{I}_a$. In this case, the entire internal voltage $\mathbf{E}$ is dropped across the internal impedance of the machine. In mathematical terms,

$$\mathbf{E} = \mathbf{I}_a \mathbf{Z}_s$$

and $\mathbf{Z}_s$ is known as the *synchronous impedance*. One portion of $\mathbf{Z}_s$ is R_a and the other a reactance, X_s, known as synchronous reactance; that is,

$$\mathbf{Z}_s = R_a + jX_s \tag{6-13}$$

If the generator operates at a terminal voltage $\mathbf{V}_t$ while supplying a load corresponding to an armature current I_a, then

$$\mathbf{E} = \mathbf{V}_t + \mathbf{I}_a (R_a + jX_s) \tag{6-14}$$

where X_s, the *synchronous reactance,* which exists by virtue of the current-carrying armature windings.

In an actual synchronous machine, except in very small ones, we almost always have $X_s >> R_a$, in which case $\mathbf{Z}_s \simeq jX_s$. We will use this restriction in most of the analysis. Among the steady-state characteristics of a synchronous generator, its voltage regulation and power-angle characteristics are the most important ones. As for a transformer and a dc generator, we define the voltage regulation of a synchronous generator at a given load as

$$\text{percent voltage regulation} = \frac{E - V_t}{V_t} \times 100 \tag{6-15}$$

where V_t is the terminal voltage on load and E is the no-load terminal voltage. Clearly, for a given V_t, we can find E from (6-14) and hence the voltage regulation, as illustrated by the following examples.

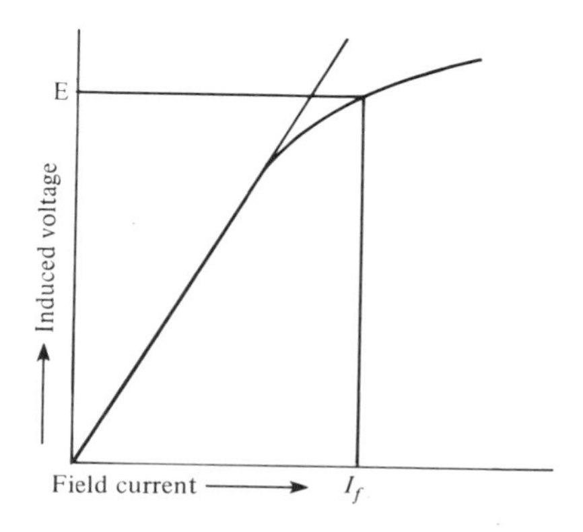

FIGURE 6-15. Open-circuit characteristics of a synchronous generator.

EXAMPLE 6-5

Calculate the percent voltage regulation for a three-phase wye-connected 2500-kVA 6600-V turboalternator operating at full-load and 0.8 lagging power factor. The per phase synchronous reactance and the armature resistance are 10.4 Ω and 0.071 Ω, respectively.

SOLUTION

Clearly, we have $X_s >> R_a$. The phasor diagram for the lagging power factor, neglecting the effect or R_a, is shown in Fig. 6-16(a). The numerical values are as follows:

$$V_t = \frac{6600}{\sqrt{3}} = 3810 \text{ V}$$

$$I_a = \frac{2500 \times 1000}{\sqrt{3} \times 6600} = 218.7 \text{ A}$$

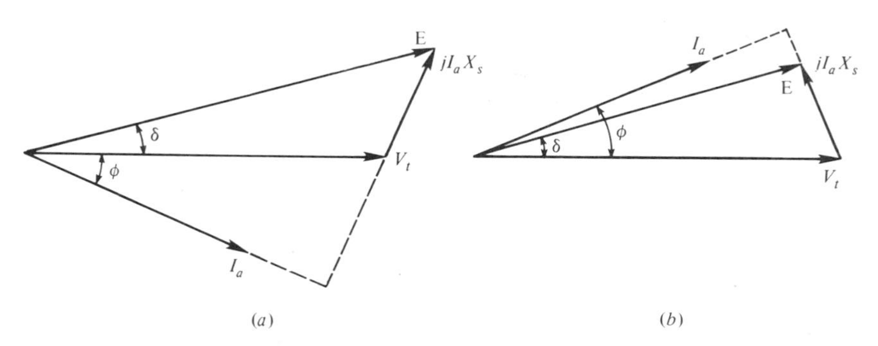

FIGURE 6-16. Phasor diagrams: (a) lagging power factor; (b) leading power factor.

From (6-14) we have

$$\mathbf{E} = 3810 + 218.7(0.8 - j0.6)j10.4 = 5485 \angle 19.3^\circ$$

and, from (6-15),

$$\text{percent regulation} = \frac{5485 - 3810}{3810} \times 100 = 44\% \quad \blacksquare$$

EXAMPLE 6-6

Repeat the preceding calculations with 0.8 leading power factor.

SOLUTION

In this case we have the phasor diagram shown in Fig. 6-16(b), from which we get

$$\mathbf{E} = 3810 + 218.7(0.8 + j0.6)j10.4 = 3048 \angle 36.6^\circ$$

and

$$\text{percent voltage regulation} = \frac{3048 - 3810}{3810} \times 100 = -20\% \quad \blacksquare$$

We observe from these examples that the voltage regulation is dependent on the power factor of the load. Unlike what happens in a dc generator, the voltage regulation for a synchronous generator may even become negative. The angle between $\mathbf{E}$ and $\mathbf{V}_t$ is defined as the *power angle*, δ. Notice that the power angle,

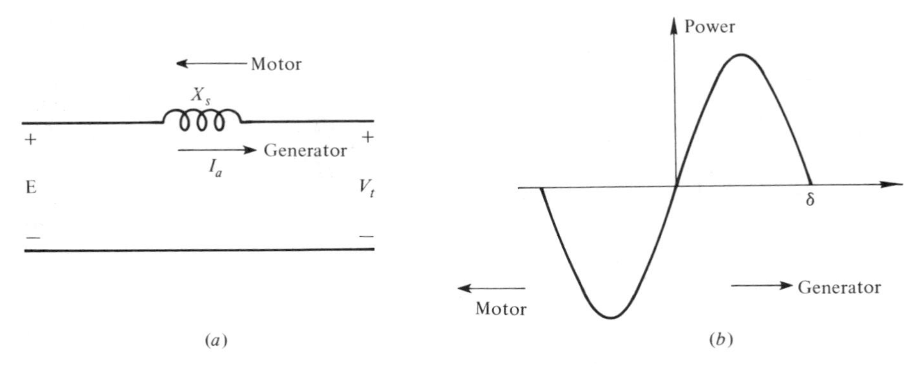

FIGURE 6-17. (a) An approximate equivalent circuit; (b) power–angle characteristics of a round-rotor synchronous machine.

δ, is not the same as the power factor angle, φ. To justify this definition, we consider Fig. 6-17(a), from which we obtain

$$I_a X_s \cos \varphi = E \sin \delta \tag{6-16}$$

Now, from the approximate equivalent circuit (assuming that $X_s >> R_a$), the power delivered by the generator = power developed, $P_d = V_t I_a \cos \varphi$ [which follows from Fig. 6-17(a) also]. Hence, in conjunction with (6-16), we get

$$P_d = \frac{EV_t}{X_s} \sin \delta \tag{6-17}$$

which shows that the internal power of the machine is proportional to sin δ. Equation (6-17) is often said to represent the *power-angle characteristic* of a synchronous machine. A plot of (6-17) is shown in Fig. 6-17(b), which shows that for a negative δ, the machine will operate as a motor, as discussed next.

6-7 SYNCHRONOUS MOTOR OPERATION

We know (from Section 6-3) that the stator of a three-phase synchronous machine, carrying a three-phase excitation, produces a rotating magnetic field in the air gap of the machine. Referring to Fig. 6-18, we will have a rotating magnetic field in the air gap of the salient pole machine when its stator (or armature) windings are fed from a three-phase source. Let the rotor (or field) winding be unexcited. The rotor will have a tendency to align with the rotating field at all times in order to present the path of least reluctance. Thus if the field is rotating, the rotor will tend to rotate with the field. From Fig. 6-13(a), we see that a round rotor will not tend to follow the rotating magnetic field because the uniform air gap presents the same reluctance all around the air gap and the rotor does not have any preferred direction of alignment with the magnetic field. This torque, which we have in the machine of Fig. 6-18 but not in the machine of Fig. 6-13(a), is called the *reluctance torque*. It is present by virtue of the variation of the reluctance around the periphery of the machine.

Next, let the field winding [Fig. 6-13(a) or 6-18] be fed by a dc source that produces the rotor magnetic field of definite polarities, and the rotor will tend to align with the stator field and will tend to rotate with the rotating magnetic field. We observe that for an excited rotor, a round rotor or a salient rotor will both tend to rotate with the rotating magnetic field, although the salient rotor will have an additional reluctance torque because of the saliency. In a later section we derive expressions for the electromagnetic torque in a synchronous machine attributable to field excitation and to saliency.

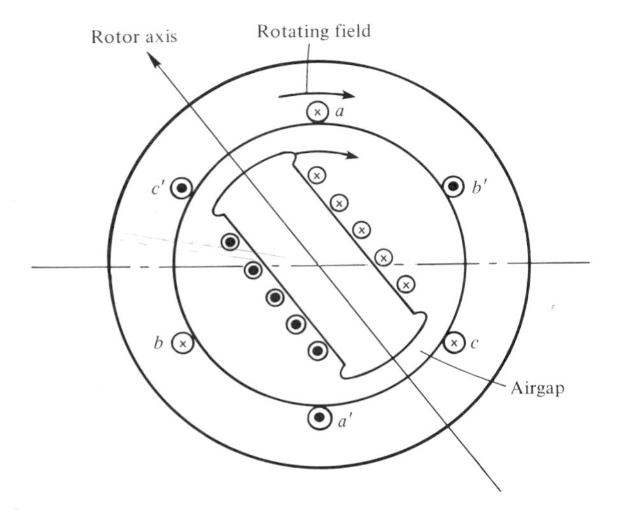

FIGURE 6-18. A salient-rotor machine.

So far we have indicated the mechanism of torque production in round-rotor and salient-rotor machines. To recapitulate, we might say that the stator rotating magnetic field has a tendency to "drag" the rotor along, as if a north pole on the stator "locks in" with a south pole of the rotor. However, if the rotor is at a standstill, the stator poles will tend to make the rotor rotate in one direction and then in the other as they rapidly rotate and sweep across the rotor poles. Therefore, a synchronous motor is not self-starting. In practice, as mentioned earlier, the rotor carries damper bars that act like the cage of an induction motor and thereby provide a starting torque. The mechanism of torque production by the damper bars is similar to the torque production in an induction motor, discussed in Chapter 7. Once the rotor starts running and almost reaches the synchronous speed, it locks into position with the stator poles. The rotor pulls into step with the rotating magnetic field and runs at the synchronous speed; the damper bars go out of action. Any departure from the synchronous speed results in induced currents in the damper bars, which tend to restore the synchronous speed. Machines without damper bars, or very large machines with damper bars, may be started by an auxiliary motor. We discuss the operating characteristics of synchronous motors in the next section.

6-8

PERFORMANCE OF A ROUND-ROTOR SYNCHRONOUS MOTOR

Except for some precise calculations, we may neglect the armature resistance as compared to the synchronous reactance. Therefore, the steady-state per phase equivalent circuit of a synchronous machine simplifies to the one shown in Fig. 6-17(a). Notice that this circuit is similar to that of a dc machine, where the dc armature resistance has been replaced by the synchronous reactance. In Fig. 6-17(a) we have shown the terminal voltage V_t, the internal excitation voltage E, and the armature current I_a going "into" the machine or "out of" it, depending on the mode of operation—"into" for motor and "out of" for generator. With the help of this circuit and (6-17) we will study some of the steady-state operating characteristics of a synchronous motor. In Fig. 6-17(b) we show the power-angle characteristics as given by (6-17). Here positive power and positive δ imply the generator operation, while a negative δ corresponds to a motor operation. Because δ is the angle between $\mathbf{E}$ and $\mathbf{V}_t$, $\mathbf{E}$ is ahead of $\mathbf{V}_t$ in a generator, whereas in a motor, $\mathbf{V}_t$ is ahead of $\mathbf{E}$. The voltage-balance equation for a motor is, from Fig. 6-17(a),

$$\mathbf{V}_t = \mathbf{E} + j\mathbf{I}_a\mathbf{X}_s$$

If the motor operates at a constant power, then (6-16) and (6-17) require that

$$E \sin \delta = I_a X_s \cos \varphi = \text{constant} \tag{6-18}$$

We recall that E depends on the field current, I_f (see Fig. 6-15). Consider two cases: (1) when I_f is adjusted so that $E < V_t$ and the machine is underexcited, and (2) when I_f is increased to a point that $E > V_t$ and the machine becomes overexcited. The voltage–current relationships for the two cases are shown in Fig. 6-19(a). For $E > V_t$ at constant power, δ is greater than the δ for $E < V_t$, as governed by (6-14). Notice that an underexcited motor operates at a lagging power factor (I_a lagging V_t), whereas an overexcited motor operates at a leading power factor. In both cases the terminal voltage and the load on the motor are the same. Thus we observe that the operating power factor of the motor is controlled by varying the field excitation, hence altering E. This is a very important property of synchronous motors. The locus of the armature current at a constant load, as given by (6-18), for varying field current is also shown in Fig. 6-19(a). From this we can obtain the variations of the armature current I_a with the field current, I_f (corresponding to E), and this can be done for different loads, as shown in Fig. 6-19(b). These curves are known as the *V curves* of the synchronous motor. One of the applications of a synchronous motor is in power factor correction, as demonstrated by the following examples. In addition to the V curves, we have also shown the curves for constant power factors. These curves are known as *compounding curves*.

In the preceding paragraph, we have discussed the effect of change in the field current on the synchronous machine power factor. However, the load supplied by a synchronous machine cannot be varied by changing the power factor. Rather, the load on the machine is varied by instantaneously changing the speed

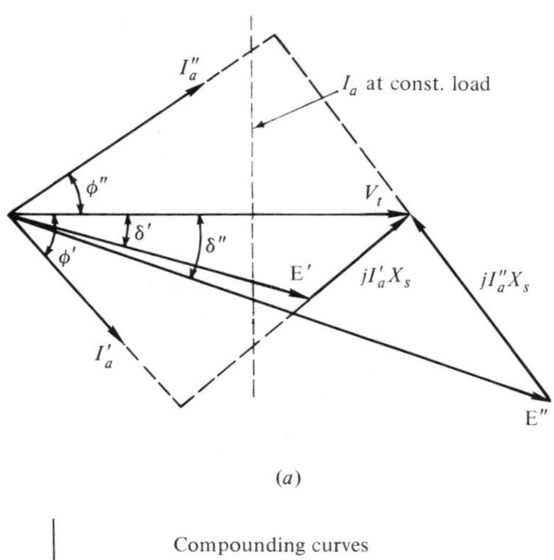

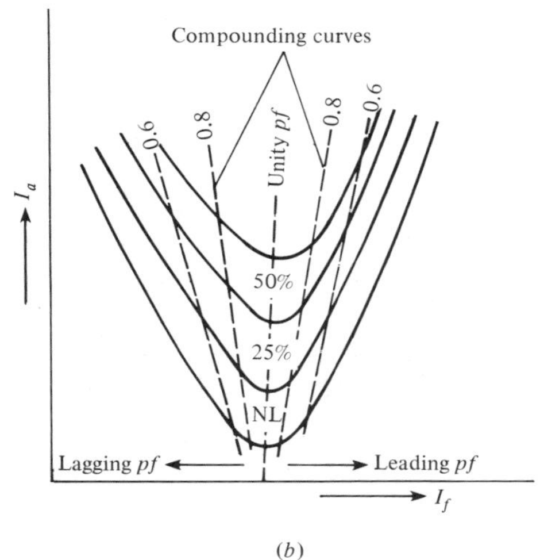

FIGURE 6-19. (a) Phasor diagram for motor operation (E′, $I_a{}'$, φ′, and δ′) correspond to underexcited operation. (E″, $I_a{}'$, φ″, and δ″) correspond to overexcited operation. (b) *V*-curves of a synchronous motor.

(in case of a generator, by supplying additional power by the prime mover), and thus changing the power angle corresponding to the new load. In a synchronous motor, a load change results in a change in the power angle.

EXAMPLE 6-7

A three-phase wye-connected load takes 50 A of current at 0.707 lagging power factor at 220 V between the lines. A three-phase wye-connected round-rotor synchronous motor, having a synchronous reactance of 1.27 Ω per phase, is connected in parallel with the load. The power developed by the motor is 33 kW at a power angle, δ, of 30°. Neglecting the armature resistance, calculate (a) the reactive kilovoltamperes (kvar) of the motor and (b) the overall power factor of the motor and the load.

SOLUTION

(a) The circuit and the phasor diagram on a per phase basis are shown in Fig. 6-20. From (6-17) we have

$$P_d = \frac{1}{3} \times 33{,}000 = \frac{220}{\sqrt{3}} \frac{E}{1.27} \sin 30^\circ$$

which yields $E = 220$ V. From the phasor diagram, $I_a X_s = 127$ or $I_a = 127/1.27 = 100$ A and $\varphi_a = 30^\circ$. The reactive kilovolt-amperes of the motor $= \sqrt{3} \times V_t I_a \sin \varphi_a = \sqrt{3} \times 220/1000 \times 100 \times \sin 30 =$ 19 kvar.

(b) Notice that φ_a, the power-factor angle of the motor, φ_L, the power-factor angle of the load, and φ, the overall power-factor angle, are shown in Fig. 6-20(b). The power angle, δ, is also shown in this phasor diagram, from which

$$\mathbf{I} = \mathbf{I}_L + \mathbf{I}_a$$

Or algebraically adding the real and reactive components of the currents, we obtain

$$I_{\text{real}} = I_a \cos \varphi_a + I_L \cos \varphi_L$$

$$I_{\text{reactive}} = I_a \sin \varphi_{a'} - I_L \sin \varphi_L$$

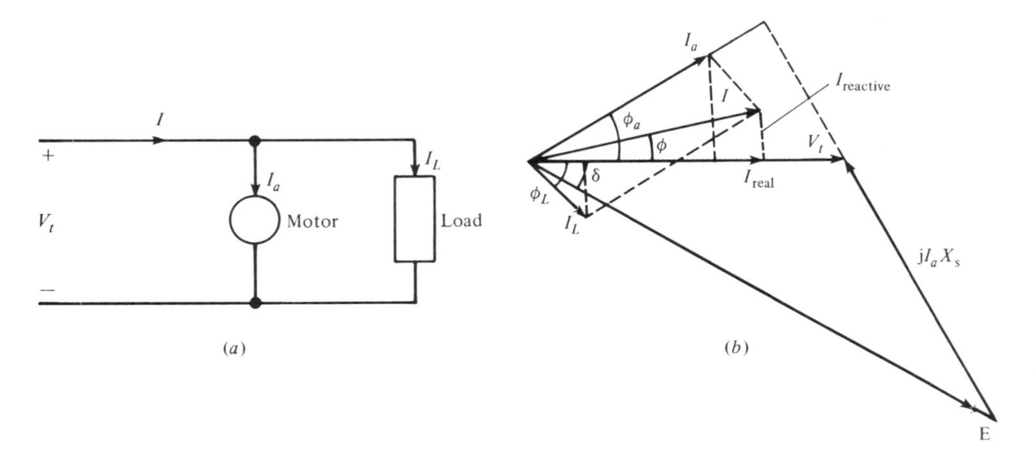

FIGURE 6-20. (a) Circuit diagram; (b) phasor diagram.

The overall power factor angle, φ, is thus given by

$$\tan \varphi = \frac{I_a \sin \varphi_a - I_L \sin \varphi_L}{I_a \cos \varphi_a + I_L \cos \varphi_L} = 0.122$$

or $\varphi = 7°$ and $\cos \varphi = 0.992$ leading. ■

EXAMPLE 6-8

For the generator of Example 6-5 calculate the power factor for zero voltage regulation on full load.

SOLUTION

Let φ be the power factor angle. Then

$$\mathbf{I}_a \mathbf{Z}_s = 218.7 \times 10.4 \underline{/\varphi + 89.6} = 2274.48 \underline{/\varphi + 89.6} \text{ V}$$

For voltage regulation to be zero, $|\mathbf{E}| = |\mathbf{V}_t|$, where $\mathbf{E}$ and $\mathbf{V}_t$ are related by

$$\mathbf{E} = \mathbf{V}_t + \mathbf{I}_a \mathbf{Z}_s$$

Rewriting the right-hand side in complex form and substituting the numerical values yields

$$\mathbf{E} = 3810 + j0 + 2274.48 \cos(\varphi + 89.6) + j2274.48 \sin(\varphi + 89.6)$$

For zero voltage regulation, $|\mathbf{E}| = 3810$. Hence

$$3810^2 = [3810 + 2274.48 \cos(\varphi + 89.6)]^2 + [2274.48 \sin(\varphi + 89.6)]^2$$

from which

$$\varphi = 17.76 \qquad \text{and} \qquad \cos \varphi = 0.95 \text{ leading}$$ ■

6-9 SALIENT-POLE SYNCHRONOUS MACHINES

In the preceding discussion we have analyzed the round-rotor machine and made extensive use of the machine parameter, which we defined as synchronous reactance. Because of saliency, the reactance measured at the terminals of a salient-rotor machine will vary as a function of the rotor position. This is not so in a round-rotor machine. Thus a simple definition of the synchronous reactance for a salient-rotor machine is not immediately forthcoming.

To overcome this difficulty, we use the two-reaction theory proposed by André Blondel. The theory proposes to resolve the given armature mmf's into two mutually perpendicular components, with one located along the axis of the rotor salient pole, known as the direct (or d) axis and with the other in quadrature and known as the quadrature (or q) axis. Correspondingly, we may define the d-axis and q-axis synchronous reactances, X_d and X_q, for a salient-pole synchronous machine. Thus, for generator operation, we draw the phasor diagram of Fig. 6-21. Notice that I_a has been resolved into its d- and q-axis (fictitious) components, I_d and I_q. With the help of this phasor diagram, we obtain

$$I_d = I_a \sin(\delta + \varphi) \qquad I_q = I_a \cos(\delta + \varphi)$$

$$V_t \sin\delta = I_q X_q = I_a X_q \cos(\delta + \varphi)$$

From these we get

$$V_t \sin\delta = I_a X_q \cos\delta \cos\varphi - I_a X_q \sin\delta \sin\varphi$$

or

$$(V_t + I_a X_q \sin\varphi)\sin\delta = I_a X_q \cos\delta \cos\varphi$$

Dividing both sides by cos δ and solving for tan δ yields

$$\tan\delta = \frac{I_a X_q \cos\varphi}{V_t + I_a X_q \sin\varphi} \tag{6-19}$$

With δ known (in terms of φ), the voltage regulation may be computed from

$$E = V_t \cos\delta + I_d X_d$$

$$\text{percent regulation} = \frac{E - V_t}{V_t} \times 100\%$$

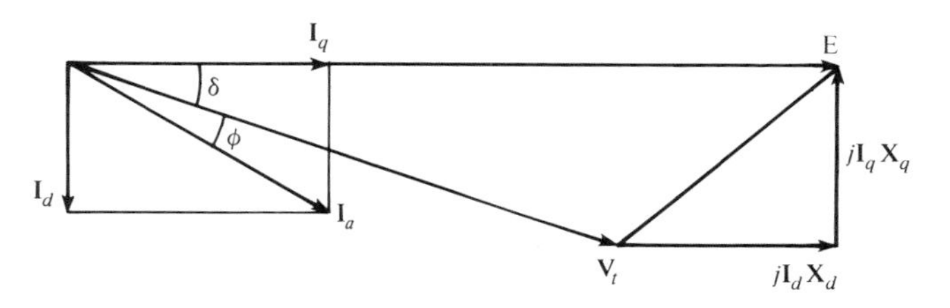

FIGURE 6-21. Phasor diagram of a salient-pole generator.

In fact, the phasor diagram depicts the complete performance characteristics of the machine.

Let us now use Fig. 6-21 to derive the power-angle characteristics of a salient-pole generator. If armature resistance is neglected, $P_d = V_t I_a \cos \varphi$. Now, from Fig. 6-21, the projection of I_a on V_t is

$$\frac{P_d}{V_t} = I_a \cos \varphi = I_q \cos \delta + I_d \sin \delta \tag{6-20}$$

Solving

$$I_q X_q = V_t \sin \delta \qquad \text{and} \qquad I_d X_d = E - V_t \cos \delta$$

for I_q and I_d, and substituting in (6-13), gives

$$P_d = \frac{EV_t}{X_d} \sin \delta + \frac{V_t^2}{2}\left(\frac{1}{X_q} - \frac{1}{X_d}\right) \sin 2\delta \tag{6-21}$$

Equation (6-21) can also be established for a salient-pole motor ($\delta < 0$); the graph of (6-21) is given in Fig. 6-22. Observe that for $X_d = X_q = X_s$, (6-21) reduces to the round-rotor equation, (6-17).

EXAMPLE 6-9

A 20-kVA 220-V 60-Hz wye-connected three-phase salient-pole synchronous generator supplies rated load at 0.707 lagging power factor. The phase constants

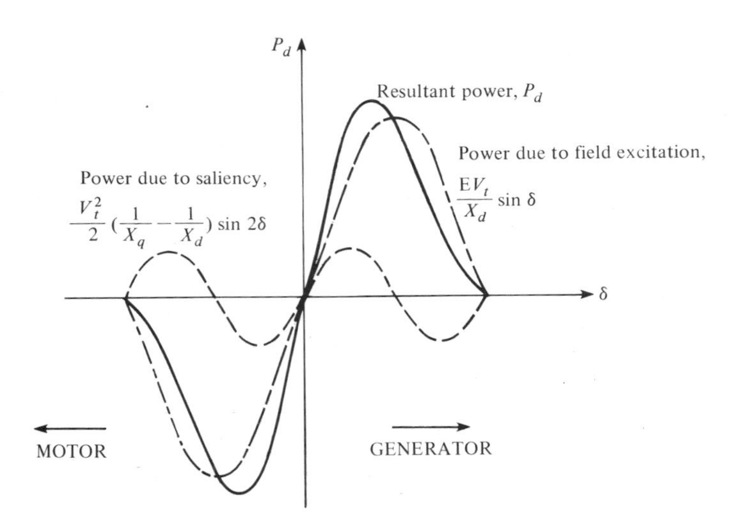

FIGURE 6-22. Power–angle characteristics of salient-pole machines.

of the machine are $R_a = 0.5\ \Omega$ and $X_d = 2X_q = 4.0\ \Omega$. Calculate the voltage regulation at the specified load.

SOLUTION

$$V_t = \frac{220}{\sqrt{3}} = 127\ \text{V}$$

$$I_a = \frac{20{,}000}{\sqrt{3} \times 120} = 52.5\ \text{A}$$

$$\varphi = \cos^{-1} 0.707 = 45^\circ$$

From (6-19),

$$\tan \delta = \frac{I_a X_q \cos \varphi}{V_t + I_a X_q \sin \varphi}$$

$$= \frac{52.5 \times 2 \times 0.707}{127 + 52.5 \times 2 \times 0.707} = 0.37$$

or

$$\delta = 20.6^\circ$$

$$I_d = 52.5 \sin (20.6 + 45) = 47.5\ \text{A}$$

$$I_d X_d = 47.5 \times 4 = 190.0\ \text{V}$$

$$E = V_t \cos \delta + I_d X_d$$

$$= 127 \cos 20.6 + 190 = 308\ \text{V}$$

and

$$\text{percent regulation} = \frac{E - V_t}{V_t} \times 100\% = \frac{308 - 127}{127} \times 100 = 142\%$$ ■

6-10 PARALLEL OPERATION

An electric power station often has several synchronous generators operating in parallel with each other. Some of the advantages of parallel operation are:

1. In the absence of one of the several machines, for maintenance or some other reason, the power station can function with the remaining units.

2. Depending on the load, generators may be brought on line, or taken off, and thus result in the most efficient and economical operation of the station.
3. For future expansion, units may be added on and operate in parallel.

In order that a synchronous generator may be connected in parallel with a system (or bus), the following conditions must be fulfilled:

1. The frequency of the incoming generator must be the same as the frequency of the power system to which the generator is to be connected.
2. The magnitude of the voltage of the incoming generator must be the same as the system terminal voltage.
3. With respect to an external circuit, the voltage of the incoming generator must be in the same phase as system voltage at the terminals.
4. In a three-phase system, the generator must have the same phase sequence as that of the bus.

The process of properly connecting a synchronous generator in parallel with a system is known as *synchronizing*. Two generators can be synchronized either by using a synchroscope or lamps. Figure 6-23 shows a circuit diagram showing lamps as well as synchroscope. The potential transformers (PTs) are used to reduce the voltage for instrumentation. Let the generator G_1 be already in operation with its switch S_{g1} closed. Other switches—S_{g2}, S_1, and S_2—are all open. After the generator G_2 is started and brought up to approximately synchronous speed, S_2 is closed. Subsequently, the lamps—L_a, L_b, and L_c—begin to flicker at a frequency equal to the difference of the frequencies of G_1 and G_2. The equality of the voltages of the two generators is ascertained by the voltmeter V, connected by the double-pole double-throw switch S. Now, if the voltages and frequencies of the two generators are the same, but there is a phase difference between the two voltages, the lamps will glow steadily. The speed of G_2 is then slowly adjusted until the lamps remain permanently dark (because they are connected such that two voltages through them are in opposition). Next, S_{g2} is closed and S_2 may be opened.

In the discussion above, it has been assumed that G_1 and G_2 both have the same phase rotation. On the other hand, let the phase sequence of G_1 be *abc* counterclockwise and that of G_2 be $a'b'c'$ clockwise. At the synchronous speed of G_1, a and a' may be coincident. This will be indicated by a dark L_a. But L_b and L_c will have equal brightness. When G_2 runs at a speed slightly less than the synchronous speed, the lamps will be dark and bright in the cyclical order L_a, L_b, and L_c. If either of the two conditions prevail, the phase rotation of G_1 must be reversed. This process of testing the phase sequence is known as *phasing out*.

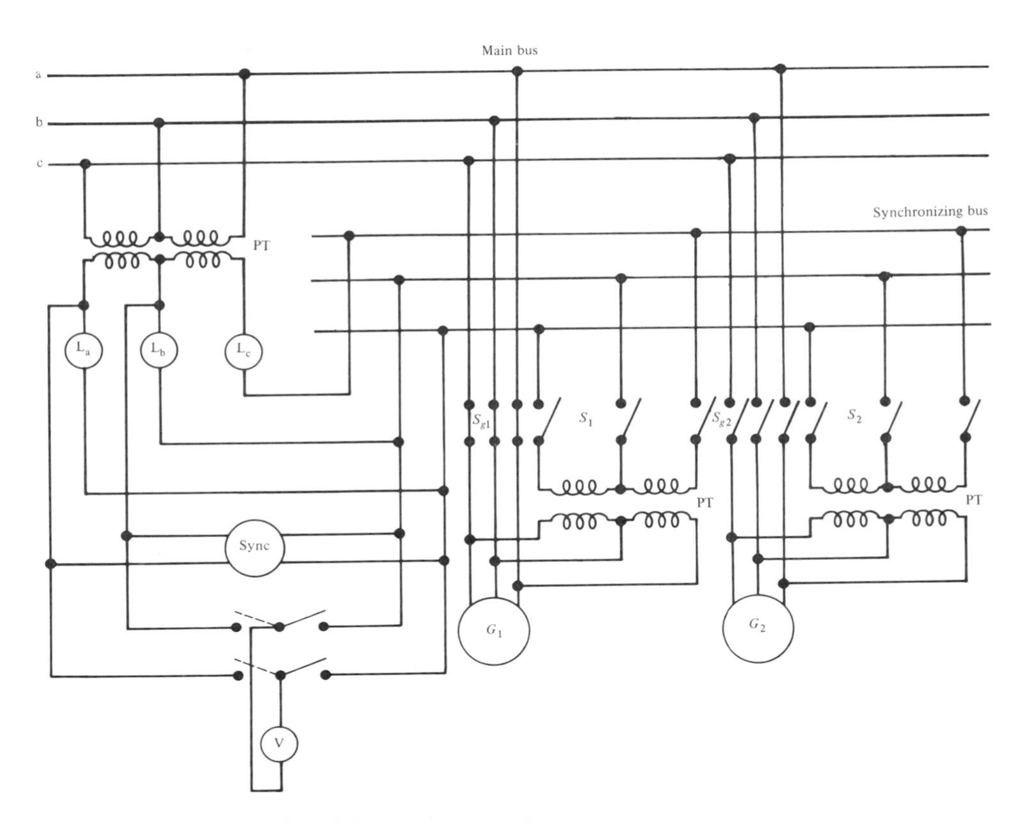

FIGURE 6-23. Synchronizing two generators.

A synchroscope is often used to synchronize two generators which have previously been phased out. A synchroscope is an instrument having a rotating pointer, which indicates whether the incoming machine is slow or fast. One type of synchroscope is shown schematically in Fig. 6-24. It consists of a field coil, F, connected to the main busbars through a large resistance R_F to ensure that the field current is almost in phase with the busbar voltage, V. The rotor consists of two windings R and X, in space quadrature, connected in parallel to each other and across the incoming generator. The windings R and X are so designed that their respective currents are approximately in phase and 90° behind the terminal voltage, V_i, of the incoming generator. The rotor will align itself so that the axes of R and F are inclined at an angle equal to the phase displacement between V and V_i. If there is a difference between the frequencies of V and V_i, the pointer will rotate at a speed proportional to this difference. The direction of rotation of the pointer will determine if the incoming generator is running below or above synchronism. At synchronism, the pointer will remain stationary at the index.

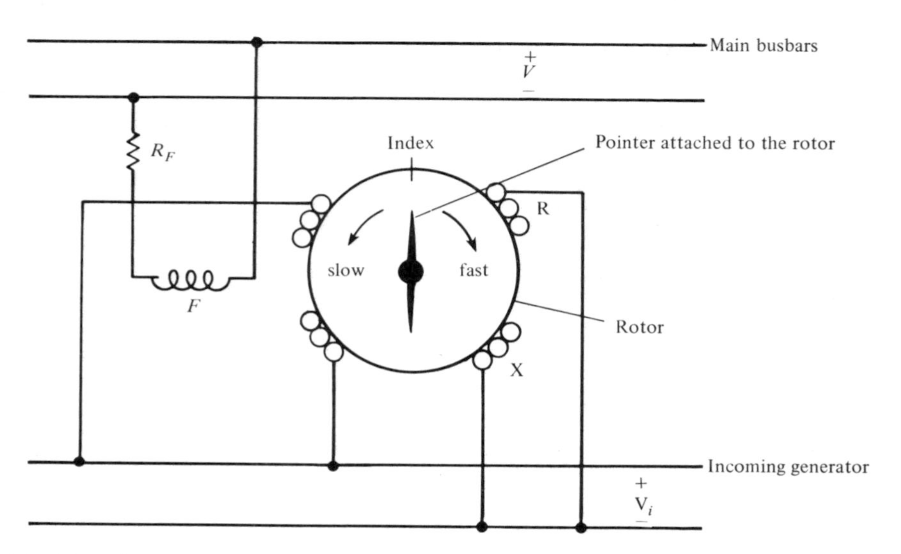

FIGURE 6-24. A synchroscope.

In modern power stations, automatic synchronizers are used. But a discussion of these is beyond the scope of this book.

Circulating Current and Load Sharing. At the time of synchronizing (that is, when S_2 of Fig. 6-23 is closed), if G_2 is running at a speed slightly less than that of G_1, the phase relationships of their terminal voltages with respect to the local circuit are as shown in Fig. 6-25(a). The resultant voltage V_c acts in the local circuit to set up a circulating current I_c lagging V_c by a phase angle φ_c. For simplification, if we assume the generators to be identical, then

$$\tan \varphi_c = \frac{R_a}{X_s} \tag{6-22}$$

and

$$I_c = \frac{V_c}{2Z_s} \tag{6-23}$$

where $R_a + jX_s = \mathbf{Z}_s$ = synchronous impedance R_a = armature resistance and X_s = synchronous reactance.

Notice from Fig. 6-25(a) that $\mathbf{I}_c$ has a component in phase with $\mathbf{V}_1$, and thus acts as a load on G_1 and tends to slow it down. The component of $\mathbf{I}_c$ in phase opposition to $\mathbf{V}_2$ aids G_2 to operate as a motor and thereby G_2 picks up speed. On the other hand, if G_2 was running faster than G_1 at the instant of synchro-

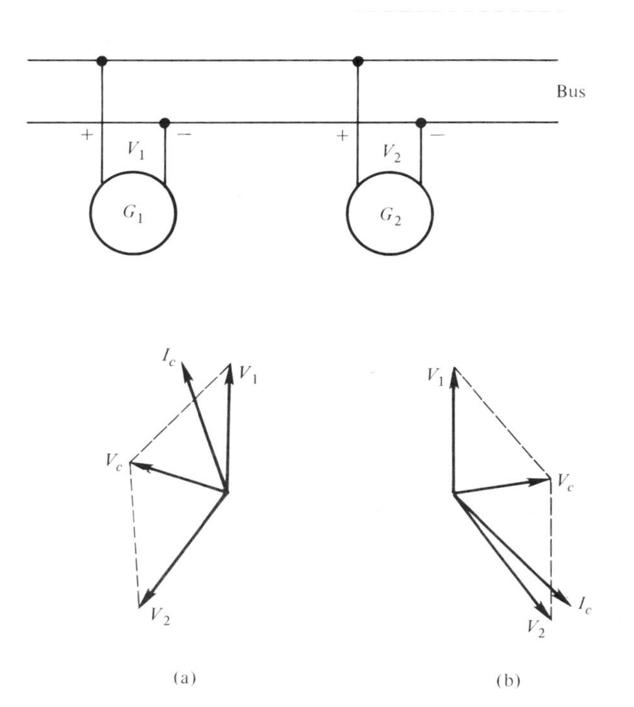

FIGURE 6-25. Circulating currents between two generators.

nization, the phase relationships of the voltages and the circulating current become as shown in Fig. 6-25(b). Consequently, G_2 will function as a generator and will tend to slow down; and while acting as a motor, G_1 will pick up speed. Thus there is an inherent synchronizing action which aids the machines to stay in synchronism.

We now recall from Section 6-6 that the power developed by a synchronous machine is given by (6-17). Observe that V_t is the terminal voltage, which is the same as the system busbar voltage. The voltage E is the internal voltage of the generator and is determined by the field excitation. As we have discussed earlier, a change in the field excitation merely controls the power factor and the circulation current at which the synchronous machine operates. The power developed by the machine depends on the power angle δ. For G_2 to share the load, (6-17) must be satisfied. Thus, for a given V_t and E, the power angle must be increased by increasing the prime-mover power. The load sharing between two synchronous generators is illustrated by the following examples.

EXAMPLE 6-10

Two identical three-phase wye-connected synchronous generators share equally a load of 10 MW at 33 kV and 0.8 lagging power factor. The synchronous reactance of each machine is 6 Ω per phase and the armature resistance is negligible. If one of the machines has its field excitation adjusted to carry 125 A of lagging current, what is the current supplied by the second machine? The prime mover inputs to both machines are equal.

SOLUTION

The phasor diagram of current division is shown in Fig. 6-26, wherein $\mathbf{I}_1 = 125$ A. Because the machines are identical and the prime-mover inputs to both machines are equal, each machine supplies the same true power:

$$I_1 \cos \varphi_1 = I_2 \cos \varphi_2 = \tfrac{1}{2} I \cos \varphi$$

Now

$$I = \frac{10 \times 10^6}{\sqrt{3}(33 \times 10^3)(0.8)} = 218.7 \text{ A}$$

whence

$$I_1 \cos \varphi_1 = I_2 \cos \varphi_2 = \tfrac{1}{2}(218.7)(0.8) = 87.5 \text{ A}$$

The reactive current of the first machine is therefore

$$I_1 |\sin \varphi_1| = \sqrt{(125)^2 - (87.5)^2} = 89.3 \text{ A}$$

and since the total reactive current is

$$I |\sin \varphi| = (218.7)(0.6) = 131.2 \text{ A}$$

the reactive current of the second machine is

$$I_2 |\sin \varphi_2| = 131.2 - 89.3 = 41.9 \text{ A}$$

Hence

$$I_2 = \sqrt{(87.5)^2 + (41.9)^2} = 97 \text{ A}$$

■

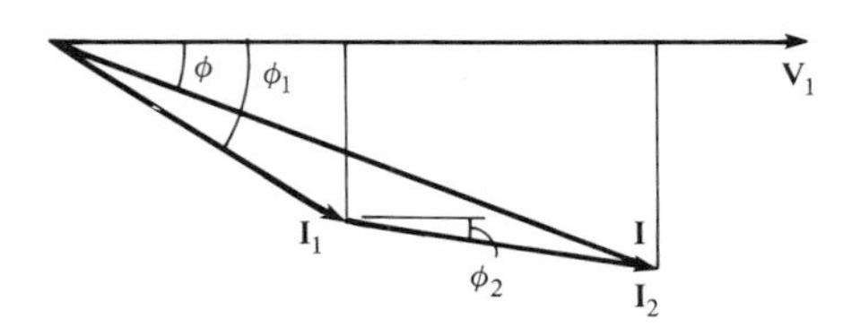

FIGURE 6-26. Example 6-10.

EXAMPLE 6-11

Consider the two machines of Example 6-10. If the power factor of the first machine is 0.9 lagging and the load is shared equally by the two machines, what are the power factor and current of the second machine?

SOLUTION

Load:

$$\text{power} = 10{,}000 \text{ kW}$$

$$\text{apparent power} = 12{,}500 \text{ kVA}$$

$$\text{reactive power} = -7500 \text{ kvar}$$

First machine:

$$\text{power} = 5000 \text{ kW}$$

$$\varphi_1 = \cos^{-1} 0.9 = -25.8°$$

$$\text{reactive power} = 5000 \tan \varphi_1 = -2422 \text{ kvar}$$

Second machine:

$$\text{power} = 5000 \text{ kW}$$

$$\text{reactive power} = -7500 - (-2422) = -5078 \text{ kvar}$$

$$\tan \varphi_2 = \frac{-5078}{5000} = -1.02$$

$$\cos \varphi_2 = 0.7$$

$$I_2 = \frac{5000}{\sqrt{3}(33)(0.7)} = 124.7 \text{ A}$$

■

6-11 DETERMINATION OF MACHINE CONSTANTS

The synchronous reactance, X_s, of a round-rotor synchronous machine can be obtained from open-circuit and short-circuit tests on the machine. The no-load or open-circuit voltage characteristic of a synchronous generator is similar to that of a dc generator. Figure 6-27 shows such a characteristic, with the effect of magnetic saturation included. Now, if the terminals of the generator are short-circuited, the induced voltage is dropped internally within the generator. The

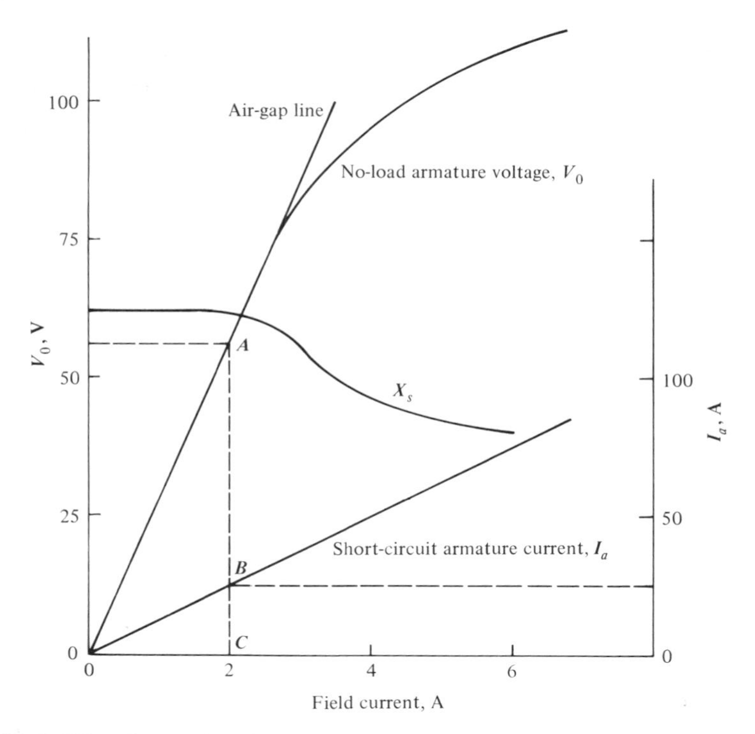

FIGURE 6-27. Open-circuit and short-circuit characteristics of a synchronous generator.

short-circuit current characteristic is also shown in Fig. 6-27. Expressed mathematically (on a per phase basis),

$$E = I_a Z_s = I_a(R_a + jX_s) \tag{6-24}$$

In (6-24), E is the no-load armature voltage at a certain field current, and I_a is the short-circuit armature current at the same value of the field current. The impedance Z_s is known as the *synchronous impedance*; R_a is the armature resistance and X_s is defined as the *synchronous reactance*. The synchronous reactance is readily measured for a round-rotor generator, since it is independent of the rotor position in such a machine. In salient-pole generators, however, the synchronous reactance depends on the rotor position.

In most synchronous machines, $R_a << X_s$, so that in terms of Fig. 6-26,

$$X_s \approx Z_s = \frac{\overline{AC}}{\overline{BC}}$$

Thus X_s varies with field current as indicated by the falling (because of saturation) curve in Fig. 6-27. However, for most calculations, we shall use the linear (constant) value of X_s.

Sudden Short-Circuit at the Armature Terminals. Consider a three-phase generator, on no-load, running at its synchronous speed and carrying a constant field current. Suddenly, the three phases are short-circuited. Symmetrical short-circuit armature current is graphed in Fig. 6-28. Notice that for the first few cycles the current, i_a, decays very rapidly; we term this duration the *subtransient period*. During the next several cycles, the current decreases somewhat slowly, and this range is called the *transient period*. Finally, the current reaches its steady-state value. These currents are limited, respectively, by the *subtransient reactance*, x_d''; the *transient reactance*, x_d'; and the synchronous reactance, X_d (or X_s). The subtransient reactance is essentially due to the presence of damper bars, the transient reactance accounts for the field winding, and the synchronous reactance is reactance due to the armature windings. It can be shown that the envelope of the instantaneous armature current (dashed curves in Fig. 6-28) is given by

$$i_a^* = \pm E\left[\left(\frac{1}{x_d''} - \frac{1}{x_d'}\right)e^{-t/\tau_d''} + \left(\frac{1}{x_d'} - \frac{1}{X_d}\right)e^{-t/\tau_d'} + \frac{1}{X_d}\right]$$

where τ_d'' = subtransient time constant

τ_d' = transient time constant

E = open-circuit armature phase voltage

The upper branch of the envelope is shown separately in Fig. 6-29.

The reactances and the time constants can be determined from design data, but the details are extremely cumbersome. On the other hand, these may be

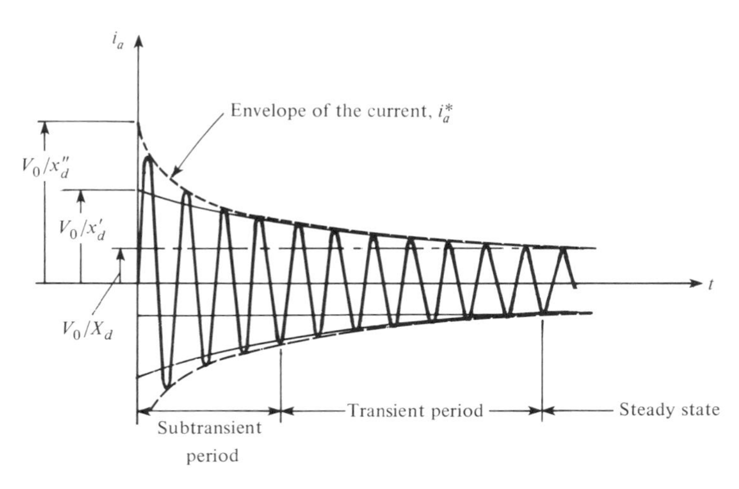

FIGURE 6-28. Armature current under sudden short-circuit.

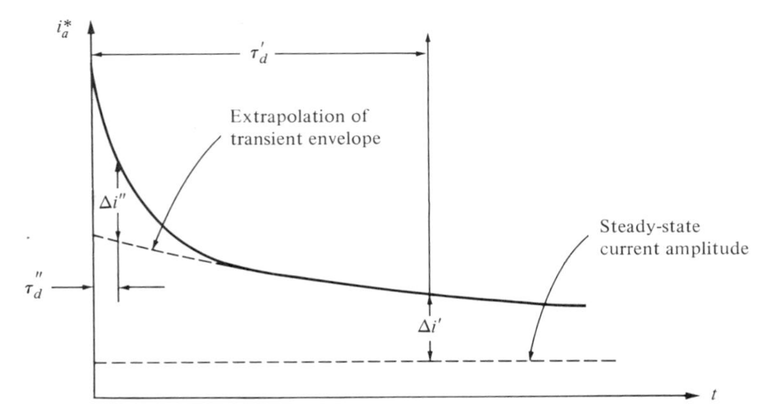

FIGURE 6-29. Envelope of the transient current.

determined from test data and Figs. 6-28 and 6-29, as illustrated by Example 6-12. In Fig. 6-29,

$$\Delta i'' = 0.368(i_d'' - i_d') \qquad \Delta i' = 0.368(i_d' - i_d)$$

where $i_d'' \equiv V_0/x_d''$, $i_d' \equiv V_0/x_d'$, and $i_d \equiv V_0/X_d$. Table 6-1 gives typical values of synchronous machine constants; the per unit values are based on the machine rating.

EXAMPLE 6-12

A three-phase short-circuit test is performed on a synchronous generator for which the envelope of the armature current is shown in Fig. 6-28. Given E = 231 V, $\Delta i''$ = 113 A, and $\Delta i'$ = 117 A, the steady-state short-circuit current is 144 A. Determine (a) X_d, (b) x_d', and (c) x_d''.

SOLUTION

(a)

$$X_d = \frac{E}{i_d} = \frac{231}{144} = 1.6\ \Omega$$

(b)

$$\Delta i' = 0.368(i_d' - i_d)$$

$$117 = 0.368(i_d' - 144)$$

$$i_d' = 462\ \text{A}$$

$$x_d' = \frac{E}{i_d'} = \frac{231}{462} = 0.5\ \Omega$$

TABLE 6-1 Per Unit Synchronous Machine Reactances and Time Constants

Constant	Salient-Pole Machine	Round-Rotor Machine
X_d	1.0–1.25	1.0–1.2
X_q	0.65–0.80	1.0–1.2
x'_d	0.35–0.40	0.15–0.25
x''_d	0.20–0.30	0.10–0.15
τ_d	0.15	0.15
τ'_d	0.9–1.1	1.4–2.0
τ''_d	0.03–0.04	0.03–0.04

(c)

$$\Delta i'' = 0.368(i''_d - i'_d)$$

$$113 = 0.368(i''_d - 462)$$

$$i''_d = 769 \text{ A}$$

$$x''_d = \frac{E}{i''_d} = \frac{231}{769} = 0.3\ \Omega$$

■

PROBLEMS

6-1 At what speed must a six-pole synchronous generator run to generate 50-Hz voltage?

6-2 A three-phase four-pole 60-Hz synchronous generator has 24 slots. The coil pitch is five slots. Calculate (a) the pitch factor and (b) the distribution factor.

6-3 A three-phase eight-pole 60-Hz synchronous generator has a 60-mWb flux per pole. The winding is full-pitched, and the distribution factor is 0.96. The armature has 120 turns per phase. Calculate the induced voltage per phase.

6-4 The armature of a three-phase eight-pole 900-r/min synchronous generator is wye-connected, and has 72 slots with 10 conductors per slot. The winding factor is 0.96. If 2400 V is measured across the line, determine the flux per pole.

6-5 The armature of a four-pole three-phase 1800 r/min machine is wye-connected and has 48 slots with four conductors per slot wound in two layers. The coil pitch is 150°, and the flux per pole is 80 mWb. Determine the open-circuit line-to-line voltage.

6-6 The open-circuit voltage of a 60-Hz generator is 11,000 V at a field current of 5 A. Calculate the open-circuit voltage at 50 Hz and 2.5 A of field current. Neglect saturation.

6-7 A 60-kVA three-phase wye-connected 440-V 60-Hz synchronous generator has a resistance of 0.15 Ω and a synchronous reactance of 3.5 Ω per phase. At rated load and unity power factor, calculate the percent voltage regulation.

6-8 A 1000-kVA 11-kV three-phase wye-connected synchronous generator supplies a 600-kW 0.8 leading power factor load. The synchronous reactance is 24 Ω per phase and the armature resistance is negligible. Calculate (a) the power angle and (b) the voltage regulation.

6-9 A synchronous generator produces 50 A of short circuit armature current per phase at a field current of 2.3 A. At this field current the open-circuit voltage is 250 V per phase. If the armature resistance is 0.9 Ω per phase, and the generator supplies a purely resistive load of 3 Ω per phase at a 130-V phase voltage and 50 A of armature current, determine the voltage regulation.

6-10 A 1000-kVA 11-kV three-phase wye-connected synchronous generator has an armature resistance of 0.5 Ω and a synchronous reactance of 5 Ω. At a certain field current the generator delivers rated load at 0.9 lagging power factor at 11 kV. For the same excitation, what is the terminal voltage at 0.9 leading power factor full-load?

6-11 An 11-kV three-phase wye-connected generator has a synchronous impedance of 6 Ω per phase, and negligible armature resistance. For a given field current, the open-circuit voltage is 12 kV. Calculate the maximum power developed by the generator. Determine the armature current and power factor for the maximum power condition.

6-12 A 400-V three-phase wye-connected synchronous motor delivers 12 hp at the shaft and operates at 0.866 lagging power factor. The total iron, friction, and field copper losses are 1200 W. If the armature resistance is 0.75 Ω per phase, determine the efficiency of the motor.

6-13 The motor of Problem 6-12 has a synchronous reactance of 6 Ω per phase, and operates at 0.9 leading power factor while taking an armature current of 20 A. Calculate the induced voltage.

6-14 A 1000-kVA 11-kV three-phase wye-connected synchronous motor has a 10-Ω synchronous reactance and a negligible armature resistance. Calculate the induced voltage for (a) 0.8 lagging power factor, (b) unity power factor, and (c) 0.8 leading power factor, when the motor takes 1000 kVA (in each case).

6-15 The per phase induced voltage of a synchronous motor is 2500 V. It lags behind the terminal voltage by 30°. If the terminal voltage is 2200 V per phase, determine the operating power factor. The per phase armature reactance is 6 Ω. Neglect armature resistance.

6-16 The per phase synchronous reactance of a synchronous motor is 8 Ω, and its armature resistance is negligible. The per phase input power is 400 kW and the induced voltage is 5200 V per phase. If the terminal voltage is 3800 V per phase, determine (a) the power factor and (b) the armature current.

6-17 A 2200-V three-phase 60-Hz four-pole wye-connected synchronous motor has a synchronous reactance of 4 Ω and an armature resistance of 0.1 Ω. The excitation is so adjusted that the induced voltage is 2200 V (line to line). If the line current is 220 A at a certain load, calculate (a) the input power, (b) the developed torque, and (c) the power angle.

6-18 An overexcited synchronous motor is connected across a 150-kVA inductive load of 0.7 lagging power factor. The motor takes 12 kW while running on no-load. Calculate the kVA rating of the motor if it is desired to bring the overall power factor of the motor-inductive load combination to unity.

6-19 Repeat Problem 6-18 if the synchronous motor is used to supply a 100-hp load at an efficiency of 90 percent.

6-20 A 60-kVA 400-V three-phase wye-connected salient-pole synchronous generator runs at 75 percent full-load at 0.9 leading power factor. The per phase direct- and quadrature-axis reactances are 1.4 Ω and 0.8 Ω, respectively. Neglecting the armature resistance, calculate (a) the power angle and (b) the developed power.

6-21 Two identical three-phase wye-connected synchronous generators share equally a load of 2500 kW at 33 kV and 0.866 lagging power factor. Each machine has a 5-Ω synchronous reactance per phase and a negligible armature resistance. One of the generators carries 75 A of current at a lagging power factor. What is the current supplied by the second machine?

6-22 The two machines of Problem 6-21 share the load equally. If the power factor of one machine is 0.9 lagging, determine (a) the power factor and (b) the current of the second machine.

CHAPTER 7

Induction Machines

7-1 INTRODUCTION

The induction motor is the most commonly used electric motor. It is considered to be the workhorse of the industry. Like the dc machine and the synchronous machine, an induction machine consists of a stator and a rotor mounted on bearings and separated from the stator by an air gap. Electromagnetically, the stator consists of a core made up of punchings (or laminations) carrying slot-embedded conductors. These conductors are interconnected in a predetermined fashion and constitute the armature windings. These windings are similar to those shown in Fig. 6-4.

Alternating current is supplied to the stator windings, and the currents in the rotor windings are induced by the stator currents. The rotor of the induction machine is cylindrical and carries either (1) conducting bars short-circuited at both ends, as in a *cage-type* machine, or (2) a polyphase winding with terminals brought out to slip rings for external connections, as in a *wound-rotor* machine. A wound-rotor winding is similar to that of the stator. Sometimes the cage-type machine is also called a *brushless* machine and the wound-rotor machine a *slip-ring* machine. The stator and the rotor, in its three different stages of pro-

FIGURE 7-1. Rotor for a 2500-kW two-pole 400-Hz induction motor in different stages of production. (Courtesy Brown Boveri Company)

duction, are shown in Fig. 7-1. The motor is rated at 2500 kW, 3 kV, 575 A, two-pole, and 400 Hz. A finished cage-type rotor of a 3400-kW 6-kV motor is shown in Fig. 7-2, and Fig. 7-3 shows the wound rotor of a three-phase slip-ring 15,200-kW four-pole induction motor. A cutaway view of a completely assembled motor, with a cage-type rotor, is shown in Fig. 7-4. The rotor is housed within the stator and is free to rotate therein.

An induction machine operates on the basis of interaction of induced rotor currents and the air-gap fields. If the rotor is allowed to run under the torque developed by this interaction, the machine will operate as a motor. On the other hand, the rotor may be driven by an external source beyond a speed such that the machine begins to deliver electrical power and operates as an induction generator (instead of as an induction motor, which absorbs electrical power). Thus we see that the induction machine is capable of functioning as a motor as well as a generator. In practice, applications of the induction machine as a generator are less common than motor applications. We will first study the motor operation, then develop the equivalent circuit of an induction motor, and subsequently we will show that the complete characteristics of an induction machine, operating either as a motor or as a generator, are obtainable from the equivalent circuit.

FIGURE 7-2. Complete rotor of a 3400 kW 6 kV 990 rpm induction motor. (Courtesy of Brown Boveri Company)

FIGURE 7-3. Rotor of a 15,200-kW 2.4kV three-phase slip-ring induction motor.

FIGURE 7-4. Cutaway of a three-phase cage-type induction motor. (Courtesy of General Electric Company)

7-2

OPERATION OF A THREE-PHASE INDUCTION MOTOR

The key to the operation of an induction motor is the production of the rotating magnetic field. We established in Section 6-3 that a three-phase stator excitation produces a rotating magnetic field in the air gap of the machine, and the field rotates at a synchronous speed given by (6-6). As the magnetic field rotates, it "cuts" the rotor conductors. By this process, voltages are induced in the conductors. The induced voltages give rise to rotor currents, which interact with the air-gap field to produce a torque. The torque is maintained as long as the rotating magnetic field and the induced rotor current exist. Consequently, the rotor starts rotating in the direction of the rotating field. The rotor will achieve a steady-state speed, n, such that $n < n_s$. Clearly, when $n = n_s$, there will be no induced currents and hence no torque. The condition $n > n_s$ corresponds to the generator mode, as we shall see later.

An alternative approach to explaining the operation of the polyphase induction motor is by considering the interaction of the (excited) stator magnetic field with the (induced) rotor magnetic field. The stator excitation produces a rotating magnetic field, which rotates in the air gap at a synchronous speed. The field induces polyphase currents in the rotor, thereby giving rise to another rotating magnetic field, which also rotates at the same synchronous speed as that of the stator and with respect to the stator. Thus we have two rotating magnetic fields, rotating at a synchronous speed with respect to the stator but stationary with respect to each other. Consequently, according to the principle of alignment of magnetic fields, the rotor experiences a torque. The rotor rotates in the direction of the rotating field of the stator.

7-3 SLIP

The actual mechanical speed, n, of the rotor is often expressed as a fraction of the synchronous speed, n_s, as related by *slip*, s, defined as

$$s = \frac{n_s - n}{n_s} \tag{7-1}$$

where n_s is as given in (6-7), which is repeated below for convenience:

$$n_s = \frac{120f}{P} \tag{7-2}$$

The slip may also be expressed as percent slip, as follows:

$$\text{percent slip} = \frac{n_s - n}{n_s} \times 100 \tag{7-3}$$

At standstill, the rotating magnetic field produced by the stator has the same relative speed with respect to the rotor windings as with respect to the stator windings. Thus the frequency of the rotor currents, f_r, is the same as the frequency of stator currents, f. At synchronous speed, there is no relative motion between the rotating field and the rotor, and the frequency of rotor current is zero. At other speeds, the rotor frequency is proportional to the slip; that is,

$$f_r = sf \tag{7-4}$$

where f_r is the frequency of rotor currents and f is the frequency of stator input current (or voltage).

EXAMPLE 7-1

A six-pole three-phase 60-Hz induction runs at 4 percent slip at a certain load. Determine (a) the synchronous speed, (b) the rotor speed, (c) the frequency of rotor currents, (d) the speed of the rotor rotating field with respect to the stator, and (e) the speed of the rotor rotating field with respect to the stator rotating field.

SOLUTION

(a) From (7-2),

$$n_s = \frac{120 \times 60}{6} = 1200 \text{ r/min}$$

(b) From (7-1),

$$n = (1 - s)n_s = (1 - 0.04) \times 1200 = 1152 \text{ r/min}$$

(c) From (7-4),

$$f_r = 0.04 \times 60 = 2.4 \text{ Hz}$$

(d) The six poles on the stator induce six poles on the rotor. The rotating field produced by the rotor rotates at a corresponding synchronous speed, n_r, relative to the rotor such that

$$n_r = \frac{120 f_r}{P} = \frac{120 f}{P} s = s n_s$$

But the speed of the rotor with respect to the stator is

$$n = (1 - s)n_s$$

Hence the speed of the rotor field with respect to the stator is

$$n'_s = n_r + n = s n_s + (1 - s)n_s = 1200 \text{ rpm}$$

(e) The speed of the rotor field with respect to the stator field is

$$n'_s - n_s = n_s - n_s = 0$$

■

7-4 DEVELOPMENT OF EQUIVALENT CIRCUITS

In order to develop an equivalent circuit of an induction motor, we consider the similarities between a transformer and an induction motor (on a per phase basis).

If we consider the primary of a transformer to be similar to the stator of the induction motor, its rotor corresponds to the secondary of the transformer. From this analogy, it follows that the stator and rotor each have their respective resistances and leakage reactances. Because the stator and the rotor are magnetically coupled, we must have a magnetizing reactance, just as in a transformer. The air gap in an induction motor makes its magnetic circuit relatively poor and the corresponding magnetizing reactance will be relatively smaller, compared to that of a transformer. The hysteresis and eddy-current losses in an induction motor can be represented by a shunt resistance, as was done for the transformer. Up to this point, we have mentioned the similarities between a transformer and an induction motor. A major difference between the two, however, is introduced because of the rotation of the rotor. Consequently, the frequency of rotor currents is different from the frequency of the stator currents [see (7-4)]. Keeping these facts in mind, we now proceed to represent a three-phase induction motor by a stationary equivalent circuit.

Considering the rotor first and recognizing that the frequency of rotor currents is the slip frequency, we may express the per phase rotor leakage reactance, x_2, at a slip s, in terms of the standstill per phase reactance X_2:

$$x_2 = sX_2 \tag{7-5}$$

Next we observe that the magnitude of the voltage induced in the rotor circuit is also proportional to the slip.

A justification of this statement follows from transformer theory because we may view the induction rotor at standstill as a transformer with an air gap. For the transformer we know that the induced voltage, say E_2, is given by

$$E_2 = 4.44fN\varphi_m \tag{7-6}$$

But at a slip s, the frequency becomes sf. Substituting this value of frequency into (7-6) yields the voltage e_2 at a slip s as

$$e_2 = 4.44sfN\varphi_m = sE_2$$

We conclude, therefore, that if E_2 is the per phase voltage induced in the rotor at standstill, the voltage e_2 at a slip s is given by

$$e_2 = sE_2 \tag{7-7}$$

Using (7-5) and (7-7), we obtain the rotor equivalent circuit shown in Fig. 7-5(a). The rotor current I_2 is given by

$$I_2 = \frac{sE_2}{\sqrt{R_2^2 + (sX_2)^2}}$$

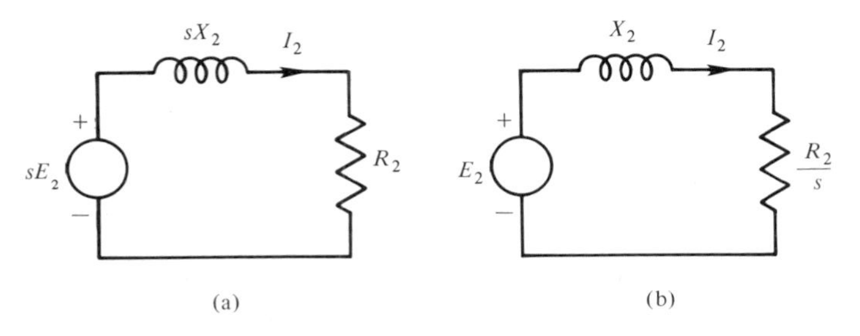

FIGURE 7-5. Two forms of rotor equivalent circuit.

which may be rewritten as

$$I_2 = \frac{E_2}{\sqrt{\left(\frac{R_2}{s}\right)^2 + X_2^2}} \tag{7-8}$$

resulting in the alternative form of the equivalent circuit shown in Fig. 7-5(b). Notice that these circuits are drawn on a per phase basis. To this circuit we may now add the per phase stator equivalent circuit to obtain the complete equivalent circuit of the induction motor.

In an induction motor, only the stator is connected to the ac source. The rotor is not generally connected to an external source, and rotor voltage and current are produced by induction. In this regard, as mentioned earlier, the induction motor may be viewed as a transformer with an air gap, having a variable resistance in the secondary. Thus we may consider that the primary of the transformer corresponds to the stator of the induction motor, whereas the secondary corresponds to the rotor on a per phase basis. Because of the air gap, however, the value of the magnetizing reactance, X_m, tends to be relatively low compared with that of a transformer. As in a transformer, we have a mutual flux linking both the stator and rotor, represented by the magnetizing reactance and various leakage fluxes. For instance, the total rotor leakage flux is denoted by X_2 in Fig. 7-5. Now considering that the rotor is coupled to the stator as the secondary of

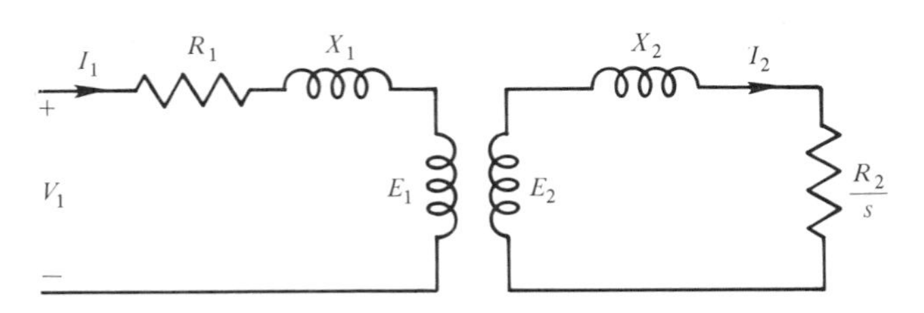

FIGURE 7-6. Stator and rotor as coupled circuits.

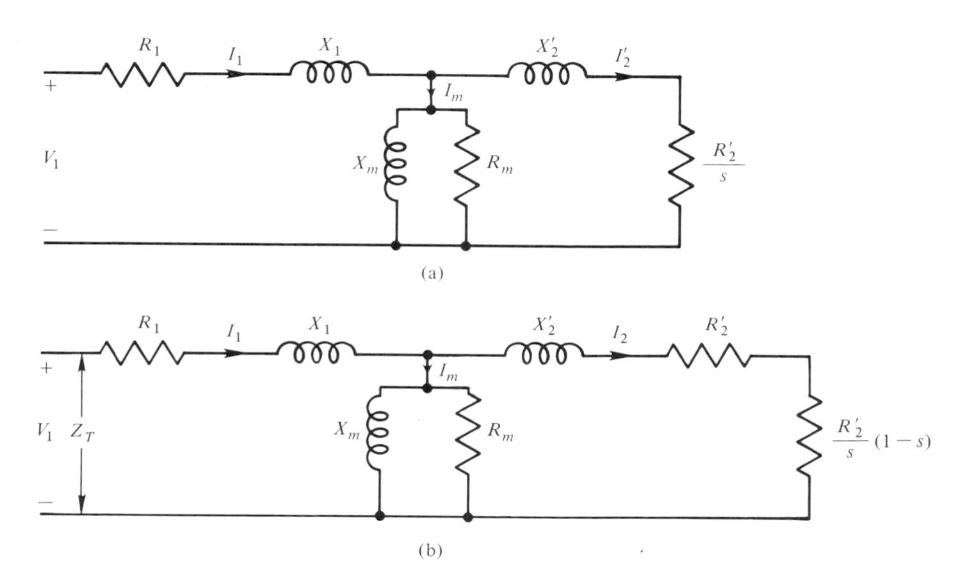

FIGURE 7-7. Two forms of equivalent circuits of an induction motor.

a transformer is coupled to its primary, we may draw the circuit shown in Fig. 7-6. To develop this circuit further, we need to express the rotor quantities as referred to the stator, as was done for the transformer. The pertinent details are cumbersome and are not considered here. However, having referred the rotor quantities to the stator, we obtain the exact equivalent circuit (per phase) shown in Fig. 7-7 from the circuit given in Fig. 7-6. For reasons that will become immediately clear, we split R'_2/s as

$$\frac{R'_2}{s} = R'_2 + \frac{R'_2}{s}(1 - s)$$

to obtain the circuit shown in Fig. 7-7. Here R'_2 is simply the per phase standstill rotor resistance referred to the stator and $R'_2(1 - s)/s$ is a dynamic resistance that depends on the rotor speed and corresponds to the load on the motor. Notice that all the parameters shown in Fig. 7-7 are standstill values and the circuit is the per phase exact equivalent circuit referred to the stator.

7-5 PERFORMANCE CALCULATIONS

We will now show the usefulness of equivalent circuit in determining the motor performance. To illustrate the procedure we refer to Fig. 7-7. We make an approximation in Fig. 7-7, and neglect R_m. We draw the approximate circuit in

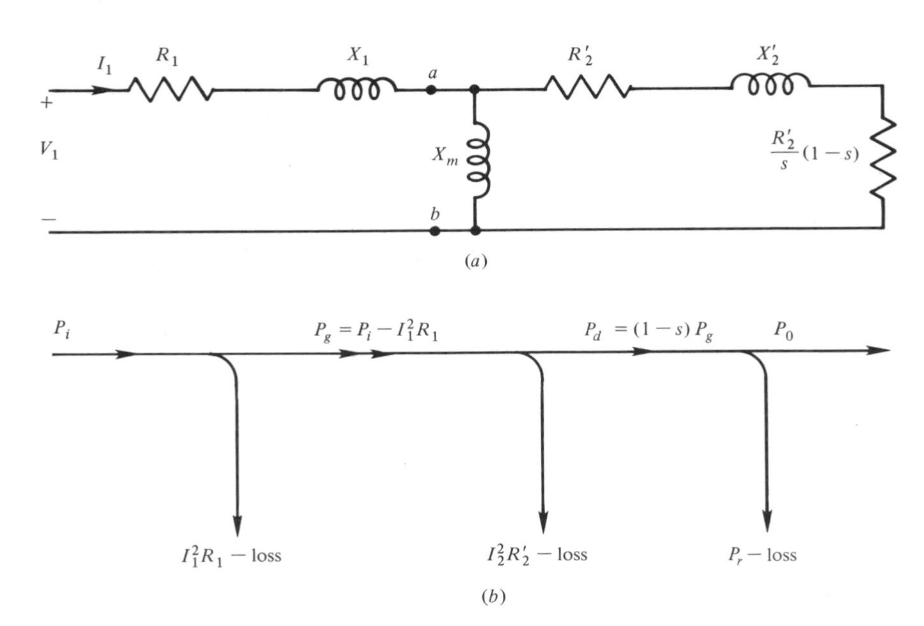

FIGURE 7-8. (a) An approximate equivalent circuit of an induction motor; (b) power flow in an induction motor.

Fig. 7-8, where we also show approximately the power flow and various power losses in one phase of the machine. From Fig. 7-8 we obtain the following relationships on a per phase basis:

$$\text{stator } I^2R \text{ loss} = I_1^2R_1 \tag{7-8}$$

$$\text{power crossing the air gap} = P_g = P_i - I_1^2R_1 \tag{7-9}$$

Since this power, P_g, is dissipated in R'_2/s [of Fig. 7-7(a)], we also have

$$P_g = \frac{I_2^2R'_2}{s} \tag{7-10}$$

Subtracting $I_2^2R'_2$ loss from P_g yields the developed electromagnetic power, P_d. Thus

$$P_d = P_g - I_2^2R'_2 \tag{7-11}$$

From (7-10) and (7-11) we get

$$P_d = (1 - s)P_g \tag{7-12}$$

This power appears across the resistance $R_2'(1 - s)/s$ (of Fig. 7-8) corresponding to the load. Subtracting the mechanical rotational power, P_r, from P_d gives the output power, P_o. Hence

$$P_o = P_d - P_r \tag{7-13}$$

and

$$\text{efficiency} = \frac{P_o}{P_i} \tag{7-14}$$

Torque calculations can be made from the power calculations. Thus, to determine the electromagnetic torque, T_e, developed by the motor, at a speed ω_m (rad/s), we write

$$T_e \omega_m = P_d \tag{7-15}$$

But

$$\omega_m = (1 - s)\omega_s \tag{7-16}$$

where ω_s is the synchronous speed in rad/s. From (7-12), (7-15), and (7-16) we obtain

$$T_e = \frac{P_g}{\omega_s} \tag{7-17}$$

which gives the torque at a slip s. At standstill $s = 1$; hence the standstill torque developed by the motor is given by

$$(T_e)_{\text{standstill}} = \frac{P_{gs}}{\omega_s} \tag{7-18}$$

Notice that in Fig. 7-8 we have neglected the core losses, most of which are in the stator. We will include core losses only in efficiency calculations. The reason for this simplification is to reduce the amount of the complex arithmetic required in numerical computations. We now illustrate the details of calculations by the following example.

EXAMPLE 7-2

The parameters of the equivalent circuit (Fig. 7-8) for a 220-V three-phase four-pole wye-connected 60-Hz induction motor are

$$R_1 = 0.2\ \Omega \qquad R'_2 = 0.1\ \Omega$$

$$X_1 = 0.5\ \Omega \qquad X'_2 = 0.2\ \Omega$$

$$X_m = 20.0\ \Omega$$

The total iron and mechanical losses are 350 W. For a slip of 2.5 percent, calculate input current, output power, output torque, and efficiency.

SOLUTION

From Fig. 7-8, the total impedance is

$$\mathbf{Z}_1 = R_1 + jX_1 + \frac{jX_m\left(\dfrac{R'_2}{s} + jX'_2\right)}{\dfrac{R'_2}{s} + j(X_m + X'_2)}$$

$$= 0.2 + j0.5 + \frac{j20(4 + j0.2)}{4 + j(20 + 0.2)}$$

$$= (0.2 + j0.5) + (3.77 + j0.95) = 4.23\,\underline{/20^\circ}$$

$$V_1 = \text{phase voltage} = \frac{220}{\sqrt{3}} = 127\ \text{V}$$

$$I_1 = \text{input current} = \frac{127}{4.23} = 30\ \text{A}$$

$$\cos\varphi = \text{power factor} = \cos 20^\circ = 0.94$$

$$3V_1I_1\cos\varphi = \text{total input power} = \sqrt{3} \times 220 \times 30 \times 0.94 = 10.75\ \text{kW}$$

From the equivalence of Figs. 7-8(a) and 7-9, we obtain total power across the air gap

$$P_g = 3 \times 30^2 \times 3.77 = 10.18\ \text{kW}$$

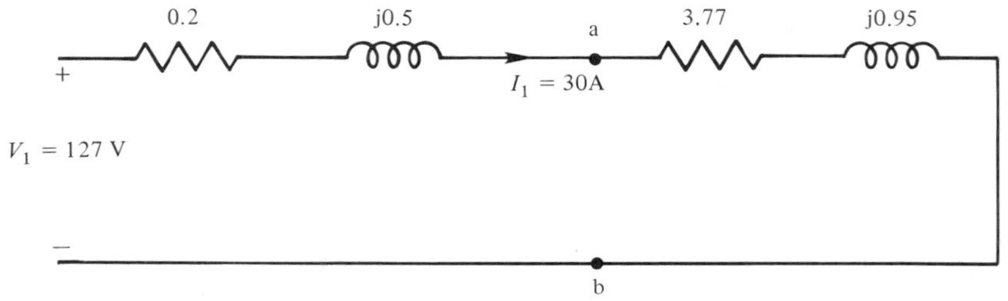

FIGURE 7-9. Example 7.2.

Notice that 3.77 Ω is the equivalent resistance between terminals a and b in the two circuits.

$$\text{total power developed, } P_d = (1 - s)P_g = 0.975 \times 10.18 = 9.93 \text{ kW}$$

$$\text{total output power} = P_d - P_{\text{core}} = 9.93 - 0.35 = 9.58 \text{ kW}$$

$$\text{total output torque} = \frac{\text{output power}}{\omega_m}$$

$$= \frac{9.58}{184} \times 1000 = 52 \text{ N-m}$$

where, from (7-6)

$$\omega_m = (1 - s)\frac{\omega}{p} = 0.075 \times 60 \times \pi = 184 \text{ rad/s}$$

$$\text{efficiency} = \frac{\text{output power}}{\text{input power}} = \frac{9.58}{10.75} = 89.1\%$$

■

Using this procedure, we can calculate the performance of the motor at other values of the slip, ranging from 0 to 1. The characteristics thus calculated are shown in Fig. 7-10.

EXAMPLE 7-3

A two-pole 60-Hz three-phase induction motor develops 25 kW of electromagnetic power at a certain speed. The rotational mechanical loss at this speed is

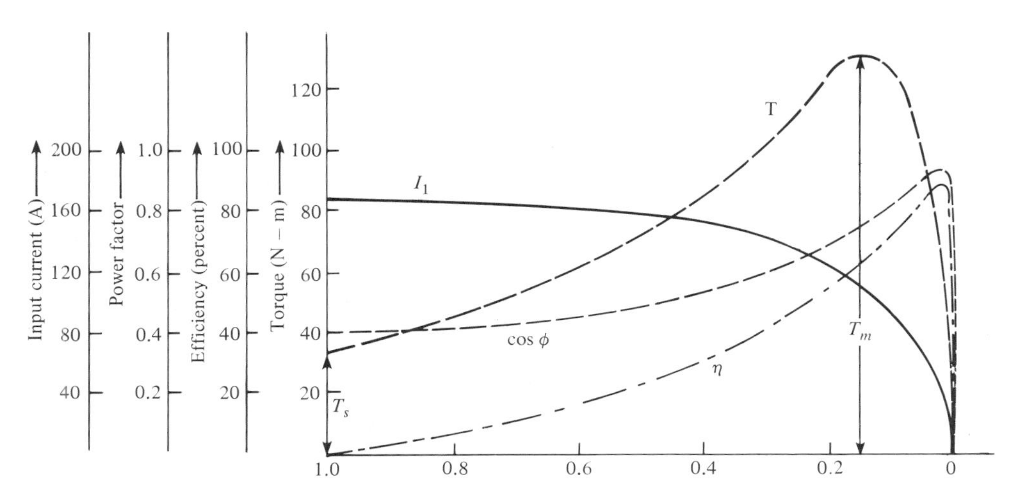

FIGURE 7-10. Characteristics of an induction motor. T_m = maximum torque; T = starting torque.

400 W. If the power crossing the air gap is 27 kW, calculate (a) the slip and (b) the output torque.

SOLUTION

(a) From (7-11) we have

$$1 - s = \frac{P_d}{P_g}$$

or

$$1 - s = \frac{25}{27}$$

Thus $s = 0.074$ (or 7.4 percent).

(b) The developed torque is given by (7-17). Substituting $\omega_s = 2\pi n_s/60 = 2\pi \times 3600/60$, $P_g = 27$ kW (given) in (7-17) gives

$$T_e = \frac{27{,}000}{2\pi \times 3600/60} = 71.62 \text{ N-m}$$

Torque lost due to mechanical rotation is found from

$$T_{\text{loss}} = \frac{P_r}{\omega_m} = \frac{400}{(1 - 0.074)2\pi \times 3600/60} = 1.15 \text{ N-m}$$

Hence

$$\text{output torque} = T_e - T_{\text{loss}} = 71.62 - 1.15 = 70.47 \text{ N-m}$$ ■

7-6 APPROXIMATE EQUIVALENT CIRCUIT FROM TEST DATA

The preceding examples illustrate the usefulness of equivalent circuits of induction motors. For most purposes, an approximate equivalent circuit is adequate. One such circuit is shown is Fig. 7-12. Obviously, in order to use this circuit for calculations, its parameters must be known. The parameters of the circuit shown in the figure can be obtained from the following tests.

No-Load Test. In this test, the motor is run on no-load. Input power and current are recorded at several voltages ranging from 25 to 125 percent of the

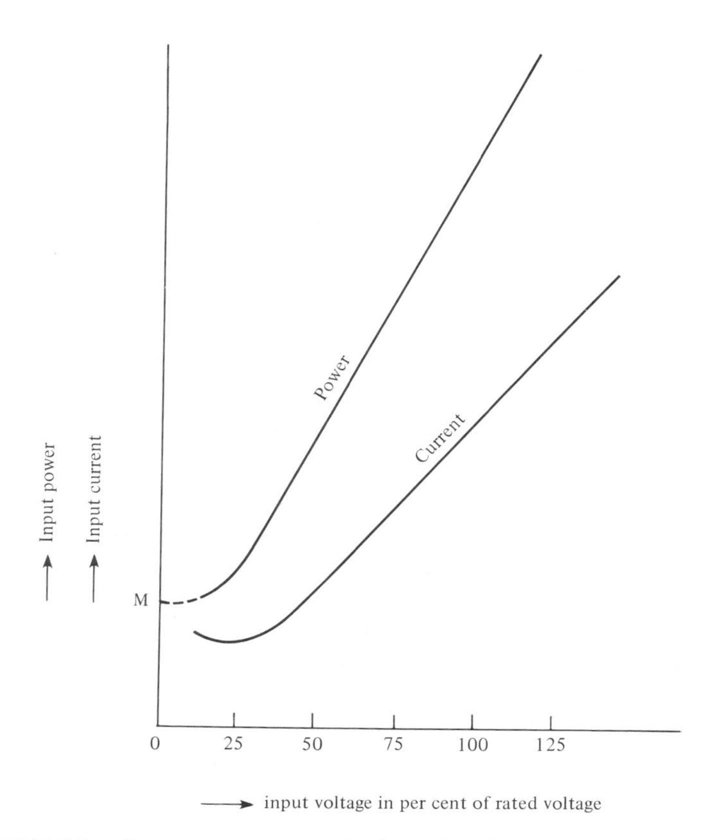

FIGURE 7-11. An approximate equivalent circuit (per phase) of an induction motor.

rated voltage. Results thus obtained are plotted in Fig. 7-11. As the voltage is reduced, the flux decreases proportionately. The power curve is approximately parabolic. At about 20 percent of the rated voltage, the magnetizing current and core loss are small, and the input power is almost entirely due to mechanical losses. By extrapolating the power curve to zero voltage, as shown in the figure,

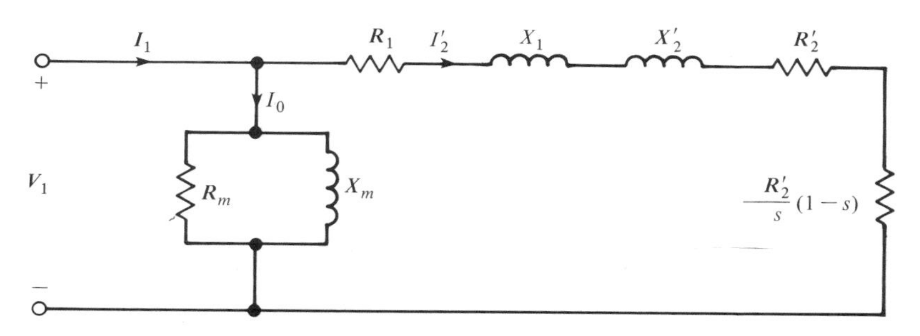

FIGURE 7-12. No-load test data.

we obtain the intercept OM which corresponds to mechanical (or friction and windage) loss. At the rated voltage, let the per phase input power (after correcting for the friction and windage loss, as mentioned above), voltage, and current be denoted by P_0, V_0, and I_0, respectively. When the machine runs on no-load, the slip is close to zero and the circuit in Fig. 7-12 to the right of the shunt branch is taken to be an open circuit. Thus the parameters R_m and X_m are found from

$$R_m = \frac{V_0^2}{P_0} \tag{7-19}$$

$$X_m = \frac{V_0^2}{\sqrt{V_0^2 I_0^2 - P_0^2}} \tag{7-20}$$

Blocked-Rotor Test. In this test, the rotor of the machine is blocked ($s = 1$), and a reduced voltage is applied to the machine so that the rated current flows through the stator windings. The input power, voltage, and current are recorded and reduced to per phase values; these are denoted by P_s, V_s, and I_s, respectively. In this test, the iron losses are assumed to be negligible and the shunt branch of the circuit shown in Fig. 7-12 is considered to be absent. The parameters are thus found from

$$R_e = R_1 + a^2 R_2 = \frac{P_s}{I_s^2} \tag{7-21}$$

$$X_e = X_1 + a^2 X_2 = \frac{\sqrt{V_s^2 I_s^2 - P_s^2}}{I_s^2} \tag{7-22}$$

In (7-21) and (7-22), the constant a is the turns ratio. The stator resistance per phase, R_1, can be directly measured, and knowing R_e from (7-21), we can determine $R_2' = a^2 R_2$, the rotor resistance referred to the stator. There is no simple method of determining X_1 and $X_2' = a^2 X_2$ separately. The total value given by (7-22) is sometimes equally divided between X_1 and X_2'.

EXAMPLE 7-4

The results of no-load and blocked-rotor tests on a three-phase wye-connected induction motor are as follows:

No-load test:

line-to-line voltage = 400 V

input power = 1770 W

input current = 18.5 A

friction and windage loss = 600 W

Blocked-rotor test:

$$\text{line-to-line voltage} = 45 \text{ V}$$

$$\text{input power} = 2700 \text{ W}$$

$$\text{input current} = 63 \text{ A}$$

Determine the parameters of the approximate equivalent circuit (Fig. 7-12).

SOLUTION

From no-load test data:

$$V_0 = \frac{400}{\sqrt{3}} = 231 \text{ V}, \quad P_0 = \tfrac{1}{3}(1770 - 600) = 390 \text{ W, and } I_0 = 18.5 \text{ A}$$

Then, by (7-19) and (7-20),

$$R_m = \frac{(231)^2}{390} = 136.8\ \Omega$$

$$X_m = \frac{(231)^2}{\sqrt{(231)^2(18.5)^2 - (390)^2}} = 12.5\ \Omega$$

From blocked-rotor test data,

$$V_s = \frac{45}{\sqrt{3}} = 25.98 \text{ V}, \quad I_s = 63 \text{ A, and } P_s = \frac{2700}{3} = 900 \text{ W}$$

Then, by (7-21) and (7-22),

$$R_e = R_1 + a^2R_2 = \frac{900}{(63)^2} = 0.23\ \Omega$$

$$X_e = X_1 + a^2X_2 = \frac{\sqrt{(25.98)^2(63)^2 - (900)^2}}{(63)^2} = 0.34\ \Omega$$

■

7-7 PERFORMANCE CRITERIA OF INDUCTION MOTORS

Example 7-2 shows the usefulness of the equivalent circuit in calculating the performance of the motor. The performance of an induction motor may be characterized by the following major factors:

1. Efficiency.
2. Power factor.

3. Starting torque.
4. Starting current.
5. Pull-out (or maximum) torque.

These characteristics are shown in Fig. 7-10. In design considerations, heating because of I^2R losses and core losses and means of heat dissipation must be included. It is not within the scope of this book to present a detailed discussion of the effects of design changes, and consequently parameter variations, on each performance characteristic. Here we summarize the results as trends. For example, the efficiency of an induction motor is proportional to $(1 - s)$. Thus the motor would be most compatible with a load running at the lowest slip. Because the efficiency is clearly dependent on I^2R losses, R_2' and R_1 must be small for a given load. To reduce core losses, the working flux density (B) must be small. But this imposes a conflicting requirement on the load current (I_2') because the torque is dependent on the product of B and I_2'. In other words, an attempt to decrease the core losses beyond a limit would result in an increase in the I^2R losses for a given load.

It may be seen from the equivalent circuits (developed in Section 7-4) that the power factor can be improved by decreasing the leakage reactances and increasing the magnetizing reactance. However, it is not wise to reduce the leakage reactances to a minimum, since the starting current of the motor is essentially limited by these reactances. Again, we notice the conflicting conditions for a high power factor and a low starting current. Also, the pull-out torque would be higher for lower leakage reactances.

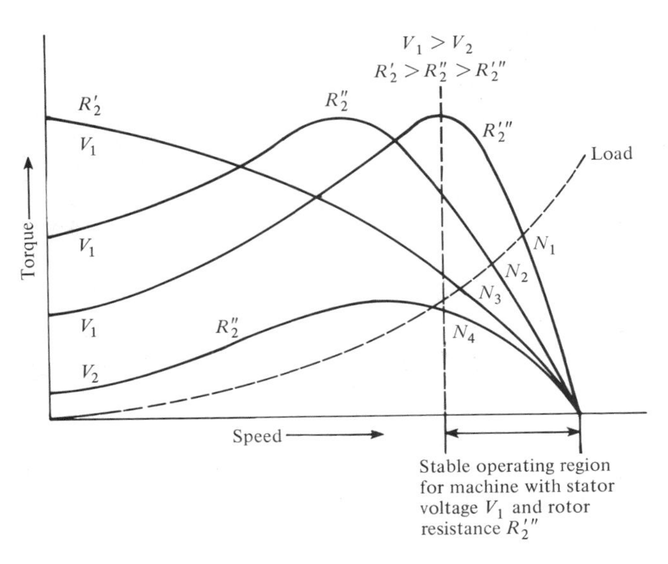

FIGURE 7-13. Effect of rotor resistance on torque–speed characteristics.

A high starting torque is produced by a high R_2'; that is, the higher the rotor resistance, the higher would be the starting torque. A high R_2' is in conflict with a high-efficiency requirement. The effect of varying rotor resistance on the motor torque–speed characteristics is shown in Fig. 7-13, which also shows three different steady-state operating speeds for three values of the rotor resistance and two stator voltages V_1 and V_2.

7-8 SPEED CONTROL OF INDUCTION MOTORS

Because of its simplicity and ruggedness, the induction motor finds numerous applications. However, it suffers from the drawback that in contrast to dc motors, its speed cannot be easily and efficiently varied continuously over a wide range of operating conditions. We will briefly review the various possible methods by which the speed of the induction motor can be varied either continuously or in discrete steps. We will not consider all these methods in detail here. Certain details are given in Chapter 11.

The speed of the induction motor can be varied (1) by varying the synchronous speed of the rotating field, or (2) varying the slip. Because the efficiency of the induction motor is approximately proportional to $(1 - s)$, any method of speed control that depends on the variation of slip is inherently inefficient. On the other hand, if the supply frequency is constant, varying the speed by changing the synchronous speed results only in discrete changes in the speed of the motor. We will now consider the governing principles of these methods of speed control.

Recall from (7-2) that the synchronous speed n_s of the rotating field in an induction machine is given by

$$n_s = \frac{120f}{P}$$

where P is the number of poles and f is the supply frequency, which indicates that n_s can be varied by (1) changing the number of poles P, or (2) changing the frequency f. Both methods have found applications, and we consider here the pertinent qualitative details.

Pole-Changing Method. In this method the stator winding of the motor is so designed that by changing the connections of the various coils (the terminals of which are brought out), the number of poles of the winding can be changed in the ratio of 2:1. Accordingly, two synchronous speeds result. We observe that only two speeds of operation are possible. Figure 7-14 shows one phase interconnection for a 2:1 pole ratio. If more independent windings (e.g., two) are provided—each arranged for pole changing—more synchronous speeds (e.g.,

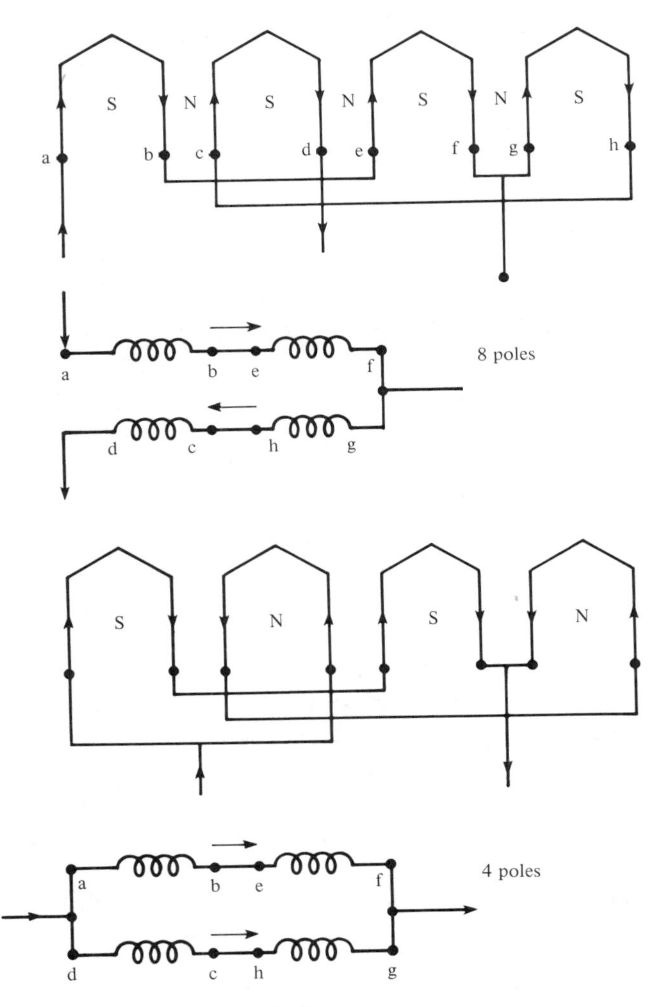

FIGURE 7-14. Pole-changing (2/1).

four) can be obtained. However, the fact remains that only discrete changes in the speed of the motor can be obtained by this technique. The method has the advantage of being efficient and reliable because the motor has a squirrel-cage rotor and no brushes.

Variable-Frequency Method. We know that the synchronous speed is directly proportional to the frequency. If it is practicable to vary the supply frequency, the synchronous speed of the motor can also be varied. The variation in speed is continuous or discrete according to continuous or discrete variation of the supply frequency. However, the maximum torque developed by the motor is inversely proportional to the synchronous speed. If we desire a constant maximum torque, both supply voltage and supply frequency should be increased if we wish to increase the synchronous speed of the motor. The inherent difficulty in the application of this method is that the supply frequency, which is commonly available, is fixed. Thus the method is applicable only if a variable-frequency supply is available. Various schemes have been proposed to obtain a variable-frequency supply. With the advent of solid-state devices with comparatively large power ratings, it is now possible to use static inverters to drive the induction motor.

Variable-Slip Method. Controlling the speed of an induction motor by changing its slip may be understood by reference to Fig. 7-13. The dashed curve shows the speed–torque characteristic of the load. The curves with solid lines are the speed–torque characteristics of the induction motor under various conditions (such as different rotor resistances—R_2', R_2'', R_3'''— or different stator voltages—V_1, V_2). We have four different torque–speed curves and, therefore, the motor can run at any one of four speeds—N_1, N_2, N_3, and N_4—for the given load. Note that the stable operating region of the motor is to the right of the peak torque. In practice, the slip of the motor can be changed by one of the following methods.

Variable Stator Voltage Methods. Since the electromagnetic torque developed by the machine is proportional to the square of the applied voltage, we obtain different torque–speed curves for different voltages applied to the motor. For a given rotor resistance, R_2, two such curves are shown in Fig. 7-13 for two applied voltages V_1 and V_2. Thus the motor can run at speeds N_2 or N_4. If the voltage can be varied continuously from V_1 to V_2, the speed of the motor can also be varied continuously between N_2 and N_4 for the given load. This method is applicable to cage-type as well as wound-rotor-type induction motors.

Variable Rotor-Resistance Method. This method is applicable only to the wound-rotor motor. The effect on the speed–torque curves of inserting external resistances in the rotor circuit is shown in Fig. 7-13 for three different rotor resistances R_2', R_2'', and R_2'''. For the given load, three speeds of operation are possible. Of course, by continuous variation of the rotor resistance, continuous variation of the speed is possible.

Control by Solid-State Switching. Other than the inverter-driven motor, the speed of the wound-rotor motor can be controlled by inserting the inverter in the rotor circuit or by controlling the stator voltage by means of solid-state switching devices such as silicon-controlled rectifiers (SCRs or thyristors). The output from the SCR feeding the motor is controlled by adjusting its firing angle. The method of doing this is similar to the variable-voltage method outlined earlier. However, it has been found that control by an SCR gives a wider range of operation and is more efficient than other slip-control methods. For details, see Chapter 11.

7-9 STARTING OF INDUCTION MOTORS

Most induction motors—large and small—are rugged enough that they could be started across the line without incurring any damage to the motor windings, although about five to seven times the rated current flows through the stator at rated voltage at standstill. However, in large induction motors, large starting currents are objectionable in two respects. First, the mains supplying the induction motor may not be of a sufficiently large capacity. Second, because of a large starting current, the voltage drops in the lines may be excessive resulting in a reduced voltage across the motor. Because the torque varies approximately as the square of the voltage, the starting torque may become so small at the reduced line voltage that the motor might not even start on load. Thus we formulate the basic requirement for starting: the line current should be limited by the capacity of the mains, but only to the extent that the motor can develop sufficient torque to start (on load, if necessary).

EXAMPLE 7-5

An induction motor is designed to run at 5 percent slip on full-load. If the motor draws six times the full-load current at starting at the rated voltage, estimate the ratio of starting torque to the full-load torque.

SOLUTION

The torque at a slip s is given by (7-17), which in conjunction with (7-10) becomes (per phase)

$$T_e = \frac{I_2^2 R_2'}{s\omega_s}$$

At full-load, with $I_2 = I_{2f}$, the torque is

$$T_{ef} = \frac{I_{2f}^2 R_2'}{0.05\omega_s}$$

At starting, $I_{2s} = 6I_{2f}$ and $s = 1$, so that

$$T_{es} = \frac{(6I_{2f})^2 R_2'}{\omega_s}$$

Hence

$$\frac{T_{es}}{T_{ef}} = \frac{(6I_{2f})^2 R_2'}{\omega_s} \frac{0.05\omega_s}{I_{2f}^2 R_2'} = 1.8$$

■

EXAMPLE 7-6

If the motor of the Example 7-5 is started at a reduced voltage to limit the line current to three times the full-load current, what is the ratio of the starting torque to the full-load torque?

SOLUTION

In this case we have

$$\frac{T_{es}}{T_{ef}} = 3^2 \times 0.05 = 0.45$$

■

Notice that the starting torque has reduced by a factor of 4, compared to the case of full-voltage starting. In many practical cases, the line current is limited to six times the full-load current and the starting torque is desired to be about 1.5 times the full-load torque.

Pushbutton Starters. There are numerous types of pushbutton starters for induction motors now commercially available. In the following, however, we will consider briefly only the principles of the two commonly used methods. We consider the current-limiting types first. Some of the most common methods of limiting the stator current while starting are:

1. *Reduced-voltage starting*. At the time of starting, a reduced voltage is applied to the stator and the voltage is increased to the rated value when the motor is within 25 percent of its final speed. This method has the obvious limitation that a variable-voltage source is needed and the starting torque drops substantially. The wye–delta method of starting is a reduced-voltage starting method. If the stator is normally connected in delta, reconnection to wye reduces the phase voltage, resulting in less current at starting. For example, at starting, if the line current is about five times the full-load current in a delta-connected stator, the current in the wye connection will be less than twice the full-load value. But at the same time, the starting torque for a wye connection would be about one-third its value for a delta connection. The advantage of wye–delta starting is that it is inexpensive and requires only a three-pole (or three single-pole) double-throw switch (or switches), as shown in Fig. 7-15.
2. *Current limiting by series resistance*. Series resistances inserted in the three lines are sometimes used to limit the starting current. These resistances are shorted out when the motor has gained speed. This method has the obvious

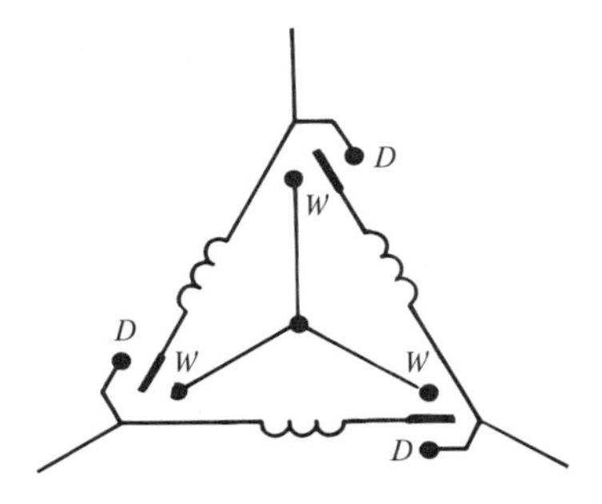

FIGURE 7-15. Wye-delta starting. Switches on *W* correspond to wye and switches on *D* correspond to the delta connection.

disadvantage of being inefficient because of the extra losses in the external resistances during the starting period.

Turning now to the starting torque, we recall from Section 7-8 that the starting torque is dependent on the rotor resistance. Thus a high rotor resistance results in a high starting torque. Therefore, in a wound-rotor machine (see Fig. 7-16), external resistance in the rotor circuit may be conveniently used. In a cage rotor, deep slots are used, where the slot depth is two or three times greater than the slot width (see Fig. 7-17). Rotor bars embedded in deep slots provide a high effective resistance and a large torque at starting. Under normal running conditions with low slips, however, the rotor resistance becomes lower and efficiency high. This characteristic of rotor bar resistance is a consequence of *skin effect*. Because of the skin effect, the current will have a tendency to concentrate at the top of the bars at starting, when the frequency of rotor currents is high. At this point, the frequency of rotor currents will be the same as the stator input

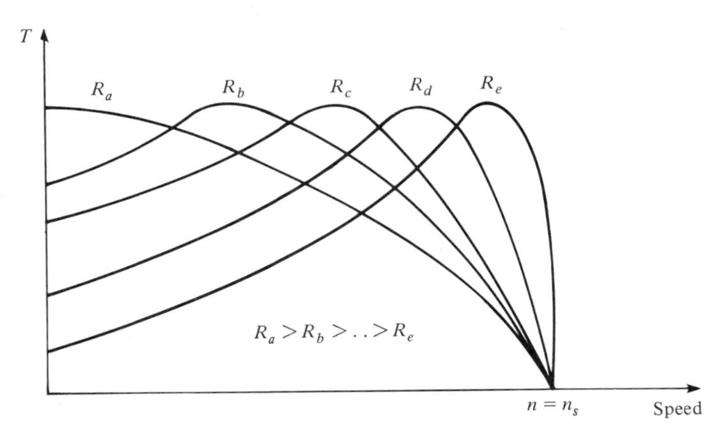

FIGURE 7-16. Effect of changing rotor resistance on the starting of a wound-rotor motor.

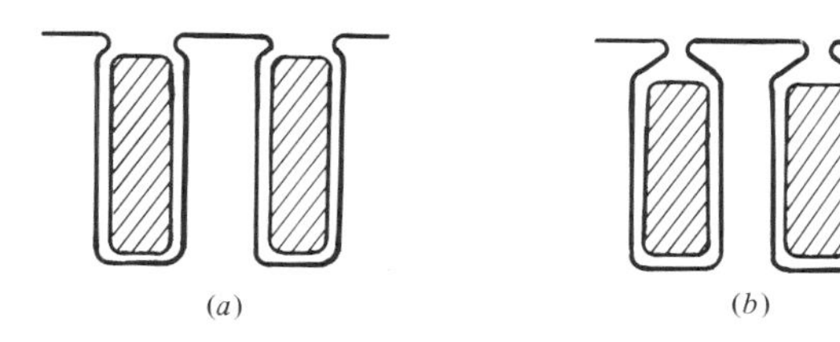

FIGURE 7-17. Deep-bar rotor slots: (a) open; (b) partially closed.

frequency (e.g., 60 Hz). While running, the frequency of rotor currents (= slip frequency = 3 Hz at 5 percent slip and 60 Hz) is much lower. At this level of operation, skin effect is negligible and the current almost uniformly distributes throughout the entire bar cross section.

The skin effect is used in an alternative form in a *double-cage* rotor (Fig. 7-18), where the inner cage is deeply embedded in iron and has low-resistance bars and has a high reactance. The outer cage has relatively high resistance bars close to the stator, and has the low reactance of a normal single-cage rotor. At starting, because of the skin effect, the influence of the outer cage dominates, thus producing a high starting torque. While running, the current penetrates to full depth into the lower cage—because of insignificant skin effect and lower reactance—which results in an efficient steady-state operation. Notice that under normal running conditions both cages carry current, thus somewhat increasing the power rating of the motor. The rotor equivalent circuit of a double-cage rotor then becomes as shown in Fig. 7-19. Approximately, the cages may be considered to develop separate torques, and the sum of these torques is the total torque. By appropriate designs, the inner- and outer-cage resistances and leakage reactances may be modified to obtain a wide range of performance characteristics. Compared to a normal single-cage motor, the inner-cage leakage reactance in a double-cage motor lowers its power factor at full-load. Furthermore, the high resistance of the outer cage increases the full-load I^2R loss and decreases the motor efficiency. But for a duty cycle of frequent starts and stops, a double-cage motor is likely to have a higher energy efficiency compared to a single-cage motor.

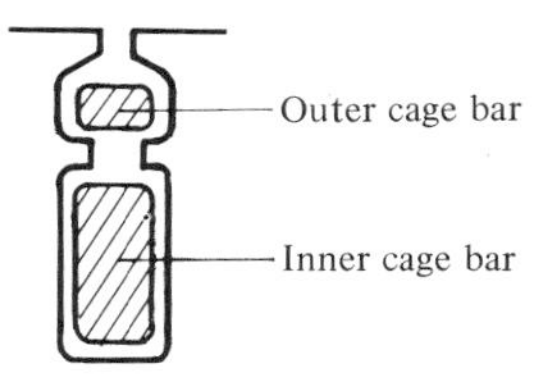

FIGURE 7-18. Form of a slot for a double-cage rotor.

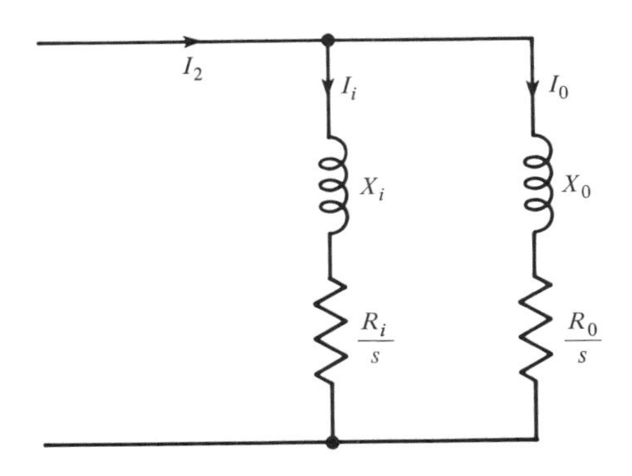

FIGURE 7-19. Equivalent circuit of a double-cage rotor.

EXAMPLE 7-7

A motor employs a wye–delta starter which connects the motor phases in wye at the time of starting and in delta when the motor is running. The full-load slip is 4 percent and the motor draws nine times the full-load current if started directly from the mains. Determine the ratio of starting torque to full-load torque.

SOLUTION

When the phases are switched to delta, the phase voltage, and hence the full-load current, is increased by a factor of $\sqrt{3}$ over the value it would have had in a wye connection. Then it follows from the last equation in Example 7-6 that

$$\frac{T_s}{T_{\mathrm{FL}}} = \left(\frac{9}{\sqrt{3}}\right)^2 (0.04) = 1.08$$

EXAMPLE 7-8

To obtain a high starting torque in a cage-type motor, a double-cage rotor is used. The forms of a slot and of the bars of the two cages are shown in Fig. 7-18. The outer cage has a higher resistance than the inner cage. At starting, because of the skin effect, the influence of the outer cage dominates, thus producing a high starting torque. An approximate equivalent circuit for such a rotor is given in Fig. 7-19. Suppose that, for a certain motor, we have the per phase values, at standstill:

$$R_i = 0.1\ \Omega \qquad R_o = 1.2\ \Omega \qquad X_i = 2\ \Omega \qquad X_o = 1\ \Omega$$

Determine the ratio of the torques provided by the two cages at (a) starting and (b) 2 percent slip.

SOLUTION

(a) From Fig. 7-17, at $s = 1$:

$$Z_i^2 = (0.1)^2 + (2)^2 = 4.01\ \Omega^2$$

$$Z_o^2 = (1.2)^2 + (1)^2 = 2.44\ \Omega^2$$

$$\text{power input to the inner cage} \equiv P_{ii} = I_i^2 R_i = 0.1 I_i^2$$

$$\text{power input to the outer cage} \equiv P_{io} = I_o^2 R_o = 1.2 I_o^2$$

$$\frac{\text{torque due to inner cage}}{\text{torque due to outer cage}} \equiv \frac{T_i}{T_o} = \frac{P_{ii}}{P_{io}} = \frac{0.1}{1.2}\left(\frac{I_i}{I_o}\right)^2 = \frac{0.1}{1.2}\left(\frac{Z_o}{Z_i}\right)^2$$

$$= \frac{0.1}{1.2}\left(\frac{2.44}{4.01}\right) = 0.05$$

(b) Similarly, at $s = 0.02$:

$$Z_i^2 = \left(\frac{0.1}{0.02}\right)^2 + (2)^2 = 29\ \Omega^2$$

$$Z_o^2 = \left(\frac{1.2}{0.02}\right)^2 + (1)^2 = 3601\ \Omega^2$$

$$\frac{T_i}{T_o} = \frac{0.1}{1.2}\left(\frac{3601}{29}\right) = 10.34$$

■

7-10

INDUCTION GENERATORS

Up to this point, we have studied the behavior of an induction machine operating as a motor. We recall from the preceding discussions that for motor operation the slip lies between zero and unity, and for this case we have a conversion of electrical power into mechanical power. If the rotor of an induction machine is driven by an auxiliary means such that the rotor speed, n, becomes greater than the synchronous speed, n_s, we have, from (7-1), a negative slip. A negative slip implies that the induction machine is now operating as an induction generator. Alternatively, we may refer to the rotor portion of the equivalent circuit, such as that of Fig. 7-11. If the slip is negative, the resistance representing the load becomes $R_2'[1 - (-s)]/(-s)$, which results in a negative value of the resistance. Because a positive resistance absorbs electrical power, a negative resistance may be considered as a source of power. Hence a negative slip corresponds to a generator operation.

To understand the generator operation, we consider a three-phase induction machine to which a prime mover is coupled mechanically. When the stator is excited, a synchronously rotating magnetic field is produced and the rotor begins to run, as in an induction motor, while drawing electrical power from the supply. The prime mover is then turned on (to rotate the rotor in the direction of the rotating field). When the rotor speed exceeds synchronous speed, the direction of electrical power reverses. The power begins to flow into the supply as the machine begins to operate as a generator. The rotating magnetic field is produced by the magnetizing current supplied to the stator winding from the three-phase source. This supply of the magnetizing current must be available as the machine operates as an induction generator. For induction generators operating in parallel

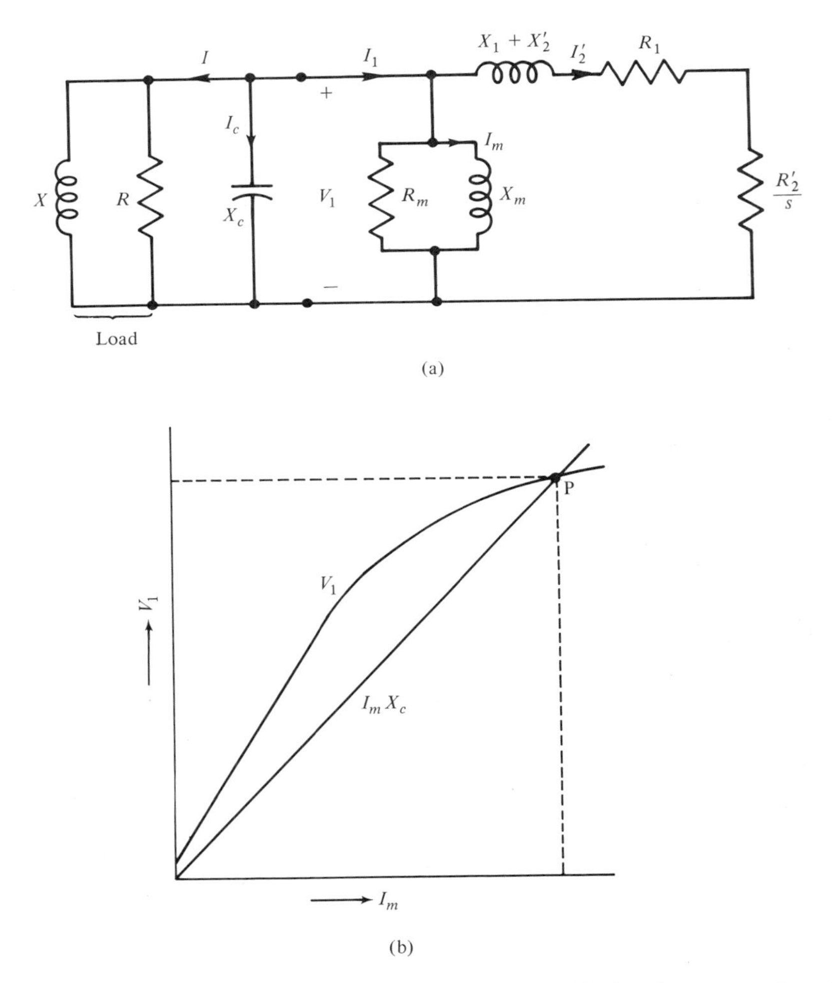

FIGURE 7-20. (a) Equivalent circuit of a self-excited induction generator; (b) determination of stable operating point.

with a three-phase source capable of supplying the necessary exciting current, the voltage and the frequency are fixed by the operating voltage and frequency of the source supplying the exciting current.

An induction generator may be self-excited by providing the magnetizing reactive power by a capacitor bank. In such a case an external ac source is not needed. The generator operating frequency and voltage are determined by the speed of the generator, its load, and the capacitor rating. As for the dc shunt generator, for the induction generator to self-excite, its rotor must have sufficient remanent flux. The operation of a self-excited induction generator may be understood by referring to Fig. 7-20(a). On no-load, the charging current of the capacitor, $I_c = V_1/X_c$, must be equal to the magnetizing current, $I_m = V_1/X_m$. Because V_1 is a function of I_m, for a stable operation the line $I_cX_c = I_mX_c$ must intersect the magnetization curve, which is a plot of V_1 versus I_m, as shown in Fig. 7-20(b). The operating point P is thus determined, and we have

$$\dot{V}_1 = I_m X_c \tag{7-23}$$

Since $X_c = 1/\omega C = 1/2\pi f C$, we rewrite (7.23) as

$$I_m = 2\pi f C V_1 \tag{7-24}$$

From (7-24) the operating frequency is given by

$$f = \frac{I_m}{2\pi C V_1} \tag{7-25}$$

On load, the generated power $V_1 I_2' \cos \varphi_2'$ provides for the power loss in R_m and the power utilized by the load R. The reactive currents are related to each other by

$$\frac{V_1}{X_c} = \frac{V_1}{X} + \frac{V_1}{X_m} + I_2' \sin \varphi_2' \tag{7-26}$$

which determines the capacitance for a given load.

Operating characteristics of an induction machine, for generator and motor modes, are shown in Fig. 7-21. Unlike in a synchronous generator, for a given load, the output current and the power factor are determined by the generator parameters. Therefore, when an induction generator delivers a certain power, it also supplies a certain in-phase current and a certain quadrature current. However, the quadrature component of the generator current generally does not have a definite relationship to the quadrature component of the load current. The quadrature current must be supplied by the synchronous generators operating in parallel with the induction generator.

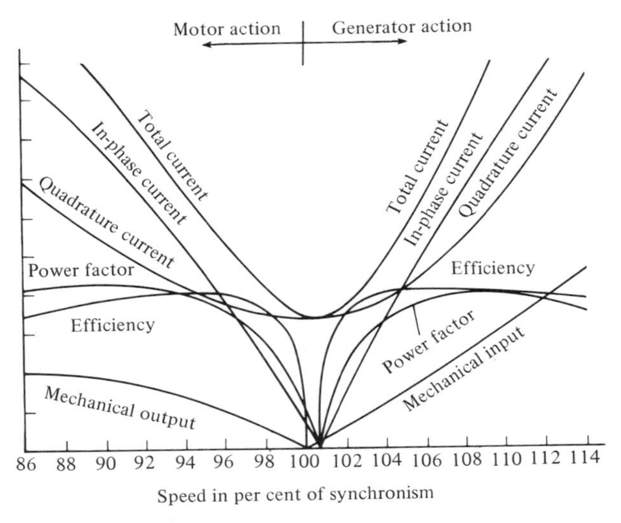

FIGURE 7-21. Motor and generator characteristics of an induction machine.

Induction generators are not suitable for supplying loads having low lagging power factors. In the past, induction generators have been used in variable-speed constant-frequency generating systems. Large induction generators have found applications in hydroelectric power stations. Induction generators are promising for windmill applications and are rapidly finding applications in this area.

7-11 ENERGY-EFFICIENT INDUCTION MOTORS

Over the last 10 years the cost of electrical energy has more than doubled. For instance, it has been reported that the annual energy cost to operate a 10-hp induction motor 4000 h per year has increased from $850 in 1972 to $1950 in 1980. The escalation of oil prices in the mid-1970s led the manufacturers of electric motors to seek methods to improve motor efficiencies. In order to improve motor efficiency, a motor's loss distribution must be studied. For a typical standard three-phase 50-hp motor, the loss distribution at full-load is given in Table 7-1. In this table we also show the average loss distribution in percent of total losses for standard induction motors. The per unit loss in Table 7-1 is defined as loss/(hp $\times$ 746).

In improving the efficiency of the motor, we must design to achieve a balance among the various losses and at the same time, meet other specifications, such as breakdown torque, locked-rotor current and torque, and power factor. For the motor designer, a clear understanding of the loss distribution is very important. Loss reductions can be made by increasing the amount of the material

TABLE 7-1 Loss Distribution in Standard Induction Motors

Loss Distribution	50-hp Motor Watts	Percent Loss	Per Unit Loss	Average Percent Loss for Standard Motors
Stator I^2R loss	1,540	38	0.04	37
Rotor I^2R loss	860	22	0.02	18
Magnetic core loss	765	20	0.02	20
Friction and windage loss	300	8	0.01	9
Stray load loss	452	12	0.01	16
Total losses	3,917	100	0.10	
Output (W)	37,300			
Input (W)	41,217			
Efficiency (%)	90.5			

in the motor. Without making other major design changes, a loss reduction of about 10 percent at full load can be achieved. Improving the magnetic circuit design using lower-loss electrical grade laminations can result in a further reduction of losses by about 10 percent. The cost of improving the motor efficiency increases with output rating (hp) of the motor. Based on the improvements just mentioned to increase the motor efficiency, Fig. 7-22 shows a comparison of the efficiencies of energy-efficient motors with those of standard motors.

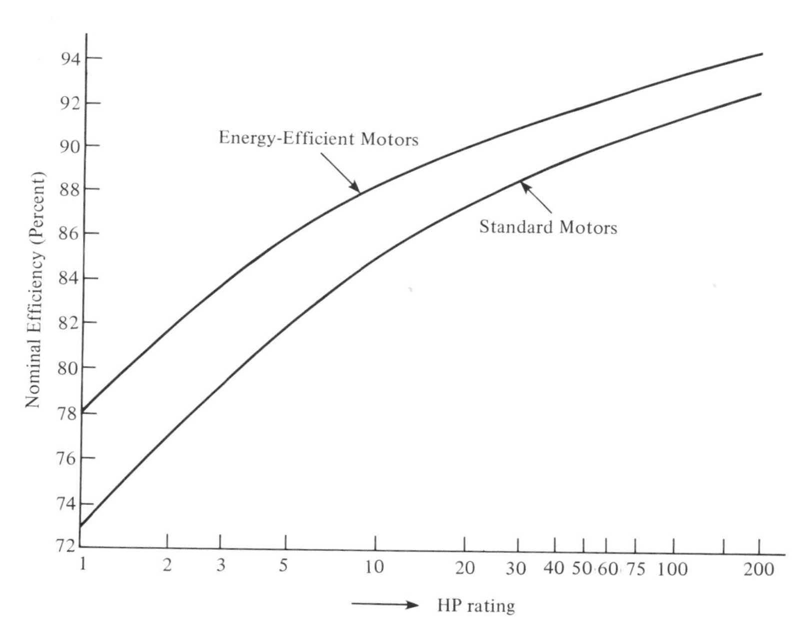

FIGURE 7-22. Comparison of nominal efficiencies of energy-efficient and standard motors.

Several of the major manufacturers of induction motors have developed product lines of energy-efficient motors. These motors are identified by their trade names, which include

E-Plus (Gould Inc.).
Energy Saver (General Electric).
XE–Energy Efficient (Reliance Electric).
Mac II High Efficiency (Westinghouse).

Because energy-efficient motors use more material, they are bigger than standard motors.

PROBLEMS

√**7-1** A six-pole 60-Hz induction motor runs at 1152 r/min. Determine the synchronous speed and the percent slip.

7-2 A six-pole induction motor is supplied by a synchronous generator having four poles and running at 1500 r/min. If the speed of the induction motor is 750 r/min, what is the frequency of rotor current?

√**7-3** A three-phase six-pole 60-Hz induction motor develops a maximum torque of 180 N·m at 800 rpm. If the rotor resistance is 0.2 Ω per phase, determine the developed torque at 1000 r/min.

7-4 A 400-V six-pole three-phase 60-Hz induction motor has a wye-connected rotor having an impedance of $(0.1 + j0.5)$ Ω per phase. How much additional resistance must be inserted in the rotor circuit for the motor to develop the maximum starting torque? The effective stator-to-rotor turns ratio is 1.

7-5 For the motor of Problem 7-4, what is the motor speed corresponding to the maximum developed torque without any external resistance in the rotor circuit?

7-6 A two-pole 60-Hz induction motor develops a maximum torque of twice the full-load torque. The starting torque is equal to the full-load torque. Determine the full-load speed.

7-7 The input to the rotor circuit of a four-pole 60-Hz induction motor, running at 1000 r/min, is 3 kW. What is the rotor copper loss?

7-8 The stator current of a 400-V three-phase wye-connected four-pole 60-Hz induction motor running at a 6 percent slip is 60 A at 0.866 power factor.

The stator copper loss is 2700 W and the total iron and rotational losses are 3600 W. Calculate the motor efficiency.

7-9 An induction motor has an output of 30 kW at 86 percent efficiency. For this operating condition, stator copper loss = rotor copper loss = core losses = mechanical rotational losses. Determine the slip.

√**7-10** A six-pole 60-Hz three-phase wye-connected induction motor has a mechanical rotational loss of 500 W. At 5 percent slip, the motor delivers 30 hp at the shaft. Calculate (a) the rotor input, (b) the output torque, and (c) the developed torque.

$P_o = 30\,HP$ $\quad P_o = P_d - P_r$ $\quad P_d = (1-s)P_g$ $\quad P_o = P_d - P_r$

7-11 A wound-rotor six-pole 60-Hz induction motor has a rotor resistance of 0.8 Ω and runs at 1150 r/min at a given load. The load on the motor is such that the torque remains constant at all speeds. How much resistance must be inserted in the rotor circuit to bring the motor speed down to 950 r/min? Neglect rotor leakage reactance.

7-12 A 400-V three-phase wye-connected induction motor has a stator impedance of $(0.6 + j1.2)$ Ω per phase. The rotor impedance referred to the stator is $(0.5 + j1.3)$ Ω per phase. Using the approximate equivalent circuit, determine the maximum electromagnetic power developed by the motor.

7-13 The motor of Problem 7-12 has a magnetizing reactance of 35 Ω. Neglecting the iron losses, at 3 percent slip calculate (a) the input current, (b) the power factor, and (c) the efficiency of the motor. Use the approximate equivalent circuit.

√**7-14** On no-load a three-phase delta-connected induction motor takes 6.8 A and 390 W at 220 V. The stator resistance is 0.1 Ω per phase. The friction and windage loss is 120 W. Determine the values of the parameters X_m and R_m of the equivalent circuit of the motor.

7-15 On blocked-rotor, the motor of Problem 7-14 takes 30 A and 480 W at 36 V. Using the data of Problem 7-14, determine the complete exact equivalent circuit of the motor. Assume that the per phase stator and rotor leakage reactances are equal.

7-16 Determine the parameters of the approximate equivalent circuit of a three-phase induction motor from the following data:

No-load test:	applied voltage	=	440 V
	input current	=	10 A
	input power	=	7600 W

Blocked rotor test:	applied voltage	= 180 V
	input current	= 40 A
	input power	= 6240 W

The stator resistance between any two leads is 0.8 Ω and the no-load friction and windage loss is 420 W.

√**7-17** A four-pole 400-V three-phase 60-Hz induction motor takes 150 A of current at starting and 25 A while running at full-load. The starting torque is 1.8 times the torque at full-load at 400 V. If it is desired that the starting torque be the same as the full-load torque, determine (a) the applied voltage and (b) the corresponding line current.

√**7-18** An induction motor is started by a wye–delta switch. Determine the ratio of the starting torque to the full-load torque if the starting current is five times the full-load current and the full-load slip is 5 percent.

7-19 An induction motor is started at a reduced voltage. The starting current is not to exceed four times the full-load current, and the full-load torque is four times the starting torque. If the full-load slip is 3 percent, calculate the factor by which the motor terminal voltage must be reduced at starting.

7-20 The per phase parameters of the equivalent circuit of a double-cage rotor are $R_o = 0.4\ \Omega$, $X_o = 0.2\ \Omega$, $R_i = 0.04\ \Omega$, and $X_i = 0.8\ \Omega$. At starting, determine the ratio of the torques provided by the outer and inner cages.

7-21 At what slip will the torques contributed by the outer and inner cages of the rotor of Problem 7-20 be equal?

CHAPTER 8

Small AC Motors

8-1 INTRODUCTION

By small motors we imply that such motors generally have outputs of less than 1 hp (or 746 W), although in a special case the output may be greater than 1 hp. Small motors find applications in electric tools, appliances, and office equipment. We shall consider specific applications in a later section. Whereas permanent-magnet small dc commutator motors have numerous applications, such as in toys, control devices, and so on, in the following we consider only small ac motors.

Almost invariably, small ac motors are designed for single-phase operation. The three common types of small ac motors are:

1. Single-phase induction motors.
2. Synchronous motors.
3. Ac commutator motors.

Special types of small ac motors include:

1. Ac servomotors.
2. Tachometers.
3. Stepper motors.

8-2 SINGLE-PHASE INDUCTION MOTORS

Let us consider a three-phase cage-type induction motor (discussed in Chapter 7) running light. It may be experimentally verified that if one of the supply lines is disconnected, the motor will continue to run, although at a speed lower than the speed of a three-phase motor. Such an operation of a three-phase induction motor (with one line open) may be considered to be similar to that of a single-phase induction motor. Next, let the three-phase motor be at rest and supplied by a single-phase source. Under this condition, the motor will not start because we have a pulsating magnetic field in the air gap rather than a rotating magnetic field which is required for torque production, as discussed in Chapter 7. Thus we conclude that a single-phase induction motor is not self-starting, but will continue to run if started by some means. This implies that to make it self-starting, the motor must be provided with an auxiliary means of starting. Later, we shall examine the various methods of starting the single-phase induction motor.

Not considering the starting mechanism for the present, the essential difference between the three-phase and single-phase induction motor is that the latter has a single stator winding which produces an air gap field that is stationary in space but alternating (or pulsating) in time. On the other hand, the stator of a three-phase induction motor has a three-phase winding that produces a time-invariant rotating magnetic field in the air gap. The rotor of the single-phase induction motor is generally a cage-type rotor and is similar to that of a three-phase induction motor. The rating of a single-phase motor of the same size as a three-phase motor would be smaller, as expected, and single-phase induction motors are most often rated as fractional-horsepower motors.

Performance Analysis. The operating performance of the single-phase induction motor is studied on the basis of the following theories: (1) cross-field theory, and (2) double-revolving field theory. We will use the revolving-field theory in the following in analyzing the single-phase induction motor, although both theories yield identical results. It is interesting, however, to study qualitatively the mechanism of torque production in a single-phase induction motor, as viewed by cross-field theory. Figure 8-1(a) shows a single-phase induction motor, the stator winding of which carries a single-phase excitation and the rotor

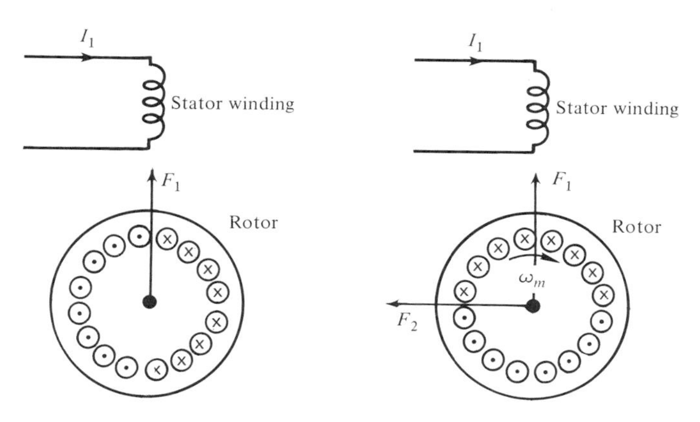

FIGURE 8-1. Representation of a single-phase induction motor: (a) standstill; (b) running.

is at standstill. The stator mmf is shown as F_1. Since the rotor is stationary, it acts like a short-circuited transformer. As a consequence, the mmf due to the rotor currents simply opposes the stator mmf. The resulting field will be stationary in space but will pulsate in magnitude. Next, let the motor be started (by some means) and let it run at some speed, as shown in Fig. 8-1(b). Rotor conductors are now rotating in the magnetic field produced by the stator. Consequently, rotational voltages will be induced in the rotor conductors. Applying the right-hand rule (see Chapter 5) yields the directions of the voltages induced, and hence the current flow, in the rotor conductors, as given in Fig. 8-1(b). We now have two mmf's: F_1 due to the stator and F_2 due to the rotation of the rotor. These mmf's produce their respective airgap fields. These mmf's are displaced from each other in space, as shown in Fig. 8-1(b). To determine the time displacement between the two mmf's, notice that the rotor-induced voltage (because of rotation) is in time phase with the stator mmf. However, the rotor circuit being highly inductive, the rotor current lags the rotor voltage by almost 90°. Hence there is a time displacement (of about 90°) between the stator and rotor mmf's. Thus we fulfill the condition for the production of a rotating magnetic field (see Chapter 6), and the rotor continues to develop a torque as long as the rotor is running in the rotating magnetic field.

We see from the above that the magnetic field produced by the stator of a single-phase motor alternates through time. The field induces a current—and consequently, an mmf—in the rotor circuit and the resultant field rotates with the rotor. A single-phase induction motor may be analyzed by considering the mmf's, fluxes, induced voltages (both rotational and transformer), and currents that are separately produced by the stator and by the rotor. Such an approach leads to the cross-field theory. However, we can also analyze the single-phase motor in a manner similar to that for the polyphase induction motor. We recall from Chapter 7 that the polyphase induction motor operates on the basis of the existence of a rotating magnetic field. This approach is based on the concept that an alternating magnetic field is equivalent to two rotating magnetic fields rotating in opposite directions. When this concept is expressed mathematically, the alternating field is of the form

$$B(\theta,t) = B_m \cos \theta \sin \omega t \tag{8-1}$$

Then (8-1) may be rewritten as

$$B_m \cos \theta \sin \omega t = \frac{1}{2} B_m \sin (\omega t - \theta) + \frac{1}{2} B_m \sin (\omega t + \theta) \tag{8-2}$$

In (8-2), the first term on the right-hand side denotes a forward rotating field, whereas the second term corresponds to a backward rotating field. The theory based on such a resolution of an alternating field into two counter-rotating fields is known as the *double-revolving field theory*. The direction of rotation of the forward rotating field is assumed to be the same as the direction of the rotation of the rotor. Thus if the rotor runs at n rpm and n_s is the synchronous speed in rpm, the slip, s_f, of the rotor with respect to the forward rotating field is the same as s, defined by (7-1), or

$$s_f = s = \frac{n_s - n}{n_s} = 1 - \frac{n}{n_s} \tag{8-3}$$

But the slip, s_b, of the rotor with respect to the backward rotating flux is given by

$$s_b = \frac{n_s - (-n)}{n_s} = 1 + \frac{n}{n_s} = 2 - s \tag{8-4}$$

We know from the operation of polyphase motors that, for $n < n_s$, (8-3) corresponds to a motor operation and (8-4) denotes the braking region. Thus the two resulting torques have an opposite influence on the rotor.

The torque relationship for the polyphase induction motor is applicable to each of the two rotating fields of the single-phase motor. We notice from (8-2) that the amplitude of the rotating fields is one-half of the alternating flux. Thus the total magnetizing and leakage reactances of the motor can be divided equally so as to correspond to the forward and backward rotating fields. The approximate equivalent circuit of a single-phase induction motor, based on the double-revolving field theory, becomes as shown in Fig. 8-2(a). The torque–speed characteristics are qualitatively shown in Fig. 8-2(b). The following example illustrates the usefulness of the circuit.

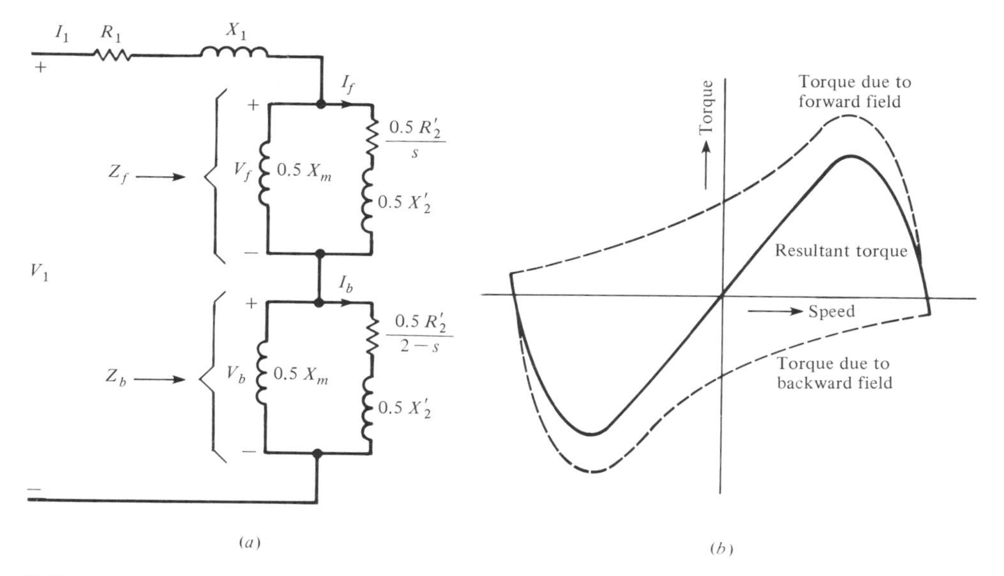

FIGURE 8-2. (a) Equivalent circuit, and (b) torque–speed characteristics, based on double-revolving field theory of the single-phase induction motor.

EXAMPLE 8-1

For a 230-V one-phase induction motor, the parameters of the equivalent circuit, Fig. 8-2(a), are: $R_1 = R_2' = 8\ \Omega$, $X_1 = X_2' = 12\ \Omega$, and $X_m = 200\ \Omega$. At a slip of 4 percent, calculate (a) the input current, (b) the input power, (c) the developed power, and (d) the developed torque (at rated voltage). The motor speed is 1728 rpm.

SOLUTION

From Fig. 8-2(a):

$$\mathbf{Z}_f = \frac{(j100)\left(\dfrac{4}{0.04} + j6\right)}{j100 + \dfrac{4}{0.04} + j6} = 47 + j50\ \Omega$$

$$\mathbf{Z}_b = \frac{(j100)\left(\dfrac{4}{1.96} + j6\right)}{j100 + \dfrac{4}{1.96} + j6} = 1.8 + j5.7\ \Omega$$

$$\mathbf{Z}_1 = R_1 + jX_1 = 8 + j12\ \Omega$$

$$\mathbf{Z}_{\text{total}} = 56.8 + j67.7 = 88.4\underline{/50^\circ}\ \Omega$$

(a) $$\text{input current} \equiv I_1 = \frac{230}{88.4} = 2.6 \text{ A}$$

$$\text{power factor} = \cos 50° = 0.64 \text{ lagging}$$

(b) $$\text{input power} = (230)(2.6)(0.64) = 382.7 \text{ W}$$

(c) Proceeding as in Example 7-2, we have

$$\begin{aligned} P_d &= [I_1^2 \text{Re}\,(\mathbf{Z}_f)](1 - s) + [I_1^2 \text{Re}\,(\mathbf{Z}_b)][1 - (2 - s)] \\ &= I_1^2[\text{Re}\,(\mathbf{Z}_f) - \text{Re}\,(\mathbf{Z}_b)](1 - s) = (2.6)^2(47 - 1.8)(1 - 0.04) \\ &= 293.3 \text{ W} \end{aligned}$$

(d) $$\text{torque} = \frac{P_d}{\omega_m} = \frac{293.3}{2\pi(1728)/60} = 1.62 \text{ N-m}$$ ■

EXAMPLE 8-2

To reduce the numerical computation, Fig. 8-2(a) is modified by neglecting $0.5X_m$ in Z_b and taking the backward-circuit rotor resistance at low slips as $0.25R_2'$. With these approximations, repeat the calculations of Example 8-1 and compare the results.

SOLUTION

$$\begin{aligned} \mathbf{Z}_f &= 47 + j50 \ \Omega \\ \mathbf{Z}_b &= 2 + j6 \ \Omega \\ \mathbf{Z}_1 &= 8 + j12 \ \Omega \\ \mathbf{Z}_{\text{total}} &= 57 + j68 = 88.7\angle 50° \ \Omega \end{aligned}$$

(a) $$I_1 = \frac{230}{88.7} = 2.6 \text{ A}$$

(b) $$\cos \varphi = 0.64 \text{ lagging}$$

$$\text{input power} = (230)(2.6)(0.64) = 382.7 \text{ W}$$

(c) $$P_d = (2.6)^2(47 - 2)(1 - 0.04) = 292.0 \text{ W}$$

(d) $$\text{torque} = \frac{292.0}{2\pi(1728)/60} = 1.61 \text{ N-m}$$ ■

Following is a comparison of the results of Examples 8-1 and 8-2:

Example Number	Input Current (A)	Input Power (W)	Power Factor	Developed Torque (N-m)
8-1	8-6	382.7	0.64	1.62
8.2	2.6	382.7	0.64	1.61

This comparison indicates that the approximation suggested in Example 8-2 is adequate for most cases.

In the next example we show the procedure for efficiency calculations for a single-phase induction motor.

EXAMPLE 8-3

A one-phase 110-V 60-Hz four-pole induction motor has the following constants in the equivalent circuit, Fig. 6.2(a): $R_1 = R_2' = 2\ \Omega$, $X_1 = X_2' = 2\ \Omega$, and $X_m = 50\ \Omega$. There is a core loss of 25 W and a friction and windage loss of 10 W. For a 10 percent slip, calculate (a) the motor input current and (b) the efficiency.

SOLUTION

$$\mathbf{Z}_f = \frac{(j25)\left(\dfrac{1}{0.1} + j1\right)}{j25 + \dfrac{1}{0.1} + j1} = 8 + j4\ \Omega$$

$$\mathbf{Z}_b = \frac{(j25)\left(\dfrac{1}{1.9} + j1\right)}{j25 + \dfrac{1}{1.9} + j1} = 0.48 + j0.96\ \Omega$$

$$\mathbf{Z}_1 = 2 + j2\ \Omega$$

$$\mathbf{Z}_{\text{total}} = 10.48 + j6.96 = 12.6\underline{/33.6^\circ}\ \Omega$$

(a)

$$I_1 = \frac{110}{12.6} = 8.73\ \text{A}$$

(b)

$$\text{developed power} = (8.73)^2(8 - 0.48)(1 - 0.10) = 516\ \text{W}$$

$$\text{output power} = 516 - 25 - 10 = 481\ \text{W}$$

$$\text{input power} = (110)(8.73)(\cos 33.6^\circ) = 800\ \text{W}$$

$$\text{efficiency} = \frac{481}{800} = 60\%$$

■

Starting of Single-Phase Induction Motors. We already know that because of the absence of a rotating magnetic field, when the rotor of a single-phase induction motor is at standstill, it is not self-starting. The two methods of starting a single-phase motor are either to introduce commutator and brushes, such as in a repulsion motor (considered later), or to produce a rotating field by means of an auxiliary winding, such as by split phasing. We consider the latter method next.

From the theory of the polyphase induction motor, we know that in order to have a rotating magnetic field, we must have at least two mmf's which are displaced from each other in space and carry currents having different time phases. Thus, in a single-phase motor, a starting winding on the stator is provided as a source of the second mmf. The first mmf arises from the main stator winding. The various methods to achieve the time and space phase shifts between the main winding and starting winding mmf's are summarized below.

Split-Phase Motors. This type of motor is represented schematically in Fig. 8-3(a), where the main winding has a relatively low resistance and a high reactance. The starting winding, however, has a high resistance and a low reactance and has a centrifugal switch as shown. The phase angle α between the two currents I_m and I_s is about 30 to 45°, and the starting torque T_s is given by

$$T_s = KI_mI_s \sin \alpha \tag{8-5}$$

where K is a constant. When the rotor reaches a certain speed (about 75 percent of its final speed), the centrifugal switch comes into action and disconnects the starting winding from the circuit. The torque–speed characteristic of the split-

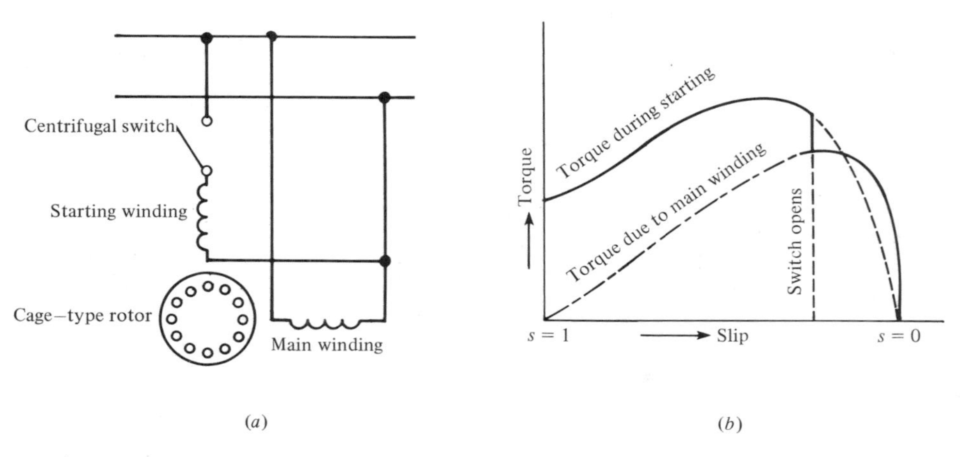

FIGURE 8-3. (a) Connections for a split-phase motor; (b) a torque–speed characteristic.

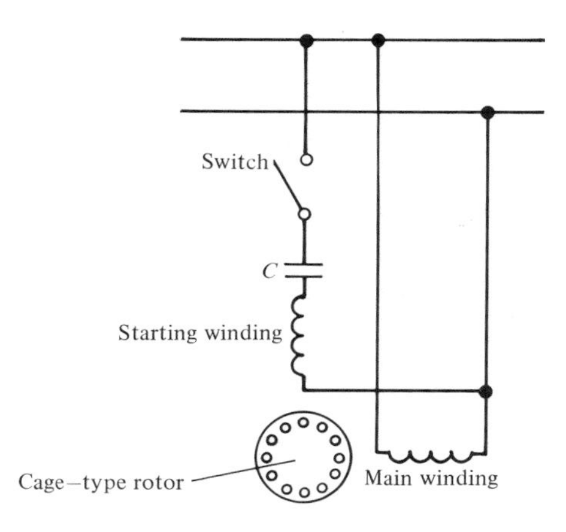

FIGURE 8-4. A capacitor-start motor.

phase motor is of the form shown in Fig. 8-3(b). Such motors find applications in fans, blowers, and so forth, and are rated up to $\frac{1}{2}$ hp.

A higher starting torque can be developed by a split-phase motor by inserting a series resistance in the starting winding. A somewhat similar effect may be obtained by inserting a series inductive reactance in the main winding. This reactance is short-circuited when the motor builds up speed.

Capacitor-Start Motors. By connecting a capacitance in series with the starting winding, as shown in Fig. 8-4, the angle α in (8-5) can be increased. The motor will develop a higher starting torque by doing this. Such motors are not restricted merely to fractional-horsepower ratings, and may be rated up to 10 hp. At 110 V, a 1-hp motor requires a capacitance of about 400 μF, whereas 70 μF is sufficient for a $\frac{1}{8}$-hp motor. The capacitors generally used are inexpensive electrolytic types and can provide a starting torque that is almost four times that of the rated torque.

As shown in Fig. 8-4, the capacitor is merely an aid to starting and is disconnected by the centrifugal switch when the motor reaches a predetermined speed. However, some motors do not have the centrifugal switch. In such a motor, the starting winding and the capacitor are meant for permanent operation and the capacitors are much smaller. For example, a 110-V $\frac{1}{2}$-hp motor requires a 15 μF capacitance.

A third kind of the capacitor motor uses two capacitors: one that is left permanently in the circuit together with the starting winding, and one that gets disconnected by a centrifugal switch. Such motors are, in effect, unbalanced two-phase induction motors.

Shaded-Pole Motors. Another method of starting very small single-phase induction motors is to use a shading band on the poles, as shown in Fig. 8-5,

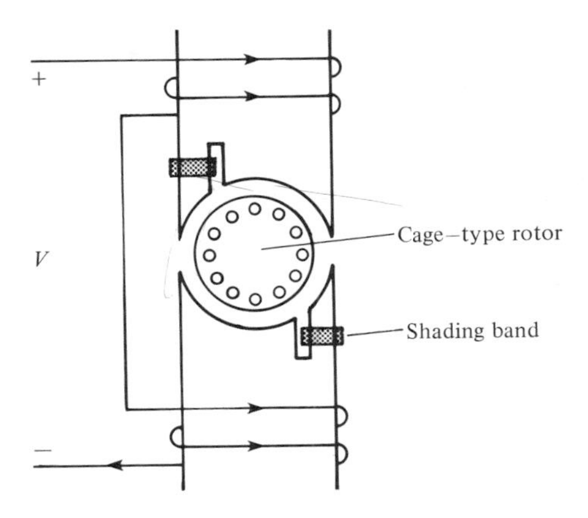

FIGURE 8-5. A shaded-pole motor.

where the main single-phase winding is also wound on the salient poles. The shading band is simply a short-circuited copper strap wound on a portion of the pole. Such a motor is known as a shaded-pole motor. The purpose of the shading band is to retard (in time) the portion of flux passing through it in relation to the flux coming out of the rest of the pole face. Thus the flux in the unshaded portion reaches its maximum before that located in the shaded portion. And we have a progressive shift of flux from the direction of the unshaded portion to shaded portion of the pole, as shown in Fig. 8-5. The effect of the progressive shift of flux is similar to that of a rotating flux, and because of it, the shading band provides a starting torque. Shaded-pole motors are the least expensive of the fractional horsepower motors and are generally rated up to $\frac{1}{20}$ hp.

8-3 SMALL SYNCHRONOUS MOTORS

The two common types of small synchronous motors are the *reluctance motor* and the *hysteresis motor*. These motors are constant-speed motors and are used in clocks, timers, turntables, and so forth.

Reluctance Motor. We are somewhat familiar with the reluctance motor from Chapter 6. We know that the torque in a reluctance motor is similar to the torque arising from saliency in a salient-pole synchronous motor. We may also recall that the time-average torque of a reluctance motor is nonzero only at one speed for a given frequency, and the power-angle characteristics of the motor are as discussed in Chapter 6. A single-phase reluctance motor is schematically shown in Fig. 8-6(a).

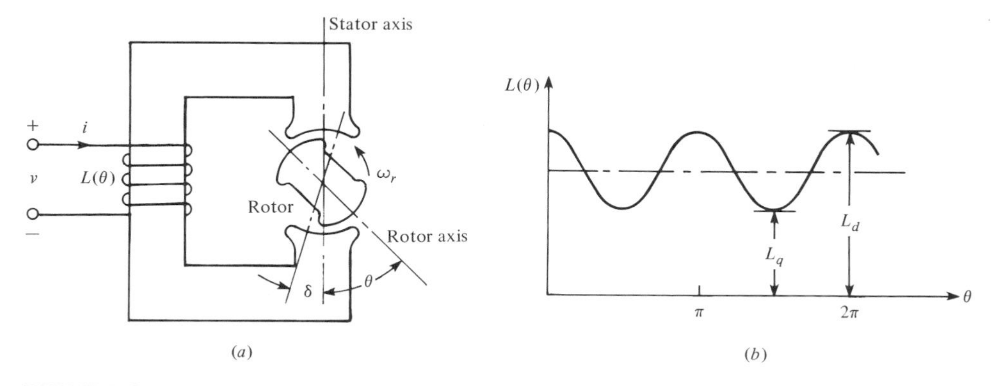

FIGURE 8-6. (a) A single-phase reluctance motor; (b) inductance of stator windings, as a function of motor position.

A reluctance motor starts as an induction motor, but normally operates as a synchronous motor. The stator of a reluctance motor is similar to that of an induction motor (single-phase or polyphase). Thus to start a single-phase motor, almost any of the methods discussed in Section 8-2 may be used. A three-phase reluctance motor is self-starting when started as an induction motor. After starting, to pull it into step and then to run it as a synchronous motor, a three-phase motor should have low rotor resistance. In addition, the combined inertia of the rotor and the load should be small. Typical construction of a four-pole rotor is shown in Fig. 8-7. Here the aluminum in the slots and in spaces where teeth have been removed serves as the rotor of an induction motor for starting.

With L_d, the maximum inductance, and L_q, the minimum inductance, as defined in Fig. 8-6(b), the average value of the torque developed by the motor, for a single-phase current $i = I_m \sin \omega t$, is found to be

$$T_e = \tfrac{1}{8} I_m^2 (L_d - L_q) \sin 2\delta \tag{8-6}$$

where δ is the power angle (defined in Chapter 6).

For a three-phase reluctance motor having a rotor of the form shown in Fig. 8-7, the torque–speed characteristic takes the form shown in Fig. 8-8. The

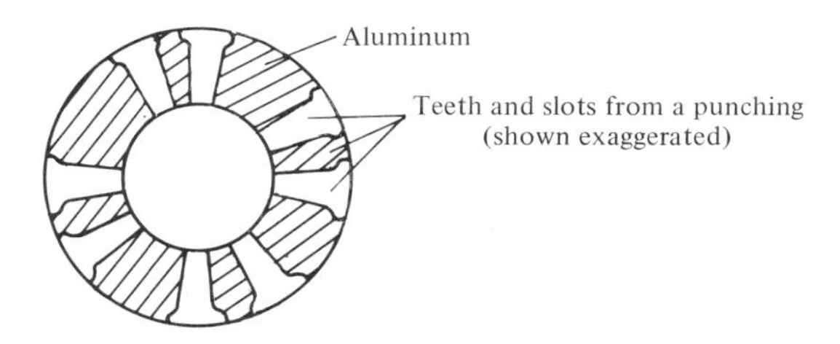

FIGURE 8-7. Rotor of a reluctance motor.

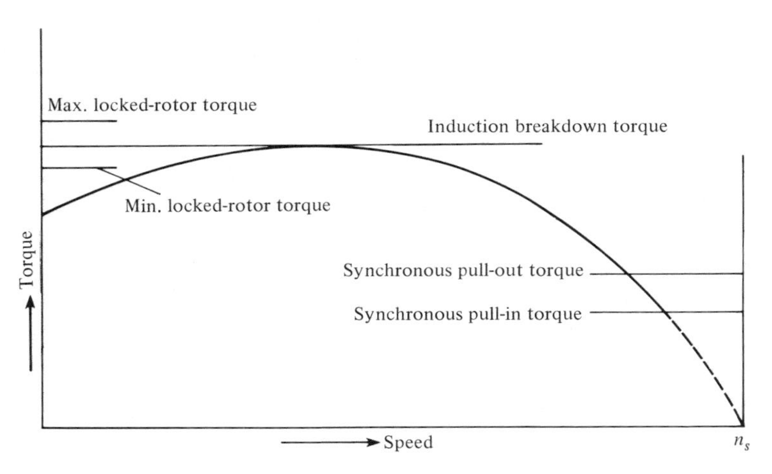

FIGURE 8-8. Torque–speed characteristic of a three-phase reluctance motor.

torques due to the induction motor action and due to reluctance are as labeled in Fig. 6.8. The power factor of a reluctance motor is poorer compared to that of an induction motor of similar rating.

EXAMPLE 8-4

A two-pole one-phase 60-Hz reluctance motor carries 5.66 A of current. The direct- and quadrature-axis inductances, L_d and L_q, respectively, are given by $L_d = 2L_q = 200$ mH. Determine (a) the rotor speed, (b) the power angle for maximum torque, and (c) the maximum value of the developed torque.

SOLUTION

(a) synchronous speed:

$$n_s = \frac{120f}{p} = \frac{120 \times 60}{2} = 3600 \text{ r/min}$$

(b) For maximum torque:

$$\sin 2\delta = 1$$

$$2\delta = 90^\circ$$

$$\delta = 45^\circ$$

(c) The maximum torque for $\delta = 45^\circ$ is obtained from (8-6), which gives

$$T_e = \tfrac{1}{8}(\sqrt{2} \times 5.66)^2(200 - 100) \times 10^{-3} = 0.8 \text{ N-m}$$ ■

Hysteresis Motor. Like the reluctance motor, a hysteresis motor does not have a dc excitation. Unlike the reluctance motor, however, the hysteresis motor does not have a salient rotor. Instead, the rotor of a hysteresis motor has a ring of special magnetic material, such as chrome, steel, or cobalt, mounted on a cylinder of aluminum or some other nonmagnetic material, as shown in Fig. 8-9. The stator of the motor is similar to that of an induction motor, and the hysteresis motor is started as an induction motor.

In order to understand the operation of the hysteresis motor, we may consider the hysteresis and eddy-current losses in the rotor. We observe that, as in an induction motor, the rotor has a certain equivalent resistance. The power dissipated in this resistance determines the electromagnetic torque developed by the motor, as discussed in Chapter 7. We may conclude that the electromagnetic torque developed by a hysteresis motor has two components: one by virtue of the eddy-current loss and the other because of the hysteresis loss. We know that the eddy-current loss can be expressed as

$$p_e = K_e f_2^2 B^2 \tag{8-7}$$

where K_e is a constant, f_2 the frequency of the eddy currents, and B the flux density. In terms of the slip s, the rotor frequency f_2 is related to the stator frequency f_1 by

$$f_2 = s f_1 \tag{8-8}$$

Thus (8-7) and (8-8) yield

$$p_e = K_e s^2 f_1^2 B^2 \tag{8-9}$$

and the torque T_e is related to p_e by (see Chapter 7)

$$T_e = \frac{p_e}{s\omega_s} \tag{8-10}$$

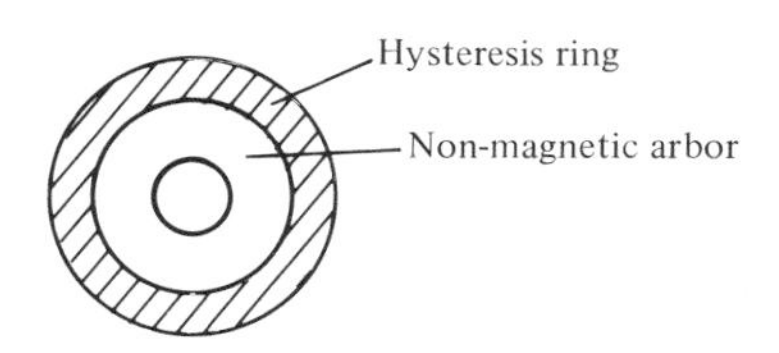

FIGURE 8-9. Rotor of a hysteresis motor.

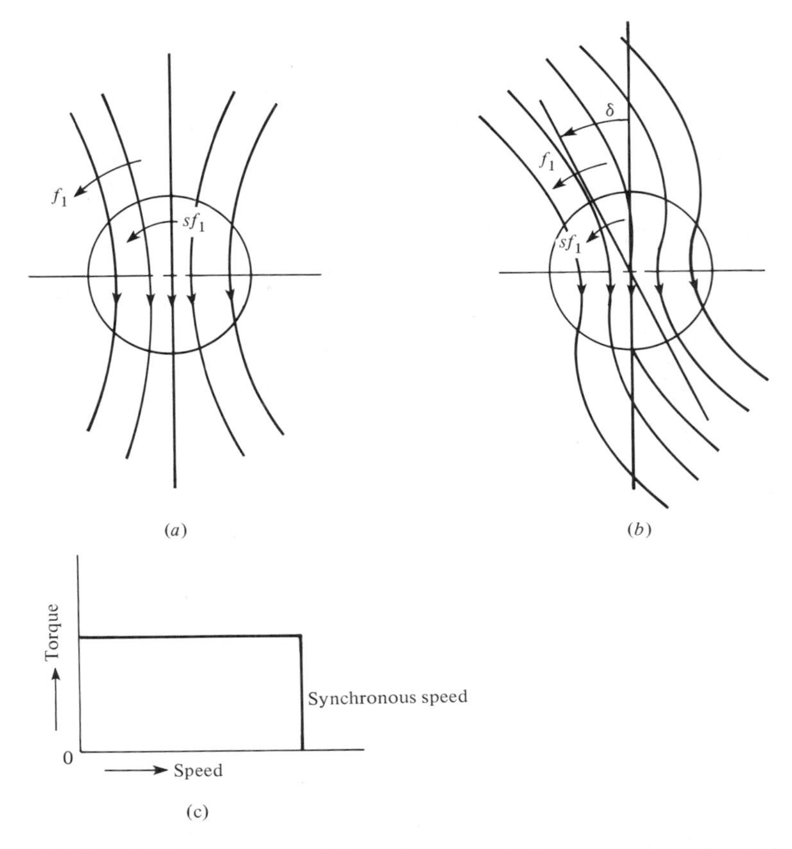

FIGURE 8-10. (a) Iron rotor, with no hysteresis in a magnetic field; (b) a rotor with hysteresis in a magnetic field; (c) torque characteristics of a hysteresis motor.

so that (8-9) and (8-10) give

$$T_e = K's \tag{8-11}$$

where $K' = K_e f_1^2 B^2/\omega_s$ = a constant.

Next, for the hysteresis loss, P_h, we have

$$p_h = K_h f_2 B^{1.6} = K_h s f_1 B^{1.6} \tag{8-12}$$

and for the corresponding torque, T_h, we obtain

$$T_h = K'' \tag{8-13}$$

where $K'' = K_h f_1 B^{1.6}/\omega_s$ = a constant.

Notice that the component T_e, as given by (8-11), is proportional to the slip and decreases as the rotor picks up speed. It is eventually zero at synchronous speed. This component of the torque aids in the starting of the motor. The second component, T_h, as given by (8-13), remains constant at all rotor speeds and is the only torque when the rotor achieves the synchronous speed. The physical basis of this torque is the hysteresis phenomenon, which causes a lag of the magnetic axis of the rotor behind that of the stator. In Fig. 8-10(a) and (b), respectively, the absence and the presence of hysteresis are shown measured by the shift of the rotor magnetic axis. The angle of lag δ, shown in Fig. 8-10(b), causes the torque arising from hysteresis. As mentioned above, this torque is independent of the rotor speed (shown in Fig. 8-10) until the breakdown torque.

8-4 AC COMMUTATOR MOTORS

In the preceding chapters we have distinguished a dc motor from an ac motor by the presence of a commutator in the dc motor. However, there exist a number of types of motors that have commutators but operate on ac. In the following discussion we consider only two types of ac commutator motors: the universal motor and the repulsion motor.

Universal Motors. A universal motor is a series motor that may be operated either on dc or on single-phase ac at approximately the same speed and power output, while supplied at the same voltage (on dc and on ac), and the frequency of the ac voltage does not exceed 60 Hz. Much of the application of universal motors is in domestic appliances and tools such as food mixers, sewing machines, vacuum cleaners, portable drills and saws, and so on. There are two types of universal motors: *uncompensated* and *compensated*. The former is less expensive and simpler in construction. It is generally used for lower power outputs and higher speeds than those of the compensated motor. Typical torque–speed characteristics of the two types of motors are shown in Fig. 8-11(a) and (b). Notice that the compensated motor has better universal characteristics in that the operation on dc and on ac at 60 Hz is not substantially different in the high-speed range. The two types of motors also differ in construction. The uncompensated motor has salient poles and a concentrated field winding, whereas the field winding of a compensated type of motor is distributed on a non-salient-pole magnetic structure. In addition, some compensated-type universal motors have a compensating winding, and some compensated motors have only one field winding, which also serves as a compensating winding.

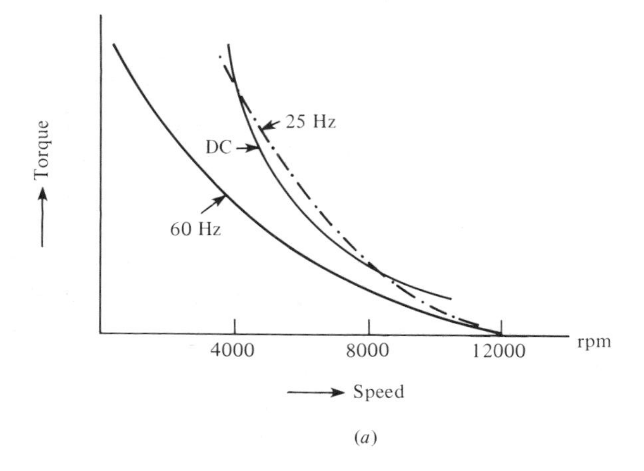

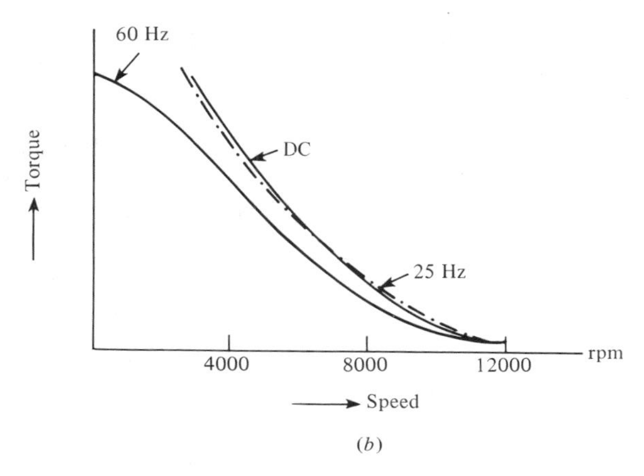

FIGURE 8-11. Characteristics of universal motors: (a) uncompensated; (b) compensated.

The principle of operation of the universal motor is similar to that of the dc series motor, the construction of the two being essentially similar. As seen from Chapter 5, the torque equation (5-10) of a dc motor may be written as

$$T_e = k_m \varphi I_a$$

where $k_m = Zp/2\pi a$ = a constant. If saturation is neglected, $\varphi = k_f I_f$, where k_f is a constant and I_f is the field current. Since $I_f = I_a = I$ in a series motor, denoting $k_f k_m = k$, the developed torque of a dc series motor is given as

$$T_e = kI^2 \tag{8-14}$$

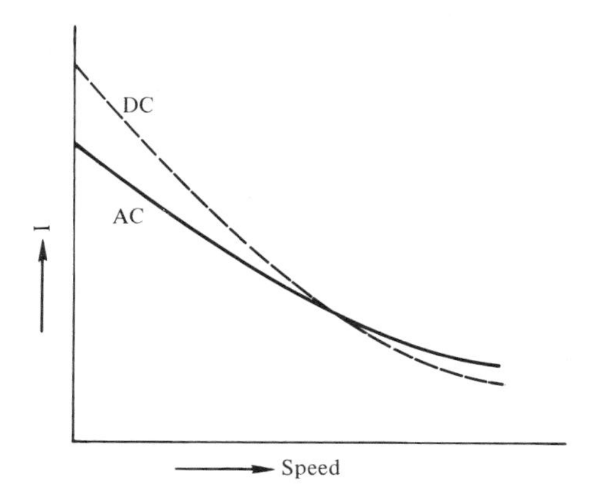

FIGURE 8-12. Current characteristics of a universal motor.

where k is a constant and I is the current through the field and the armature, the two being in series. Under ac operation, with a current $I_m \sin \omega t$, (8-14) yields

$$T_e = kI_m^2 \sin^2 \omega t \tag{8-15}$$

Since $\sin^2 \omega t = \frac{1}{2}(1 - \cos 2\omega t)$, (8-15) becomes

$$T_e = kI^2(1 - \cos 2\omega t) \tag{8-16}$$

which is an expression for the instantaneous torque. The time-average value of T_e is, therefore,

$$(T_e)_{\text{average}} = kI^2 \tag{8-17}$$

Hence the average torque is unidirectional and has the same magnitude as that with dc excitation, but the torque pulsates at twice the supply frequency.

In contrast to the solid poles of small dc series motors, the field structure of a universal motor is laminated to reduce hysteresis and eddy-current losses on ac operation. For a given voltage, the universal motor will draw less current (because of the reactance of the field and armature windings; see Fig. 8-12), develop less torque, and hence operate at a lower speed on ac than when operated on dc. Another difference between the ac and dc operations of a universal motor is that on ac the commutation is poorer because of the voltage induced, by transformer action, in the coils undergoing commutation.

A compensating winding is used to neutralize the reactance voltage of the

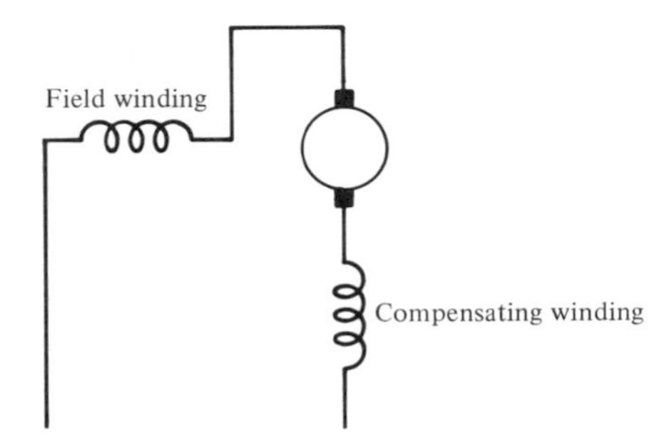

FIGURE 8-13. Conductively compensated universal motor.

armature winding. The compensating winding is connected in series with the armature, but is so arranged that the mmf of this winding opposes and neutralizes the armature mmf. Thus the compensating winding is displaced 90° (electrical) from the field winding, as shown in Fig. 8-13. The compensating winding not only neutralizes the armature reactance, but aids in commutation.

Repulsion Motors. A repulsion motor is represented schematically in Fig. 8-14(a). The rotor is similar to that of a dc machine armature. However, the brushes are short-circuited along an axis displaced by an angle ψ from the axis of a single-phase stator winding. The stator winding is similar to that of a single-phase induction motor.

In order to understand the operation of the repulsion motor, we resolve the stator winding into two components on the stator as marked by F and T in Fig. 8-14(b). Then T may be considered as the primary of a transformer with the

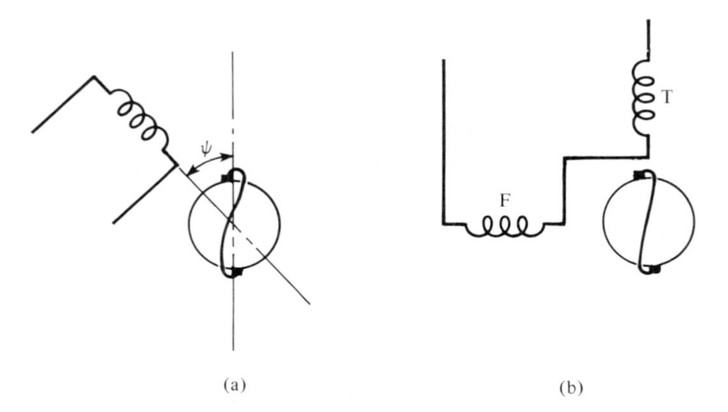

FIGURE 8-14. (a) A repulsion motor; (b) its equivalent.

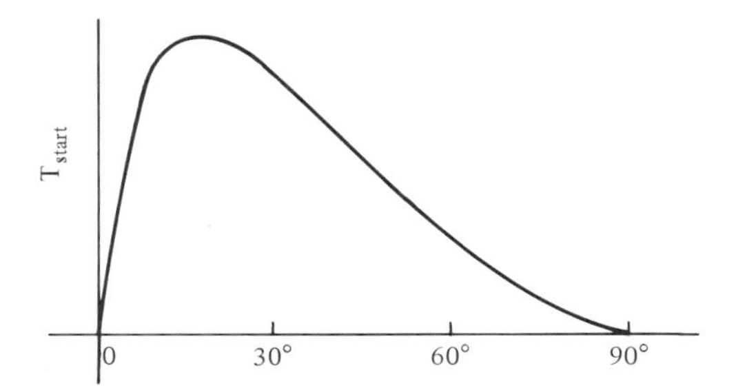

FIGURE 8-15. Starting torque versus brush position.

armature as the short-circuited secondary. When the stator is energized, the armature sets up its own mmf by induction action (via coupling through T). This mmf interacts with the stator mmf produced by F, resulting in a torque production. The current taken by the motor depends on the load. Hence the mmf due to F is directly proportional to the load current, and the motor has a variable-speed series motor characteristic. The motor is inherently compensated for armature reaction because an increase in the armature current results in an increase in the current through T and thereby increasing its mmf. This mmf is in opposition to the armature mmf. Inherent compensation is one of the advantages of a repulsion motor compared to a series motor. Low power factor and tendency to spark at the brushes are among the disadvantages of a repulsion motor. These disadvantages can be offset by modified and compensated repulsion motors. But these motors are too specialized to be considered here.

Figures 8-15 and 8-16 show the various operating characteristics of a repulsion motor. The maximum starting torque is obtained when the brush shift is about 10 to 15°. At rated load, the optimum brush shift is between 15 and 25°.

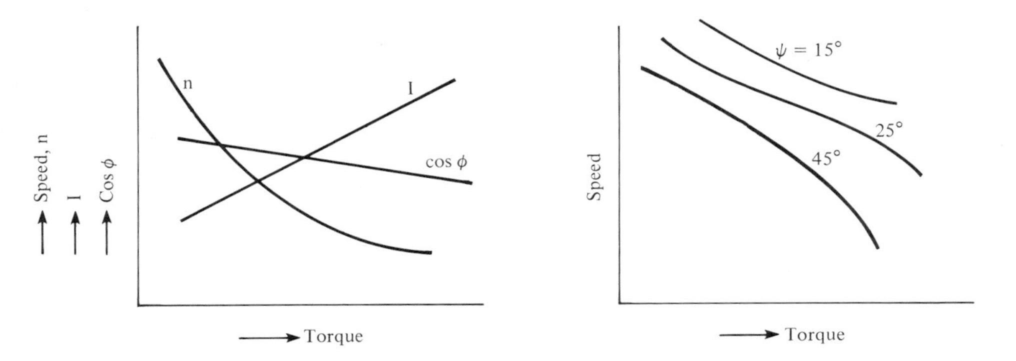

FIGURE 8-16. Characteristics of a repulsion motor.

TWO-PHASE MOTORS

Two-phase ac motors are usually used in instrumentation and control systems. Two such motors, discussed next, are ac tachometers and two-phase servomotors.

AC Tachometers. In many control applications it is desirable to express the speed of a motor as an ac voltage. A small two-phase induction motor, connected as shown in Fig. 8-17, is suitable for this purpose. The reference winding R is connected to an ac source of constant voltage and frequency. While the rotor is rotating, a voltage of the same frequency as that of the reference voltage is available at the control winding. The magnitude of this voltage is (ideally) linearly proportional to the speed of the rotor, and the phase of the voltage is fixed with respect to the reference voltage. In practice, the control winding is connected to a high-impedance amplifier and thus may be considered open-circuited for practical purposes. For a given current in the reference winding, $\mathbf{I}_r$, the control-winding voltage, $\mathbf{V}_c$, is given by

$$\mathbf{V}_c = kI_r(\mathbf{Z}_f - \mathbf{Z}_b) \tag{8-18}$$

where k is a constant and $\mathbf{Z}_f$ and $\mathbf{Z}_b$ are the forward and backward impedances of Fig. 8-2(a). Because $\mathbf{Z}_f$ and $\mathbf{Z}_b$ are functions of speed, $\mathbf{V}_c$ will measure the motor speed. Ac tachometers are commonly used in 400-Hz systems.

Two-Phase Servomotors. A two-phase servomotor, used in control applications, is very similar to the two-phase ac tachometer just described. Servomotors are used to transform a time-varying signal to a time-varying motion: for example, in an antenna-positioning system. An arrangement of a two-phase servomotor is shown in Fig. 8-18. The motor has two stator field windings

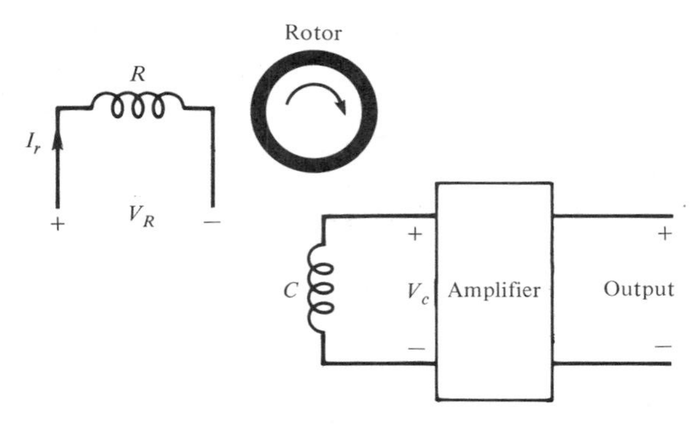

FIGURE 8-17. A two-phase tachometer.

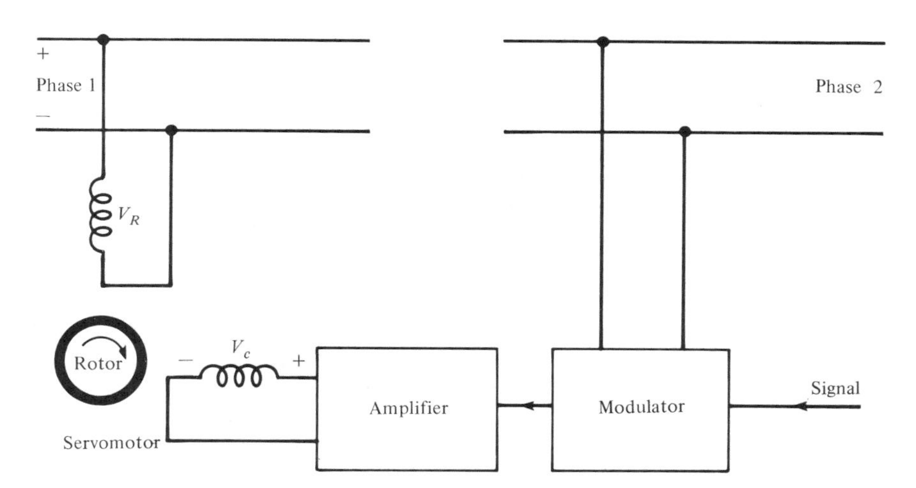

FIGURE 8-18. A two-phase servomotor.

displaced from each other by 90 electrical degrees. In a conventional two-phase induction motor, the voltages V_r and V_c (Fig. 8-18) are equal and 90° displaced from each other in time. The speed of the induction motor depends on the load (or slip). However, the speed of the servomotor must be proportional to an input signal voltage. The control voltage V_c results from the amplitude of the voltage of phase 2, but modulated by the signal. The motor speed, in turn, will depend on the signal or control voltage, V_c. In this sense, the ac tachometer is a generator, whereas the servomotor functions as a motor. The torque–speed characteristics of a typical two-phase servomotor are shown in Fig. 8-19 for different values of the signal voltage.

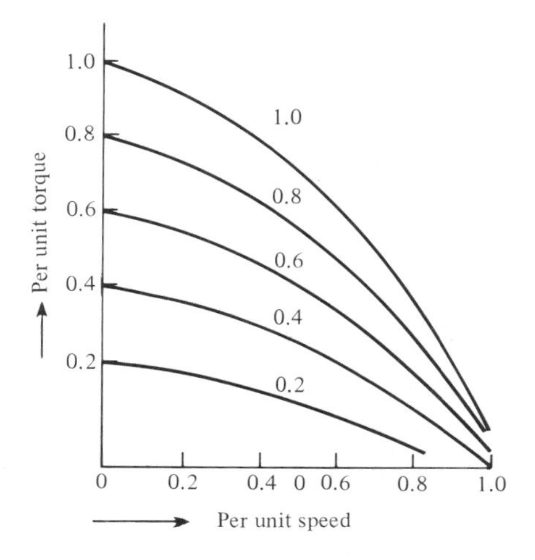

FIGURE 8-19. Speed–torque characteristics of a servomotor.

STEPPER MOTORS

The stepper motor is not operated to run continuously. Rather, it is designed to rotate in steps in response to electrical pulses received at its input from a control unit. The motor indexes in precise angular increments. The average shaft speed of the motor, n_{average}, is given by

$$n_{\text{average}} = \frac{60 \text{ (pulses per second)}}{\text{number of phases in the winding}} \quad \text{r/min}$$

Two basic types of stepper motors are variable-reluctance (VR) and permanent-magnet (PM) stepper motors. In principle, a stepper motor is a synchronous motor, and typically has three-phase or four-phase windings on the stator. The number of poles on the rotor depends on the step size (or angular displacement) per input pulse. The rotor is either reluctance type or may be made of permanent magnet. When a pulse is given to one of the phases on the stator, the rotor

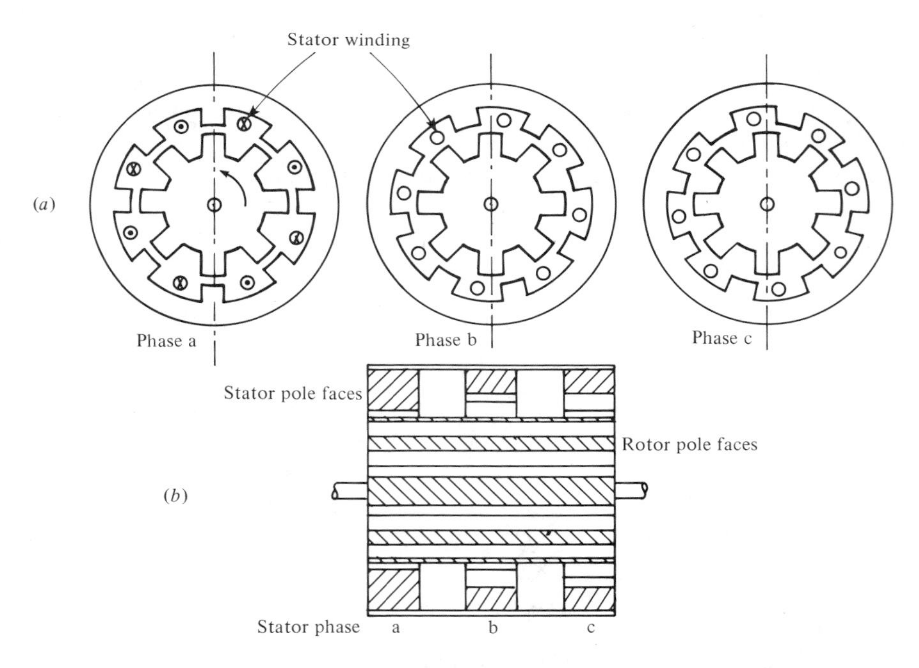

FIGURE 8-20. A variable-reluctance stepper motor: (a) phase a energized and relative displacements of stator phases b and c; (b) cross-section of the assembled motor.

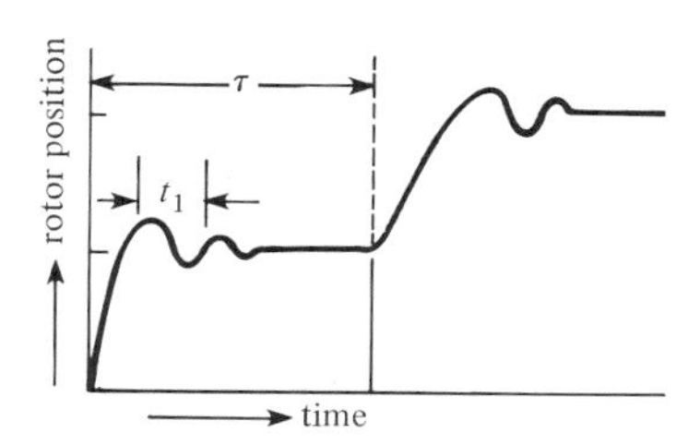

FIGURE 8-21. Undamped response of the rotor to step inputs.

tends to align with the mmf axis of the stator coil. These coils are sequentially switched and the rotor follows the stator mmf in sequence.

A variable-reluctance motor is shown in Fig. 8-20. The machine has a rotor with eight poles and three independent eight-pole stators arranged coaxially with the rotor as shown in Fig. 8-20(b). When phase *a* is energized, the motor poles align with the stator poles of phase *a*. Notice from Fig. 8-20(a) that the phase *b* stator is displaced from the phase *a* stator by 15° in the counterclockwise direction, and the phase *c* stator is further displaced from the phase *b* stator by another 15° in the counterclockwise direction. Now, when the current in phase *a* is turned off and phase *b* is energized, the rotor will rotate counterclockwise through 15°. Next, when phase *b* current is turned off and phase *c* is energized, the rotor will turn by another 15° in the counterclockwise direction. Finally, turning off the current in phase *c* and exciting phase *a* will complete one step (of 45°) in the counterclockwise direction. Additional current pulses in the sequence *abc* will produce further steps in the counterclockwise direction. Reversal of rotation is obtained by reversing the phase sequence to *acb*.

The stepping motion of the rotor of a stepper motor is typical of an undamped system, as illustrated in Fig. 8-21. The initial displacement of the rotor overshoots the final position and then gradually settles down to the final position. Some means of damping these oscillations is necessary in stepper motors. As the frequency of pulsing is increased, the period τ decreases. When τ is close to the oscillatory period t_1, the motor reaches its operating limit.

As seen from the preceding discussion, the step angle of a motor is determined by the number of poles. Typical step angles are 15°, 5°, 2°, and 0.72°. The choice of step angle depends on the angular resolution required for the application. The speed at which a stepping motor can operate is limited by the degree of damping existing in the system. Speeds up to 200 steps per second are typically attainable. A steady and continuous speed of rotation (*slewing*) greater than this value can be achieved, but the motor is then unable to stop the system in a single step.

A permanent-magnet stepper motor is shown in Fig. 8-22. The stator has a two-phase winding and the rotor has five pole pairs. For the position shown,

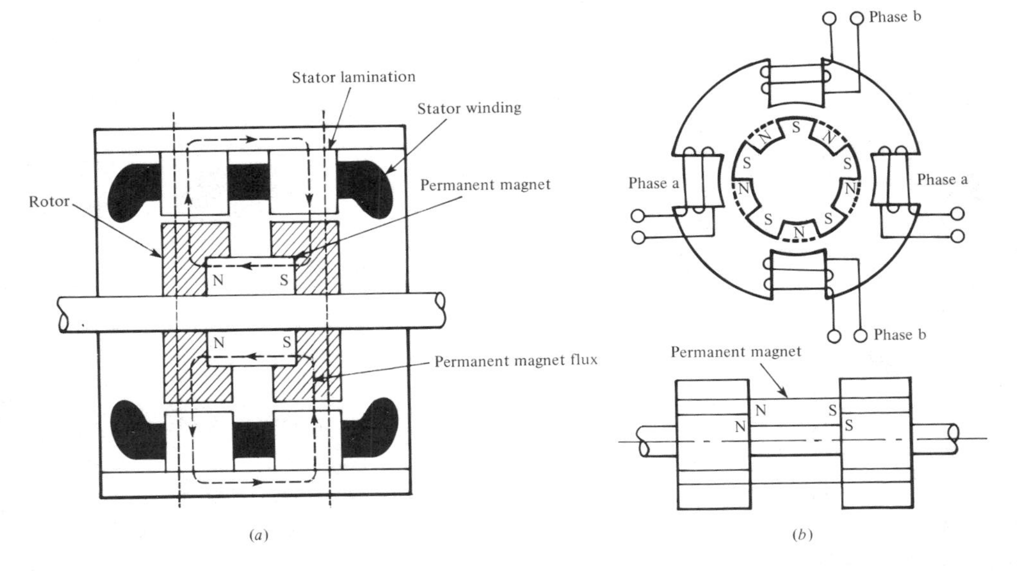

FIGURE 8-22. A permanent-magnet stepper motor: (a) axial view; (b) radial cross-section.

phase b is energized and the rotor turns by $(90° - 72°) = 18°$ to align with the pole of the phase b winding. The switching of the phases for a clockwise rotation is shown in Fig. 8-23. The rotor achieves an equilibrium position after each step. If disturbed, the rotor will tend to return to its equilibrium position.

The parameters that are important in the selection of a stepper motor are speed, torque, single-step response, holding torque, settling time, and slewing. In the slewing mode, the rotor does not come to a complete stop before the next pulse is switched on. The typical torque–steps per second characteristic of a step motor are shown in Fig. 8-24.

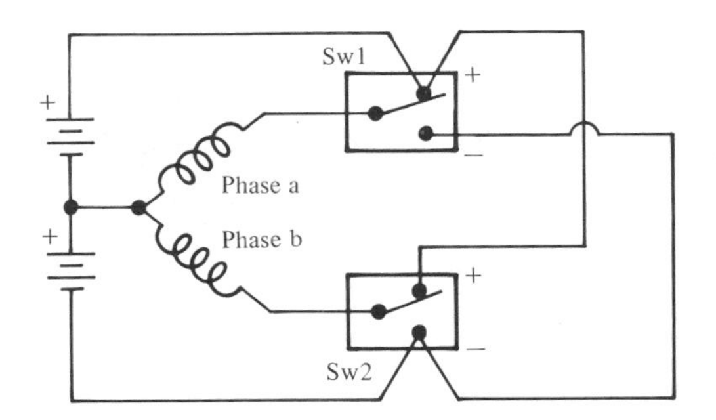

Step	CW rotation	
	Sw1	Sw2
1	off	−
2	+	off
3	off	+
4	−	off
1	off	−

FIGURE 8-23. (a) Circuit diagram and (b) switching sequence for clockwise rotation of motor shown in Fig. 6-21.

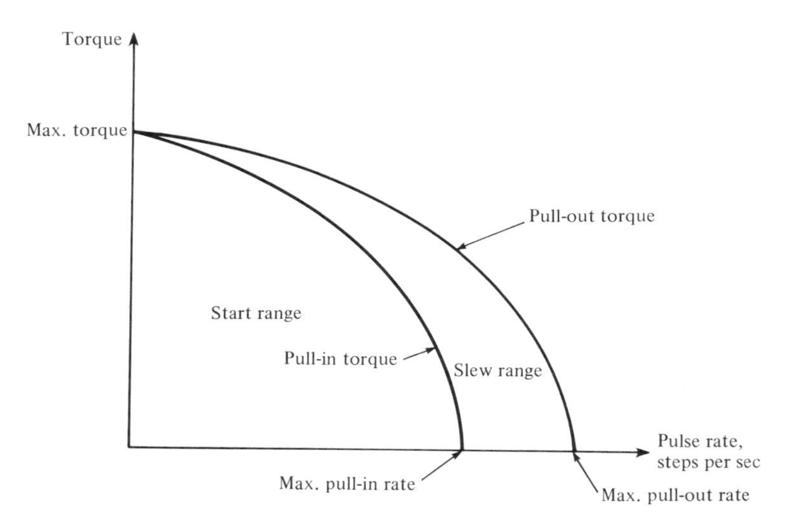

FIGURE 8-24. Torque–pulse rate characteristics.

A comparison of variable-reluctance (VR) and permanent-magnet (PM) stepper motors is given in Table 8-1.

Stepper motors have a wide range of applications, including process control, machine tools, computer peripherals, and certain medical equipment.

TABLE 8-1 Comparison Between VR and PM Stepper Motors

VR Motor	PM Motor
No torque with no excitation	Residual torque with no excitation
Higher inductance and slower electrical response	Lower inductance and faster electrical response
Low rotor inertia and faster mechanical response	High inertia and slower mechanical response

PROBLEMS

8-1 A 60-Hz single-phase induction motor has six poles and runs at 1000 r/min. Determine the slip with respect to (a) the forward rotating field and (b) the backward rotating field.

8-2 A 110-V one-phase two-pole 60-Hz induction motor is designed to run at 3420 r/min. The parameters of the motor equivalent circuit of Fig. 8.2(a) are R_1 =

$R_2' = 6\ \Omega$, $X_1 = X_2' = 10\ \Omega$, and $X_m = 80\ \Omega$. Determine (a) the input current and (b) the developed torque at 3420 r/min.

8-3 The circuit of Fig. 8-2(a) is modified by neglecting $0.5X_m$ from the backward circuit of the rotor. In this circuit the rotor resistance is replaced by $0.3R_2'$. With these approximations, repeat the calculations of Problem 8-2 and compare the results.

8-4 If the motor of Problem 8-2 has a core loss of 8 W and a friction and windage loss of 10 W, determine the motor efficiency at 5 percent slip using (a) the equivalent of Fig. 8-2(a) and (b) the modified equivalent circuit of Problem 8-3.

8-5 What is the relative amplitude of the resultant forward-rotating flux density to the resultant backward-rotating flux density for the motor of Problem 8-2 at 5 percent slip?

8-6 Draw the modified equivalent circuit suggested in Problem 8-3. Determine the parameters of this modified circuit from the following test data:

No-load test:	input voltage, 110 V; input current, 2.5 A input power, 45 W friction and windage loss, 8 W
Blocked-rotor test:	input voltage, 46 V input current, 1.8 A stator resistance, 1.2 Ω (assume that $X_1 = X_2'$)

8-7 The direct- and quadrature-axis inductances of a reluctance motor are 60 mH and 25 mH. The motor operates at a power angle of 15° while taking 2.2 A of current. Calculate the torque developed by the motor.

8-8 Determine the torque developed by the motor of Problem 8-7 if the motor is supplied by a 110-V 60-Hz source and the power angle is 15°. Neglect the motor winding resistance.

8-9 What is the operating speed of a reluctance motor if it has four poles and is operated at 400 Hz?

8-10 The hysteresis loss in the material of a two-pole 60-Hz hysteresis motor is 600 W. Determine the torque developed by the motor. If 8 W is lost because of various rotational losses, calculate the power and the torque available at the shaft.

8-11 A 110-V universal motor has an input impedance of $(30 + j40)\ \Omega$. The armature has 36 conductors and is lap-wound. The field has two poles and the flux per pole is 30 mWb. Determine the torque developed by the motor.

8-12 The motor of Problem 8-11 has a total no-load and rotational loss of 30 W at 4000 r/min. Calculate (a) the power factor and (b) the efficiency of the motor.

CHAPTER 9

General Theory of Electromechanical Energy Conversion

9-1

PHYSICAL PRINCIPLES OF OPERATIONS OF GENERATORS AND MOTORS

Whereas we have studied specific machine types in Chapters 5 through 8, in the present chapter we present a unified approach to electric generators, motors, and electromechanical transducers. Before presenting the various analytical equations, we review the physical principles governing the operations of electric machines.

Electric generators, which convert mechanical energy into electrical energy, operate on the basis of Faraday's law of electromagnetic induction. According to Faraday's law, a voltage is induced in a conductor when it "cuts" magnetic flux lines, or when the flux linking a circuit changes in time. As seen in earlier chapters, dc and ac generators operate on this principle. For future reference in this chapter, we rewrite the basic induced emf equation as

$$e = \frac{d\lambda}{dt} = \frac{d}{dt}(Li) \tag{9-1}$$

which implies that a voltage e is induced in an N-turn coil if the flux linking the coil changes in time. By definition, the flux linkage $\lambda = N\varphi = Li$, where L is the inductance of the coil and the flux φ is produced by the current i in the coil. Equation (9-1) governs the production of emfs in all electromechanical devices.

Devices which convert electrical energy into mechanical energy, operate either on the principle of (1) alignment of flux, or (2) interaction between magnetic fields and current-carrying conductors (Ampère's law). Alignment of flux provides the force that pulls ferromagnetic items to a common magnet. The interaction between magnetic fields and currents, on the other hand, provides the force that tends to expel current-carrying conductors from the presence of a magnetic field. Two examples of "alignment of flux" are illustrated in Fig. 9-1. In Fig. 9-1(a), the force on the ferromagnetic pieces causes them to align with the flux lines, thus shortening the magnetic flux path and reducing the reluctance. Figure 9-1(b) shows the alignment of two current-carrying coils. Examples of "interaction between current-carrying conductors and magnetic fields" are given in Fig. 9-2, where current-carrying conductors experience a force when placed in magnetic fields. For instance, in Fig. 9-2(b) the force is produced by the interaction of flux lines and coil current, resulting in a torque on the coil. This mechanism forms the basis of a variety of electrical measuring instruments and electric motors.

Whereas motors and generators are gross-motion devices, electromechanical transducers are incremental-motion devices in the sense that their motion is restricted to small displacements. For instance, the displacement of the diaphragm of a loudspeaker (processing electrical energy and converting it to mechanical form) is small compared to the motion of an electric motor. The loudspeaker is an incremental-motion device and may be considered as an energy processor. A motor is a gross-motion device and is an energy converter. An

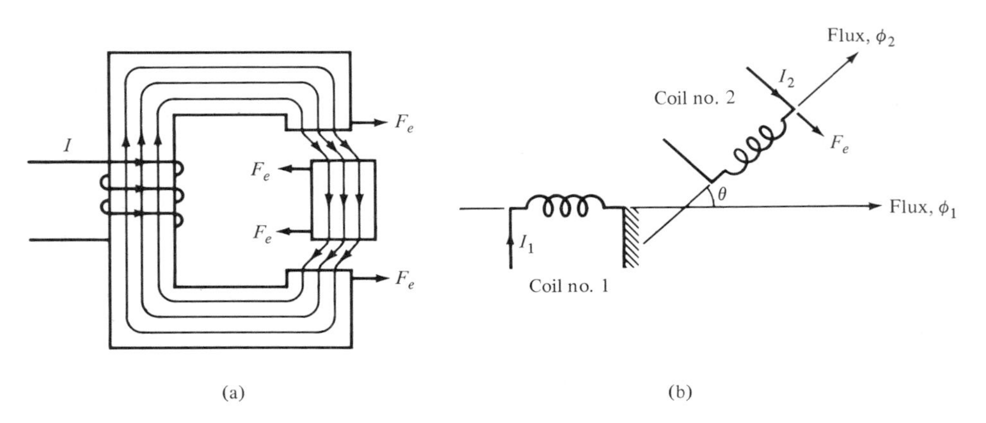

FIGURE 9-1. Two examples of alignment of flux for force production.

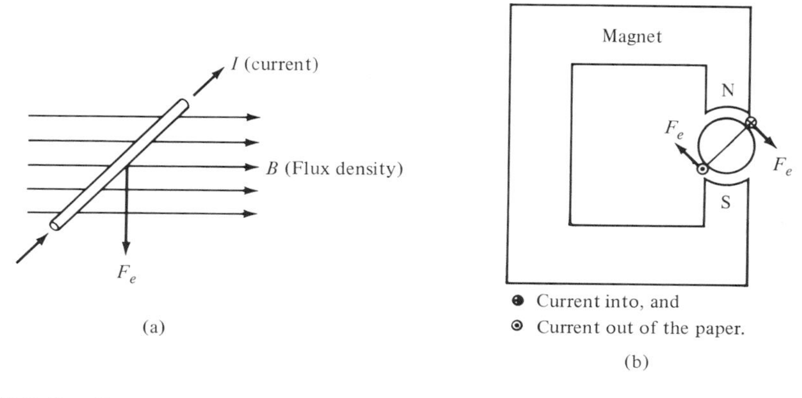

FIGURE 9-2. Forces on current-carrying conductors located in magnetic fields.

electromagnetic solenoid also belongs to the category of incremental-motion devices. Electrical forces in such devices can be determined from the principle of energy conservation. The method, although derived (in the next section) for an incremental-motion device—an electromagnet—is applicable to gross-motion devices as well.

In the preceding we have given a few examples showing how mechanical forces are produced via magnetic fields. For energy conversion (i.e., for doing work), mechanical motion is as important as mechanical force. During mechanical motion, the energy stored in the coupling magnetic field is disturbed. In Fig. 9-1(a), for instance, most of the magnetic field energy is stored in the air gap separating the two ferromagnetic pieces. The air-gap field may be termed the *coupling field*. When electromechanical energy conversion occurs, the coupling fields are disturbed such that the energy stored in the field changes with mechanical motion. We can relate the change in energy stored in the magnetic field to the mechanical work done, and this relationship will enable us to determine the magnitudes of mechanical forces arising from magnetic field effects.

The energy conservation principle in connection with electromechancal systems may be stated in a number of ways. For instance, we may say that

$$\begin{matrix}\text{input} & & \text{input} & & \text{increase} & & \text{energy} \\ \text{electrical} & + & \text{mechanical} & = & \text{in stored} & + & \text{dissipated} \\ \text{energy} & & \text{energy} & & \text{energy} & & \text{as heat}\end{matrix}$$

or

$$\begin{matrix}\text{input} & & \text{mechanical} & & \text{increase} & & \text{energy} \\ \text{electrical} & + & \text{work} & = & \text{in stored} & + & \text{dissipated} \\ \text{energy} & & \text{done} & & \text{energy} & & \text{as heat}\end{matrix}$$

Or, if only the conservative (or lossless) portion of the system is considered, we have

$$\text{sum of input energy} = \text{change in stored energy}$$

or (9-2)

$$\text{input electrical energy} = \text{mechanical work done} + \text{increase in stored energy}$$

9-2 FORCE EQUATION

Bearing in mind the foregoing expressions of the principle of energy conservation, let us consider the electromagnet shown in Fig. 9-3(a). For the present, we assume that the system is lossless. The coil shown in Fig. 9-3(a) may be excited by a current source or a voltage source. Excitation by a current source implies that the coil current is held constant at all times, whereas voltage excitation requires that the voltage across the coil, or the flux linking it, must remain constant. When the coil is excited, the core acts like an electromagnet, attracting the iron mass (armature) with a force, F_e, of electromagnetic origin. Now suppose that the armature undergoes a small motion, dx. Over the period of the motion of the armature, let dW_e be the input electrical energy and dW_m the increase in the energy stored in the magnetic field. Then (9-2) may be written as

$$dW_e = F_e\,dx + dW_m \tag{9-3}$$

Since $v = d\lambda/dt$ (from Faraday's law), dW_e may be written as

$$dW_e = vi\,dt = \frac{d\lambda}{dt}\,i\,dt = i\,d\lambda$$

and (9-3) becomes

$$i\,d\lambda = F_e\,dx + dW_m \tag{9-4}$$

We now let the current in the coil be held constant at I_0 during the period that the armature undergoes motion from position 1 to position 2, through dx. During this motion, the flux linkage changes from λ_1 to λ_2 and the electrical

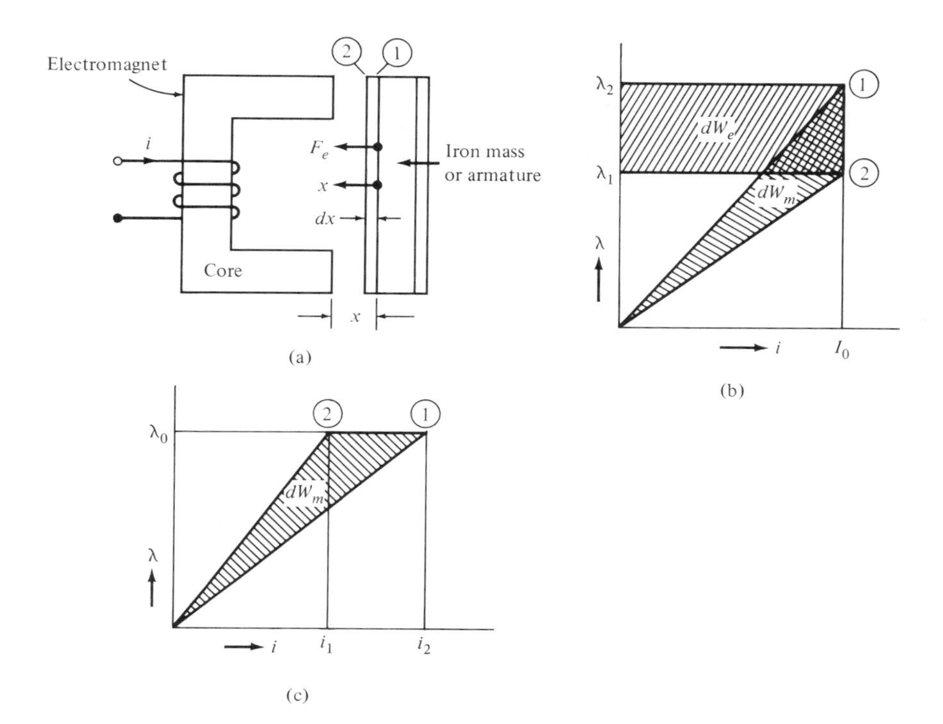

FIGURE 9-3. Energy balance in an electromechanical system: (a) a simple system; (b) constant-current operation; (c) constant-voltage (or flux linkage) operation.

energy input becomes $dW_e = I_0 (\lambda_2 - \lambda_1)$. The electrical energy comes from the current source, as shown in Fig. 9-3(b). Also, during the process, the increase in field energy is $dW_m = \frac{1}{2}I_0(\lambda_2 - \lambda_1)$. Thus, from (9-4), we have

$$I_0(\lambda_2 - \lambda_1) = F_e\, dx + \tfrac{1}{2}I_0(\lambda_2 - \lambda_1)$$

or (9-5)

$$F_e\, dx = \tfrac{1}{2}I_0(\lambda_2 - \lambda_1)$$

indicating that for a current-excited system the electrical energy input divides equally between increasing stored energy and doing mechanical work.

Next we consider the case in which the flux linkage is kept constant at λ_0 and the current is allowed to vary from i_1 to i_2 ($i_2 < i_1$) during motion, as shown in Fig. 9-3(c). In this case there is no electrical energy input from the source, as may be seen by comparing parts (b) and (c). The change in stored energy is

$$dW_m = \tfrac{1}{2}\lambda_0(i_2 - i_1)$$

which is negative. Thus, from (9-5), we get

$$-F_e\,dx = dW_m$$

or (9-6)

$$-F_e\,dx = \tfrac{1}{2}\lambda_0(i_2 - i_1)$$

indicating that the mechanical work done equals the reduction in stored energy.

The term on the right-hand side of (9-6) is simply dW_m. Hence we may write

$$F_e = -\frac{\partial W_m}{\partial x}(\lambda, x) \tag{9-7}$$

Similarly, we may rewrite (9-5) as

$$F_e = \frac{\partial W_m}{\partial x}(i, x) \tag{9-8}$$

Equations (9-7) and (9-8) are the force equations, respectively, for voltage- and current-excited systems. In (9-7) and (9-8), we have used partial derivatives because W_m is a function of more than one variable—(λ, x) in (9-7) and (i, x) in (9-8).

General Derivation of the Force Equation

Derivation leading to (9-7) and (9-8) are, obviously, carried through by considering a special case—that of the electromagnet. A general derivation now follows.

As mentioned earlier, in an electromechanical system either (i, x) or (λ, x) may be considered as independent variables. If we consider (i, x) as independent, the flux linkage λ is given by $\lambda = \lambda(i, x)$, which can be expressed in terms of small changes as

$$d\lambda = \frac{\partial \lambda}{\partial i}\,di + \frac{\partial \lambda}{\partial x}\,dx$$

Also, we have $W_m = W_m(i, x)$, so that

$$dW_m = \frac{\partial W_m}{\partial i}\,di + \frac{\partial W_m}{\partial x}\,dx \tag{9-9}$$

Equation (9-9), when substituted into (9-4), yields

$$F_e\,dx = -\frac{\partial W_m}{\partial x}\,dx - \frac{\partial W_m}{\partial i}\,di + i\,\frac{\partial \lambda}{\partial x}\,dx + i\,\frac{\partial \lambda}{\partial i}\,di$$

or (9-10)

$$F_e\,dx = \left(-\frac{\partial W_m}{\partial x} + i\,\frac{\partial \lambda}{\partial x}\right)dx + \left(-\frac{\partial W_m}{\partial i} + i\,\frac{\partial \lambda}{\partial i}\right)di$$

Because the incremental changes di and dx are arbitrary, F_e must be independent of these changes. Thus for F_e to be independent of di, its coefficient in the preceding equation must be zero. Consequently, we obtain

$$F_e = -\frac{\partial W_m}{\partial x}\,(i, x) + i\,\frac{\partial \lambda}{\partial x}\,(i, x) \tag{9-11}$$

which is the force equation and holds true if i is the independent variable. To show that the coefficient of di is always zero, that is,

$$-\frac{\partial W_m}{\partial i} + i\,\frac{\partial \lambda}{\partial i} = 0 \tag{9-12}$$

we recall that

$$W_m = \int i\,d\lambda = i\lambda - \lambda\,di \tag{9-13}$$

Substituting (9-13) into (9-12) shows that (9-12) is always satisfied. Hence (9-11) is valid.

If, on the other hand, λ is taken as the independent variable, that is, if $i = i(\lambda, x)$ and $W_m = W_m(\lambda, x)$, then

$$dW_m = \frac{\partial W_m}{\partial \lambda}\,d\lambda + \frac{\partial W_m}{\partial x}\,dx$$

which, when substituted into (9-4), gives

$$F_e\,dx = -\frac{\partial W_m}{\partial x}\,dx - \frac{\partial W_m}{\partial \lambda}\,d\lambda + i\,d\lambda \tag{9-14}$$

But (9-13) implies that $\partial W_m/\partial\lambda = i$, so (9-14) finally becomes

$$F_e = -\frac{\partial W_m}{\partial x}(\lambda, x) \qquad (9\text{-}15)$$

Observe that in view of (9-13), the coefficient of $d\lambda$ in (9-14) is zero.

Comparing (9-7) and (9-15), notice that the two force equations are identical. However, (9-8) which is different from the general force equation (9-11), is valid for the special case of a linear magnetic circuit. For such a case, from Fig. 9-3(b) or (c),

$$W_m = \tfrac{1}{2}\lambda i$$

implying that

$$\frac{\partial W_m}{\partial x} = \frac{1}{2} i \frac{\partial \lambda}{\partial x}$$

which when substituted in (9-11) yields (9-8).

We now illustrate the application of the force equation by the following examples.

EXAMPLE 9-1

A solenoid of cylindrical geometry is shown in Fig. 9-4. For the numerical values $I = 10$ A, $N = 500$ turns, $g = 5$ mm, $a = 20$ mm, $b = 2$ mm, and $l = 40$ mm, what is the magnitude of the force? Assume that $\mu_{\text{core}} = \infty$. Neglect leakage and losses. For the magnetic circuit (in terms of the symbols used in Fig. 9-4) the reluctance is

$$\mathcal{R} = \frac{g}{\mu_0 \pi c^2} + \frac{b}{\mu_0 2\pi a l} \qquad \text{where } c = a - \frac{b}{2}$$

The inductance L is then given by

$$L = \frac{N^2}{\mathcal{R}} = \frac{2\pi\mu_0 a l c^2 N^2}{2alg + bc^2} = \frac{k_1}{k_2 g + k_3}$$

where $k_1 = 2\pi\mu_0 a l c^2 N^2$, $k_2 = 2al$, and $k_3 = bc^2$. Expressing the force as in (9-10), with $W_m = \frac{1}{2}LI^2$,

$$F_e = \frac{\partial W_m}{\partial x} = \frac{1}{2} I^2 \frac{\partial L}{\partial g} = -\frac{I^2 k_1 k_2}{2(k_2 g + k_3)^2}$$

where the minus sign indicates that the force tends to decrease the air gap.

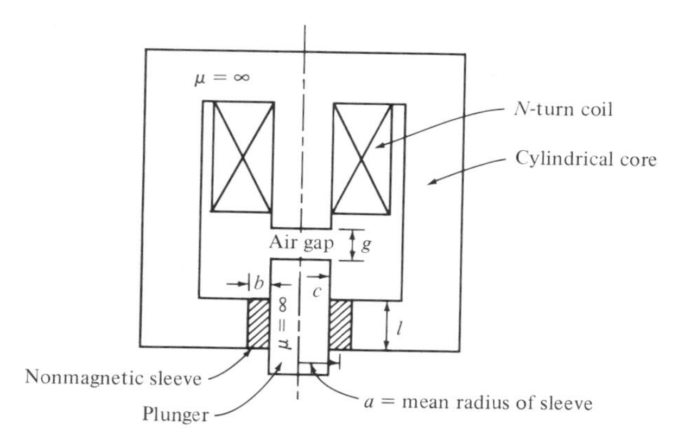

FIGURE 9-4. Example 9.1.

Substituting the numerical values in the force expression yields 600 N as the magnitude of the force. ■

EXAMPLE 9-2

The inductances of the coils shown in Fig. 9-1(b) are $L_{11} = L_{22} = 50$ mH and $L_{12} = L_{21} = 30 \cos \theta$ mH. If $I_1 = I_2 = 10$ A, determine the electromagnetic torque at $\theta = 45°$.

The magnetic energy stored in the two-coil system is given by

$$W_m = \tfrac{1}{2}L_{11}I_1^2 + \tfrac{1}{2}L_{22}I_2^2 + L_{12}I_1I_2$$

Substituting the given numerical values yields

$$W_m = 5 + 3 \cos \theta$$

The electromagnetic torque is given by

$$T_e = \frac{\partial W_m}{\partial \theta} = -3 \sin \theta \qquad \text{N-m}$$

as in (9-10). At $\theta = 45°$, $T_e = -3 \sin 45° = -3/\sqrt{2}$ N-m. The negative sign implies that T_e tends to decrease θ. ■

EXAMPLE 9-3

For the magnetic circuit of Fig. 9-3(a), the flux linkage and current are related by

$$\lambda = \frac{2(i^{1/2} + i^{1/3})}{x + 1}$$

where x is as shown. Calculate the electrical force at $x = 0$ and $i = 64$ A dc.

Because the λ–i relationship is nonlinear, we must use the general force equation, (9-11). The vanishing of the coefficient of di in (9-10) implies that

$$\frac{\partial W_m}{\partial i} = i\,\frac{\partial \lambda}{\partial i}$$

Therefore,

$$W_m = \int i\,\frac{\partial \lambda}{\partial i}\,di = \frac{2}{x+1}\int i\left(\tfrac{1}{2}i^{-1/2} + \tfrac{1}{3}i^{-2/3}\right)di = \frac{2}{x+1}\left(\tfrac{1}{3}i^{3/2} + \tfrac{1}{4}i^{4/3}\right)$$

Thus

$$-\frac{\partial W_m}{\partial x} = \frac{2}{(x+1)^2}\left(\tfrac{1}{3}i^{3/2} + \tfrac{1}{4}i^{4/3}\right) \qquad \text{and} \qquad i\,\frac{\partial \lambda}{\partial x} = \frac{-2}{(x+1)^2}\left(i^{3/2} + i^{4/3}\right)$$

and (9-11) gives

$$F_e = -\frac{2}{(x+1)^2}\left(\tfrac{4}{3}i^{3/2} + \tfrac{3}{2}i^{4/3}\right)$$

At $x = 0$ and $i = 64$ A, $F_e = -1066.67$ N, tending to decrease x. ■

9-3 GENERAL PRINCIPLES GOVERNING ROTATING MACHINES

In Section 9-1 we have alluded to the "principle of alignment" as a basis of torque (or force) production in current-carrying conductors. In fact, torque production in most common types of rotating machines can be explained and analyzed by the alignment principle. Various electric machines differ from each other depending on how the stator and rotor mmf's are kept displaced from each other at all times so that they tend to align continuously and thereby produce an average torque. In the following we present the basic operations and the fundamental governing equations pertaining to two types of most common rotating machines: dc machines and synchronous machines. At this point we are familiar with the basic elements of these machines as already discussed in earlier chapters.

In a rotating machine, electromechanical energy conversion occurs when the magnetic energy stored in the air gap changes with the rotor position, according

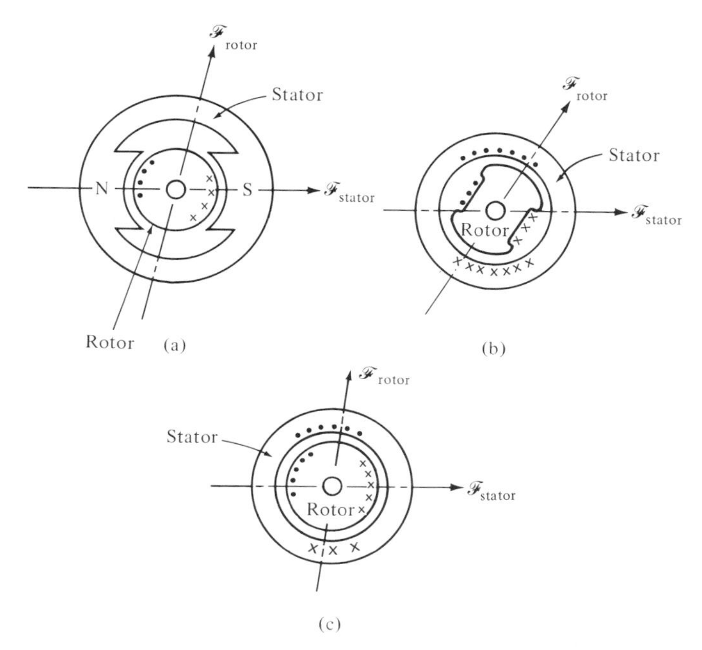

FIGURE 9-5. Configurations of common types of rotating machines: (a) saliency on stator, no saliency on rotor; (b) no saliency on stator, saliency on rotor; (c) no saliency on stator or on rotor.

to the alignment principle. For various practical reasons magnetic structures of rotating machines belong to one of the following general categories:

1. Saliency on the stator, with no saliency on the rotor (that is, the rotor is round or cylindrical), as shown in Fig. 9-5(a). Such a configuration is commonly used in dc commutator machines.
2. Saliency on the rotor, with no saliency on the stator, as represented in Fig. 9-5(b). This form of magnetic structure is employed in low-speed synchronous machines and in small reluctance machines.
3. No saliency either on stator or on rotor. This type of a configuration, shown in Fig. 9-5(c), is used in induction motors and high-speed synchronous machines.
4. Saliency on stator and rotor both. Only special machines utilize such magnetic structures, and these are not considered here.

From an analytical viewpoint, it is clear that a machine with saliency, either on the stator or on the rotor, is more general in that in various governing equations the results for the non-salient-pole machine follow by setting the terms due to saliency to zero. Thus we will consider the machine with saliency first.

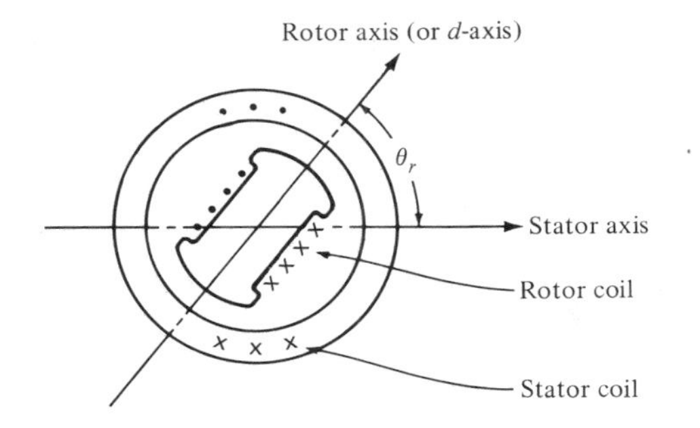

FIGURE 9-6. A salient-pole machine having one coil on stator and one on rotor.

To illustrate the analytical procedure, we assume that there is one coil on the stator and one on the rotor, as shown in Fig. 9-6, where we have shown saliency on the rotor. As mentioned in Section 9-2, to evaluate the energy conversion process quantitatively, it is convenient to express the governing equations in terms of energy storage and flux linkages. In other words, the machine inductances must be known. Referring now to Fig. 9-7, we define four inductances for the machine under consideration. These inductances are $L_{ss} \equiv$ stator self-inductance, $L_{rr} \equiv$ rotor self-inductance, $L_{sr} \equiv$ stator-to-rotor mutual inductance, and $L_{rs} \equiv$ rotor-to-stator mutual inductance. From reciprocity, $L_{rs} = L_{sr}$. The variations of various inductances with rotor positions are illustrated in Fig. 9-7 (for an idealized machine for which the variations are sinusoidal). Thus, in terms of the nomenclature shown in Fig. 9-7, the inductances may be expressed as

$$
\begin{aligned}
L_{ss} &= L_0 + L_s \cos 2\theta_r \\
L_{rr} &= L_r = \text{a constant} \\
L_{sr} &= L_{rs} = L_m \cos \theta_r
\end{aligned}
\tag{9-16}
$$

If i_s and i_r are the currents in the stator coil and the rotor coil, respectively, the corresponding flux linkages become

$$
\begin{aligned}
\lambda_s &= L_{ss} i_s + L_{sr} i_r \\
\lambda_r &= L_{sr} i_s + L_{rr} i_r
\end{aligned}
\tag{9-17}
$$

Using (9-3), (9-16), and (9-17) we may write the machine emf equations (see Problems (9-15 and 9-16).

Next, to determine the torque developed, we express the energy stored in the magnetic field as

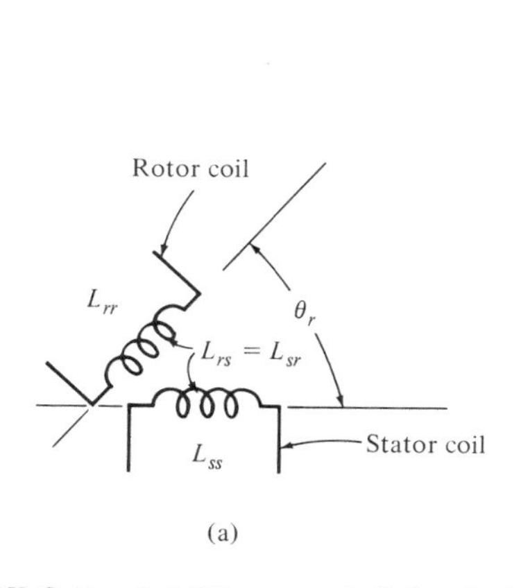

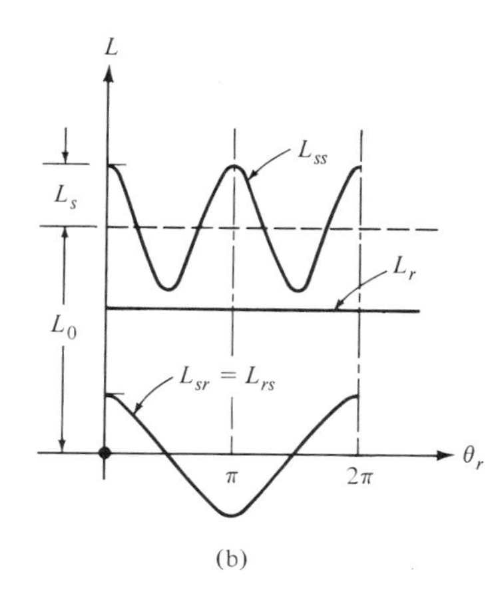

FIGURE 9-7. (a) Diagram defining inductances; (b) inductance variations.

$$W_m = \tfrac{1}{2}L_{ss}i_s^2 + \tfrac{1}{2}L_{rr}i_r^2 + L_{sr}i_si_r \tag{9-18}$$

Utilizing the procedure of Example 9-2, we may write the expression for the torque as

$$T_e = \frac{\partial W_m}{\partial \theta_r} \tag{9-19}$$

And (9-16), (9-18), and (9-19) yield the value of the instantaneous torque (see Problem 9-17).

To demonstrate the principle of alignment, let us consider the simplified case of a non-salient-pole machine. In such a machine, L_{ss} and L_{rr} are constants and $L_{sr} = L_m \cos \theta_r$. If the stator and rotor currents, respectively, are i_s and i_r, then using (9-18) and (9-19) and the given inductances, we obtain

$$T_e = -L_m i_s i_r \sin \theta_r \tag{9-20}$$

Two conclusions follow from (9-20). First, the torque tends to align the stator and rotor coils. Second, for T_e to be nonzero (for continuous rotation, the time-averaged value of T_e must not be zero), we must have a displacement between the stator and the rotor mmf's. This conclusion is consistent with the statement that we made in the beginning of this section. Based on the principle of alignment and the analytical approach discussed above, we shall now consider the operations of individual machines which are distinct from each other primarily because the displacement between the axes of stator and rotor mmf's is produced by different means.

DC MACHINES

We mentioned in Section 9-3 that in order to have a nonzero time-averaged torque (for a continuous rotation of the machine) there must be a constant angular displacement between the rotor and stator mmf axes. In a dc machine, this displacement is accomplished by the commutator. From Chapter 5 we are already familiar with the action of the commutator. Notice that the commutator segments rotate with the coil, but the brushes are fixed (with respect to the field poles). Therefore, the polarities at the brushes are independent of the rotor position. For two rotor positions the polarities of currents in the rotor conductors are shown in Fig. 9-8(a) and (b). In both positions the polarity of the rotor mmf remains unchanged. Consequently, we have a constant displacement between the stator and rotor mmf's and the rotor is capable of continuous rotation. If the rotor is rotated mechanically by an external source, the machine will operate as a generator, and the polarity of the voltage at the brushes will be independent of the rotor position.

Using the basic emf and torque equations, (9-2) and (9-19), respectively, we can now derive the voltage and torque characteristics of the dc machine. As mentioned in Section 9-3, dc machines have salient stators and cylindrical rotors. In this case, the inductances, corresponding to those given by (9-16), are

$$\begin{aligned} L_{ss} &= L_s = \text{a constant} \\ L_{rr} &= L_0 + L_r \cos 2\theta_r \\ L_{sr} &= L_{rs} = L_m \cos \theta_r \end{aligned} \tag{9-21}$$

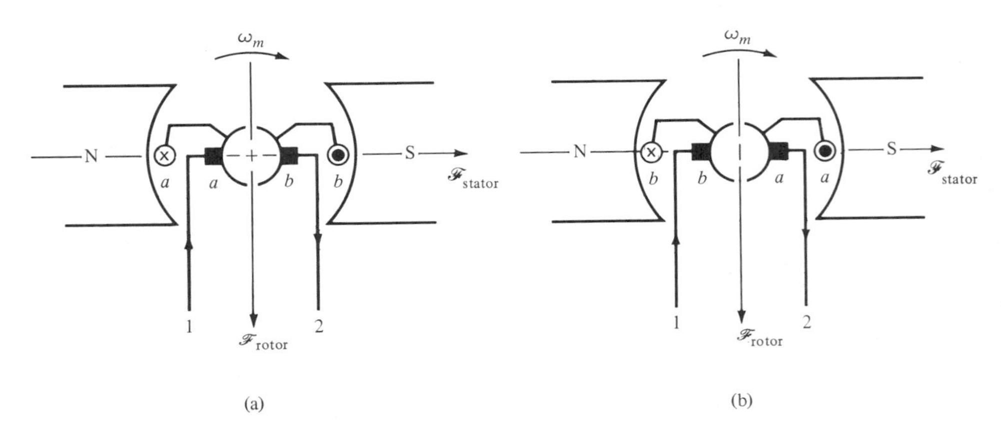

FIGURE 9-8. Conductor a connected to segment a and conductor b connected to segment b. Brushes 1 and 2 always have the same polarities in both positions (a) and (b).

where L_0, L_r, and L_m are constants and θ_r is the displacement of the rotor mmf axis from the stator mmf axis. The rotor flux linkage becomes

$$\lambda_r = L_{sr} i_s + L_{rr} i_r \tag{9-22}$$

Hence (9-2) and (9-22) yield

$$e_r = \frac{d\lambda_r}{dt} = \frac{d}{dt}(L_{sr} i_s + L_{rr} i_r) \tag{9-23}$$

If the rotor angular velocity is ω_m, then $\dot{\theta}_r = \omega_m$, and (9-21) and (9-23) yield

$$e_r = L_{sr}\frac{di_s}{dt} + L_{rr}\frac{di_r}{dt} - i_s \omega_m L_m \sin\theta_r - 2 i_r \omega_m L_r \sin 2\theta_r \tag{9-24}$$

Under steady-state conditions, let the stator and rotor currents be I_s and I_r, respectively. Since d/dt terms will then be zero, (9-24) reduces to

$$E_r = -I_s \omega_m L_m \sin\theta_r - 2 I_r \omega_m L_r \sin 2\theta_r \tag{9-25}$$

As already shown in Chapter 5, in a dc machine the brushes are located along an axis perpendicular to the stator mmf axis. Thus $\theta_r = -\pi/2$ and (9-25) becomes

$$E_r = \omega_m L_m I_s \tag{9-26}$$

This is the voltage induced in the rotor (by virtue of its rotation) because of a current I_s in the stator. If we include the effect of the rotor resistance R_r, the rotor terminal voltage, V_r, is given by

$$V_r = R_r I_r + \omega_m L_m I_s \tag{9-27}$$

Following this procedure we can easily verify that for the stator circuit, the steady-state terminal voltage, V_s is simply

$$V_s = R_s I_s \tag{9-28}$$

where R_s is the stator resistance.

Schematically, (9-27) and (9-28) can be represented by the configuration shown in Fig. 9-9. Such a machine is known as the *separately excited* machine. Now, from (9-26) and (9-1) we may write the rotor electromagnetic power as

$$E_r I_r = T_e \omega_m = \omega_m L_m I_s I_r$$

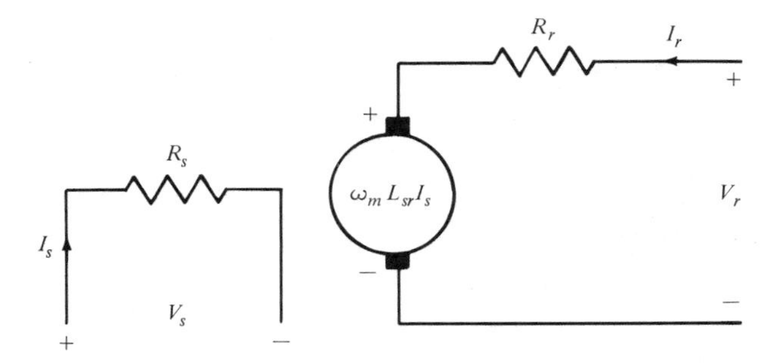

FIGURE 9-9. A separately-excited machine.

from which the electromagnetic torque, T_e, may be written as

$$T_e = L_m I_s I_r \tag{9-29}$$

A common configuration, the *shunt machine,* follows immediately from the preceding analysis. In a shunt machine, the rotor and the stator windings are connected in parallel, as shown in Fig. 9-10. This constraint can be expressed as

$$V_s = V_r = V \tag{9-30}$$

The voltage and torque equations are

$$V = R_r I_r = \omega_m L_m I_s + R_s I_s \tag{9-31}$$

$$T_e = L_m I_s I_r \tag{9-32}$$

The line current, I (Fig. 9-11), becomes

$$I = I_r \pm I_s \tag{9-33}$$

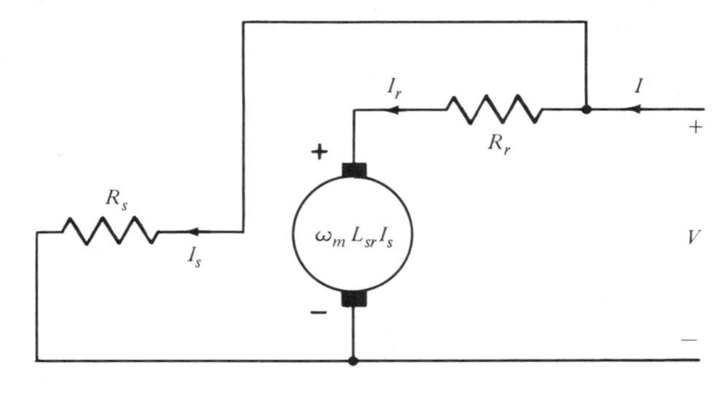

FIGURE 9-10. A shunt machine.

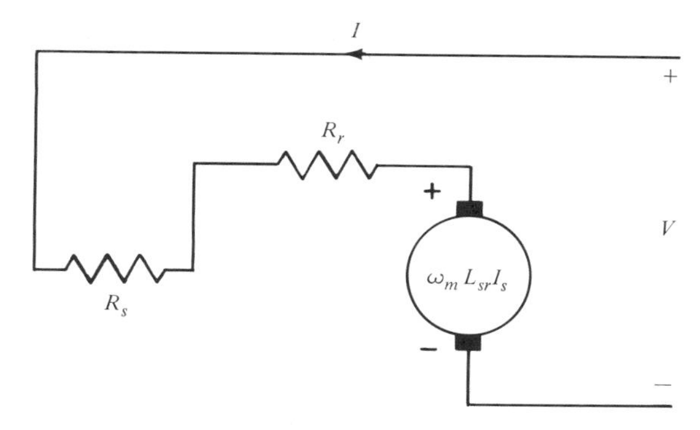

FIGURE 9-11. A series machine.

where the positive sign is for motor operation and the negative sign corresponds to generator operation.

Another configuration, the *series machine,* in which the rotor and stator windings are connected in series (Fig. 9-11). Consequently,

$$I_s = I_r = I \tag{9-34}$$

The voltage and torque equations then become, respectively,

$$V = (R_s + R_r)I + \omega_m L_{sr} I \tag{9-35}$$

and

$$T_e = L_{sr} I^2 \tag{9-36}$$

In establishing (9-21) through (9-36), we have demonstrated the general principles of electromechanical energy conversion in evolving the three common forms of dc machines.

9-5 SYNCHRONOUS MACHINES

For a simplified treatment, we consider a cylindrical-rotor machine. In such a case (9-16) becomes

$$L_{ss} = L_0 = \text{a constant}$$

$$L_{rr} = L_r = \text{a constant}$$

$$L_{sr} = L_{rs} = L_m \cos \theta_r$$

If i_s and i_r are, respectively, the stator and rotor currents, the magnetic stored energy in the system becomes

$$W_m = \tfrac{1}{2}L_0 i_s^2 + \tfrac{1}{2}L_r i_r^2 + L_m \cos \theta_r i_s i_r \tag{9-37}$$

The developed torque, from (9-37), may be written as

$$T_e = \frac{\partial W_m}{\partial \theta_r} = -L_m i_s i_r \sin \theta_r \tag{9-38}$$

We know from Chapter 6 that in a synchronous machine $i_r = I_f$, the field current, and $i_s = I_m \cos \omega t$. Furthermore, θ_r may be related to the mechanical angular velocity of the rotor by

$$\theta_r = \omega_m t + \delta$$

where δ is the constant of integration. Hence (9-38) becomes

$$T_e = I_m I_f L_m \cos \omega t \sin (\omega_m t + \delta)$$

which may also be written as

$$T_e = I_m I_f L_m (\cos \omega t \sin \omega_m t \cos \delta + \cos \omega t \cos \omega_m t \sin \delta) \tag{9-39}$$

The time-averaged value of the electromagnetic torque, as given by (9-39), is zero except when $\omega = \omega_m =$ synchronous speed. In this case (9-39) yields

$$(T_e)_{\text{average}} = \tfrac{1}{2} I_m I_f L_m \sin \delta \tag{9-40}$$

which yields the torque–angle characteristics and δ in (9-40) is indeed the power angle discussed in Chapter 6. From (9-40) it is clear that, for a nonzero average torque, the rotor must run at the synchronous speed.

PROBLEMS

9-1 An electromagnet, shown in Fig. 9P-1, is required to exert a 500-N force on the iron at an air gap of 1 mm, while the exciting coil is carrying a 5 A dc. The core cross section at the air gap is 600 mm^2 in area. Calculate the required number of turns of the exciting coil.

9-2 (a) How many turns must the exciting coil of the electromagnet of Fig. 9P-1 have to produce a 500-N (average) force if the coil is excited by a 60-Hz

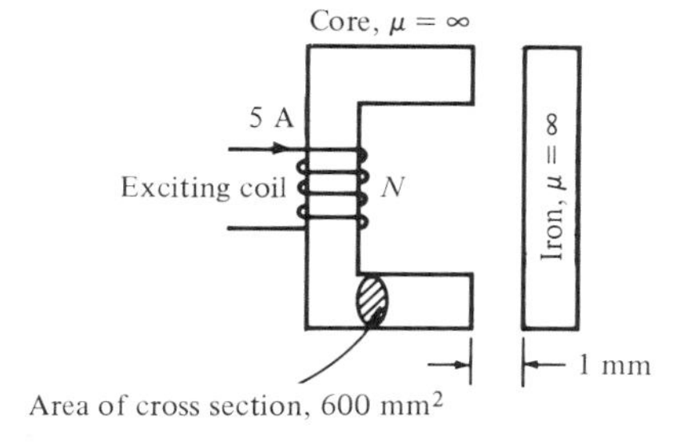

FIGURE 9-P1. PROBLEM 9-1.

alternating current having a maximum value of 7.07 A? (b) Is the average force frequency dependent?

9-3 Figure 9P-3 shows two mutually coupled cils, for which

$$L_{11} = L_{22} = 3 + \frac{2}{3x} \quad \text{mH} \qquad L_{12} = L_{21} = \frac{1}{3x} \quad \text{mH}$$

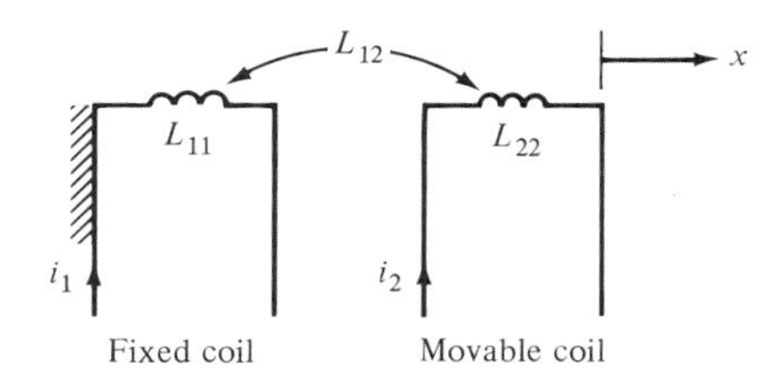

FIGURE 9-P3. PROBLEM 9-3.

where x is in meters. (a)If $i_1 = 5$ A dc and $i_2 = 0$, what is the mutual electrical force between the coils at $x = 0.01$ m? (b) If $i_1 = 5$ A dc and the second coil is open-circuited and moves in the positive x-direction at a constant speed of 20 m/s, determine the voltage across the second coil at $x = 0.01$ m.

9-4 For the two-coil system of Fig. 9P-3, if $i_1 = 7.07 \sin 377t$ A, $i_2 = 0$, and $x = 0.01$ m, determine (a) the instantaneous and (b) the time-averaged electrical force.

9-5 (a) The two coils of Fig. 9P-3 are connected in series, with 5 A dc flowing in them. Determine the electrical force between the coils at $x = 0.01$ m. Does the force tend to increase or decrease x? (b) Next, the coils are connected in parallel across a 194-V 60-Hz source. Compute the average electrical force at $x = 0.01$ m, neglecting the coil resistances.

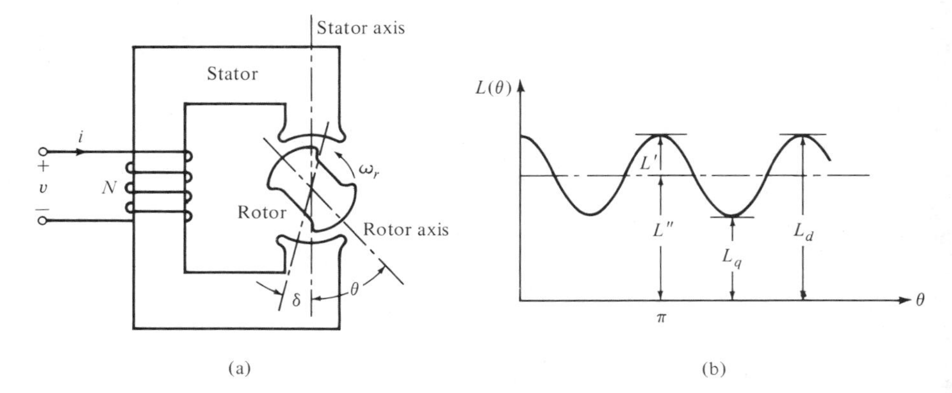

FIGURE 9-P6. PROBLEM 9-6.

9-6 The device shown in Fig. 9P-6(a) is a single-phase reluctance motor. The stator inductance varies sinusoidally with rotor position as shown in Fig. 9P-6(b). (a) If the stator is supplied with a current $i = I_m \sin \omega t$ and the rotor rotates at ω_r rad/s, derive an expression for the instantaneous torque developed by the rotor. (b) Notice from part (a) that the time-averaged torque is zero unless a certain condition is fulfilled. What is that condition? Express the nonzero average torque in terms of I_m, L_d, L_q, and the angle δ. (c) Given $L_d = 2L_q = 200$ mH and $I_m = 8$ A, what is the maximum value of the nonzero average torque?

9-7 The core of the electromagnet of Fig. 9P-1 has a saturation characteristic that may be approximated as $\lambda = 2 \times 10^{-3} \sqrt{i}/x$, where all quantities are in SI units. Calculate the force developed by the magnet if the coil current is 5 A dc and $x = 10^{-3}$ m.

9-8 A two-winding system has its inductances given by

$$L_{11} = \frac{k_1}{x} = L_{22} \qquad L_{12} = L_{21} = \frac{k_2}{x}$$

where k_1 and k_2 are constants. Neglecting the winding resistances, derive an expression for the electrical force (as a function of x) when both windings are connected to the same voltage source, $v = V_m \sin \omega t$.

9-9 Two mutually coupled coils are shown in Fig. 9P-9. The inductances of the coils are $L_{11} = A$, $L_{22} = B$, and $L_{12} = L_{21} = C \cos \theta$. Find the electrical torque for (a) $i_1 = I_0$, $i_2 = 0$; (b) $i_1 = i_2 = I_0$; (c) $i_1 = I_m \sin \omega t$, $i_2 = I_0$; (d) $i_1 = i_2 = I_m \sin \omega t$; and (e) coil 1 short-circuited and $i_2 = I_0$.

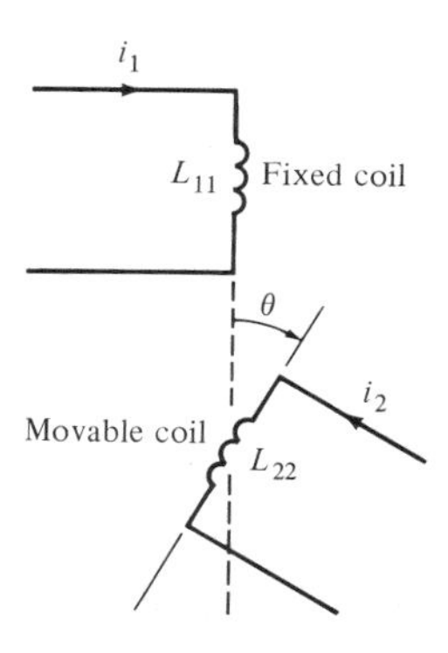

FIGURE 9-P9. PROBLEM 9-9.

9-10 Refer to Fig. 9P-6. The motor runs at 3600 r/min and develops a maximum torque of 1 N·m while being fed by a constant-current source. The motor parameters are $L_d = 2.5L_q = 160$ mH, stator winding resistance $R = 0.2\ \Omega$, and bearing friction coefficient $b = 1.1 \times 10^{-4}$ N·m-s/rad. Neglecting the core losses, compute the total losses.

9-11 A cylindrical electromagnet is shown in Fig. 9P-11. Given $a = 2$ mm, $c = 40$ mm, $l = 40$ mm, $N = 500$ turns, and $R = 3.5\ \Omega$. A 110-V 60-Hz

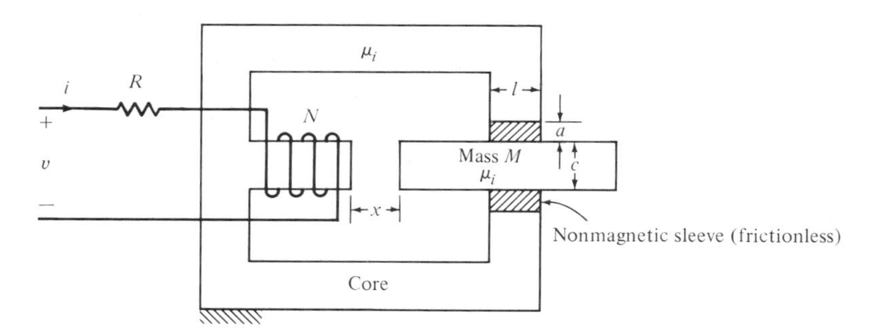

FIGURE 9-P11. PROBLEM 9-11.

source is applied across the coil. At $x = 5$ mm, determine (a) the maximum air-gap flux density and (b) the average value of the electrical force. Assume that $\mu_i >> \mu_o$.

9-12 From the data of Problem 9-11 calculate the steady-state (rms) value of (a) the coil current, (b) the input power, and (c) the power factor.

9-13 In Fig. 9P-11, initially the air gap is 5 mm; it is reduced to 2 mm while the coil is carrying a constant 4 A dc. Calculate (a) the energy supplied by the electrical source and (b) the mechanical work done in moving the mass.

9-14 The flux density in the 5-mm air gap of the electromagnet of Problem 9-11 is given as 0.65 sin $377t$ T. Determine (a) the average value of the electrical force and (b) the rms value of the induced voltage of the coil.

9-15 A two-winding machine has its inductances given by (9-16). The stator and rotor winding resistances are R_s and R_r, and the winding currents are i_s and i_r, respectively. If the rotor runs at a constant speed ω_m, write the voltage equations for the two windings.

9-16 How are the equations obtained in Problem 9-15 modified if the rotor carries a constant current of I amperes (dc)?

9-17 For the machine of Problem 9-15, we have $i_s = I_m \sin \omega t$, $i_r = I_2$ (a constant), and $\theta_r = \omega t + \delta$ (where δ is a constant). Obtain an expression for the instantaneous torque developed by the machine.

CHAPTER 10

Electric Power Transmission

10-1 INTRODUCTION

In Chapters 4 and 9 we considered two basic classes of power system components: transformers and rotating electric machines. Another major component of the power system is the transmission line. Transmission lines physically integrate the output of generating plants and the requirements of customers by providing pathways for the flow of electric power between various circuits in the system. These circuits include those between generating units, between utilities, and between generating units and substations (or load centers). For our purposes we consider a transmission line to have a sending end and a receiving end, and to have a series resistance and inductance and shunt capacitance and conductance as primary parameters. We will classify transmission lines as short, medium, and long lines. In a short line, the shunt effects (conductance and capacitance) are negligible. Often, this approximation is valid for lines up to 80 km long. In a medium line, the shunt capacitances are lumped at a few predetermined locations along the line. A medium line may be anywhere between 80 and 240 km in length. Lines longer than 240 km are considered as long lines which are represented by (uniformly) distributed parameters.

The operating voltage of a transmission line often depends on the length of the line. Some of the common operating voltages for transmission lines are 115, 138, 230, 345, 500, and 765 kV. It is estimated that by 1990 the United States will have a total length of 76,974 km (or 46,184.4 miles) of transmission lines, of which the 345-kV lines will be predominant, being 27,320 km (or 16,391.9 miles) long.

Whereas most transmission lines operate on ac, with the advent of high-power solid-state conversion equipment, high-voltage dc transmission lines are sometimes used for long distances exceeding 600 km (375 miles). We discuss dc transmission briefly toward the end of the chapter.

10-2 TRANSMISSION-LINE PARAMETERS

The first transmission-line parameter to be considered is the resistance of the conductors. Resistance is the cause of I^2R loss in the line, and also results in an IR-type voltage drop. The dc resistance R of a conductor of length l and area of cross section A is given by

$$R = \rho \frac{l}{A} \quad \Omega \tag{10-1}$$

where ρ is the *resistivity* of the conductor in Ω-m. The dc resistance is affected only by the operating temperature of the conductor, linearly increasing with the temperature. However, when operating on ac, the current-density distribution across the conductor cross section becomes nonuniform, and is a function of the ac frequency. This phenomenon is known as the *skin effect,* and as a consequence, the ac resistance is higher than the dc resistance. Approximately at 60 Hz, the ac resistance of a transmission line conductor may be 5 to 10 percent higher than its dc resistance.

The temperature dependence of a resistance is given by

$$R_2 = R_1[1 + \alpha(T_2 - T_1)] \tag{10-2}$$

where R_1 and R_2 are the resistances at temperatures T_1 and T_2, respectively, and α is defined as the temperature coefficient of resistance. The resistivities and temperature coefficients of certain materials are given in Table 10-1.

In practice, transmission-line conductors contain a stranded steel core surrounded by aluminum (for electrical conduction). Such a conductor is designated ACSR—aluminum conductor steel reinforced. Whereas it is difficult to calculate the exact ac resistance of such a conductor, measured values of resistance of

TABLE 10-1 Values of ρ and α

Material	Resistivity, ρ, at 20°C (μΩ-cm)	Temperature Coefficient, α, at 20°C
Aluminum	2.83	0.0039
Brass	6.4–8.4	0.0020
Copper		
Hard-drawn	1.77	0.00382
Annealed	1.72	0.00393
Iron	10.0	0.0050
Silver	1.59	0.0038
Steel	12–88	0.001–0.005

various types of stranded conductors are tabulated in the literature (see, for instance, Ref. 1).

The next parameter of interest is the transmission-line inductance. Considering the two-wire single-phase line first, it may be shown [1] that its inductance is given by

$$L = \frac{\mu_0}{4\pi}\left(1 + 4 \ln \frac{d}{r}\right) \qquad \text{H/m} \tag{10-3}$$

where $\mu_0 = 4\pi \times 10^{-7}$ H/m (the permeability of free space), d is the distance between the centers, and r is the radius of the conductors. For a three-phase line, it may further be shown [1] that the per phase, or line-to-neutral inductance of a line with equilaterally spaced conductors is

$$L = \frac{\mu_0}{8\pi}\left(1 + 4 \ln \frac{d}{r}\right) \qquad \text{H/m} \tag{10-4}$$

where the symbols are the same as in (10-3). In practice, the three conductors of a three-phase line are seldom equilaterally spaced. Such an unsymmetrical spacing results in unequal inductances in the three phases, leading to unequal voltage drops and an imbalance in the line. To offset this difficulty, the positions of the conductors are interchanged at regular intervals along the line. This practice is known as *transposition* and is illustrated in Fig. 10-1, which also shows the unequal spacings between the conductors. The average per phase inductance for such a case is still given by (10-4) except that the spacing d in the equation is replaced by the equivalent spacing d_e obtained from

$$d_e = (d_{ab}d_{bc}d_{ca})^{1/3} \tag{10-5}$$

where the distances d_{ab}, and so on, are shown in Fig. 10-1.

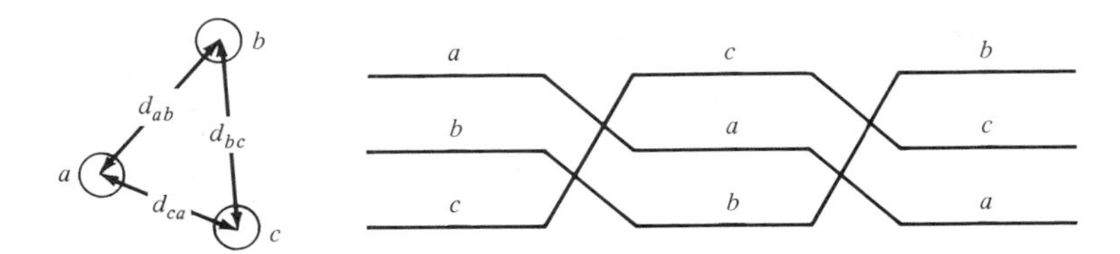

FIGURE 10-1. Transposition of unequally-spaced three-phase transmission line conductors.

The last parameter of interest is the shunt capacitance of the transmission line. The capacitance per unit length of a single-phase two-wire line is given by

$$C = \frac{\pi\varepsilon_0}{\ln (d/r)} \quad \text{F/m} \tag{10-6}$$

where ε_0 is the permittivity of free space and the other symbols are as defined in (10-3). For a three-phase line, having equilaterally spaced conductors, the per phase (or line-to-neutral) capacitance is

$$C = \frac{2\pi\varepsilon_0}{\ln (d/r)} \tag{10-7}$$

For unequal spacings between the conductors, d in (10-7) is replaced by d_e of (10-5), as was done for the case of inductance.

EXAMPLE 10-1

Determine the resistance of a 10-km-long solid cylindrical aluminum conductor, having a diameter of 250 mils, at (a) 20°C and (b) 120°C.

$$250 \text{ mils} = 0.25 \text{ in.} = 0.635 \text{ cm}$$

$$\text{area of cross section} = \frac{\pi}{4}(0.635)^2 = 0.317 \text{ cm}^2$$

from Table 10-1, $\rho = 2.83\ \mu\Omega\text{-cm}$ at 20°C.

(a) Therefore, from (10-1) at 20°C,

$$R_{20} = 2.83 \times 10^{-6} \times \frac{10 \times 10^3}{0.317} = 0.0893\ \Omega$$

(b) From (10-2) and Table 10-1, at 120°C we obtain

$$R_{120} = R_{20}[1 + \alpha(120 - 20)]$$

$$= 0.0893(1 + 0.0039 \times 100) = 0.124\ \Omega$$

■

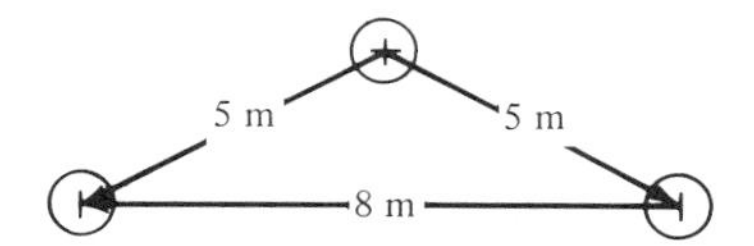

FIGURE 10-2. Example 10-2.

EXAMPLE 10-2

A single-circuit three-phase 60-Hz transmission line consists of three conductors arranged as shown in Fig. 10-2. If the conductors are the same as in Example 10-1, find the inductive and capacitive reactances per kilometer per phase.

From (10-5),

$$d_e = (5 \times 5 \times 8)^{1/3} = 5.848 \text{ m}$$

From Example 10-1,

$$r = \tfrac{1}{2} \times 0.635 \times 10^{-2} \text{ m}$$

Thus

$$\frac{d_e}{r} = \frac{5.848 \times 2 \times 10^2}{0.635} = 1841.9$$

and $\ln(d_e/r) = 7.52$. Hence, from (10-4) (with $\mu_0 = 4\pi \times 10^{-7}$ H/m),

$$L = \frac{4\pi \times 10^{-7}}{8\pi}(1 + 4 \times 7.52) \times 10^3 = 1.554 \text{ mH/km}$$

Or, inductive reactance per kilometer

$$X_L = \omega L = 377 \times 1.554 \times 10^{-3} = 0.5858\ \Omega$$

From (10-7) (with $\varepsilon_0 = 10^{-9}/36\pi$ F/m),

$$C = \frac{2\pi \times 10^{-9}/36\pi}{7.52} \times 10^3 = 7.387 \times 10^{-9} \text{ F/km}$$

Hence capacitive reactance per kilometer

$$X_c = \frac{1}{\omega C} = \frac{10^9}{377 \times 7.387} = 0.36 \times 10^6\ \Omega$$

■

In this section we have discussed the three basic parameters of transmission lines. A fourth parameter, the leakage resistance or conductance to ground, which represents the combined effect of various current flows from the line to ground, has not been considered because it is usually negligible for most calculations. However, it has been estimated [2] that on 132-kV transmission lines, leakage losses vary between 0.3 and 1.0 kW/mile.

10-3 TRANSMISSION-LINE REPRESENTATION

Returning to the three basic parameters calculated in Examples 10-1 and 10-2, we observe that the capacitance is very small for 1-km-long line. In fact, we mentioned in Section 10-1 that for a transmission line up to 80 km long, shunt effects, due to capacitance and leakage resistance, are negligible. Such a line is known as a *short transmission line,* and is represented by the lumped parameters R and L, as shown in Fig. 10-3. Notice that R is the resistance (per phase) and L is the inductance (per phase) of the entire line, although in Section 10-2 we introduced the transmission-line parameters in terms of per unit length of the line. The line is shown to have two ends: the sending end at the generator end and the receiving end at the load end. The problems of significance to be solved here are the determination of voltage regulation and of efficiency of transmission. These quantities are defined as follows:

$$\text{percent voltage regulation} = \frac{|V_{R(\text{no-load})}| - |V_{R(\text{load})}|}{|V_{R(\text{load})}|} \times 100 \qquad (10\text{-}8)$$

$$\text{efficiency of transmission} = \frac{\text{power at the receiving end}}{\text{power at the sending end}} \qquad (10\text{-}9)$$

Often, (10-8) and (10-9) are evaluated at full-load values. The calculation procedure is illustrated by the following example.

EXAMPLE 10-3

Let the transmission line of Example 10-2 be 60 km long. The line supplies a three-phase wye-connected 100-MW 0.9 lagging power factor load at 215 kV

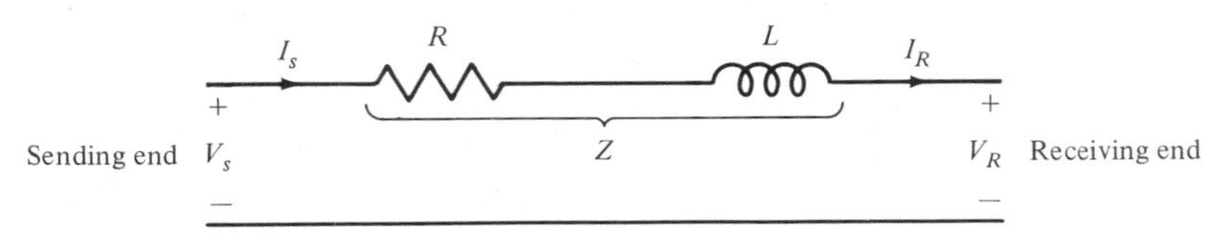

FIGURE 10-3. Representation of a short transmission line (on a per-phase basis).

line-to-line voltage. If the operating temperature of the line is 60°C, determine the regulation and efficiency of transmission in percent.

The line resistance R is given by (see Example 10-1)

$$R = R_{60}\{0.00893[1 + 0.0039(60 - 20)]\} \times 60$$
$$= 0.62\ \Omega$$

The inductance L is (from Example 10-2)

$$L = 1.554 \times 10^{-3} \times 60 = 93.24 \text{ mH}$$

and

$$X_L = \omega L = 377 \times 93.24 \times 10^{-3} = 35.15\ \Omega$$

$$\text{Line current } I\ (=I_S = I_R) = \frac{100 \times 10^6}{\sqrt{3} \times 215 \times 0.9} = 298.37 \text{ A}$$

$$\text{phase voltage at the receiving end, } V_R = \frac{215 \times 10^3}{\sqrt{3}} = 124.13 \text{ kV}$$

The phasor diagram illustrating the operating conditions is shown in Fig. 10-4, from which

$$\mathbf{V}_S = \mathbf{V}_R + \mathbf{I}(R + jX_L)$$
$$= 124.13 \times 10^3 \underline{/0^\circ} + 298.37 \underline{/-25.8}\ (0.62 + j35.15)$$
$$\simeq 124.13 \times 10^3 \underline{/0^\circ} + 298.37 \underline{/-25.8} \times 35.15 \underline{/90^\circ}$$
$$= (128.69 + j9.44) \text{ kV}$$
$$\simeq 129.04 \underline{/4.2^\circ} \text{ kV}$$

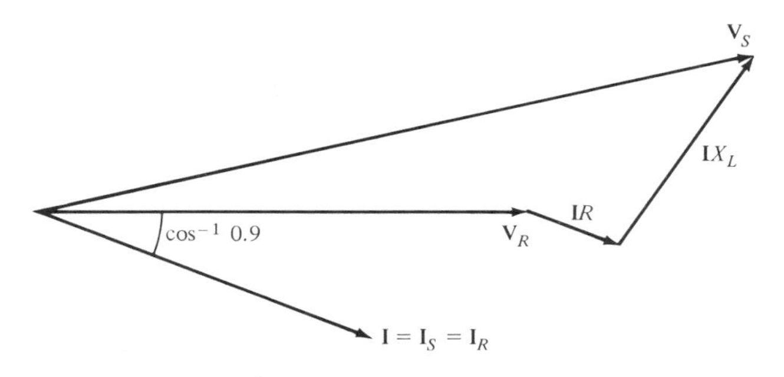

FIGURE 10-4. Example 10-3.

Hence

$$\text{percent voltage regulation} = \frac{129.04 - 124.13}{124.13} \times 100$$

$$= 3.955 \text{ percent}$$

To calculate the efficiency, we determine the loss in the line as

$$\text{line loss} = 3 \times 298.37^2 \times 0.62 = 0.166 \text{ MW}$$

$$\text{power received} = 100 \text{ MW (given)}$$

$$\text{power sent} = 100 + 0.166 = 100.166 \text{ MW}$$

$$\text{efficiency} = \frac{100}{100.166} = 99.83 \text{ percent}$$

■

The *medium-length transmission line* is considered to be up to 240 km long. In such a line the shunt effect due to the line capacitance is not negligible. Two representations for the medium-length line are shown in Figs. 10-5 and 10-6. These are known as the nominal-Π circuit and the nominal-T circuit of the transmission line, respectively. In Figs. 10-5 and 10-6, we also show the respective phasor diagrams for lagging power factor conditions. These diagrams aid in understanding the mutual relationships between currents and voltages at certain places on the line. The following examples illustrate the applications of the nominal-Π and nominal-T circuits in calculating the performances of medium-length transmission lines.

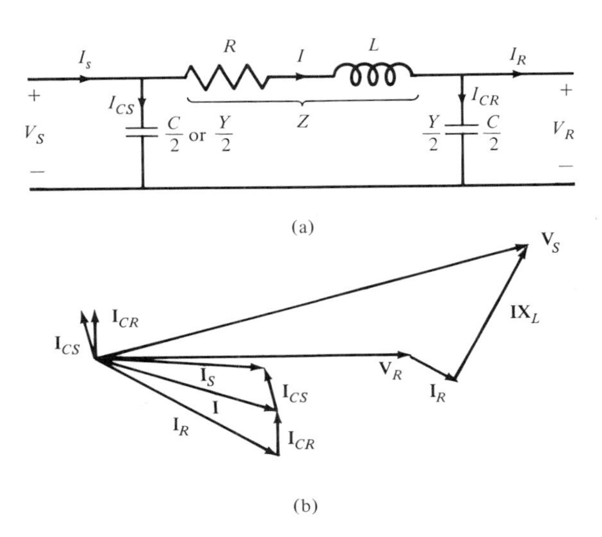

FIGURE 10-5. (a) A nominal-π circuit; (b) corresponding phasor diagrams.

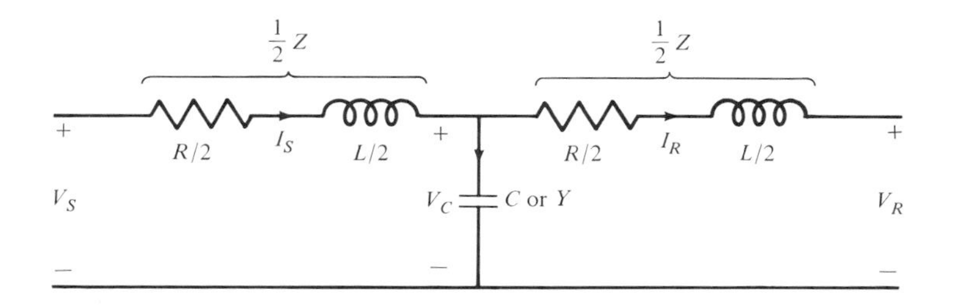

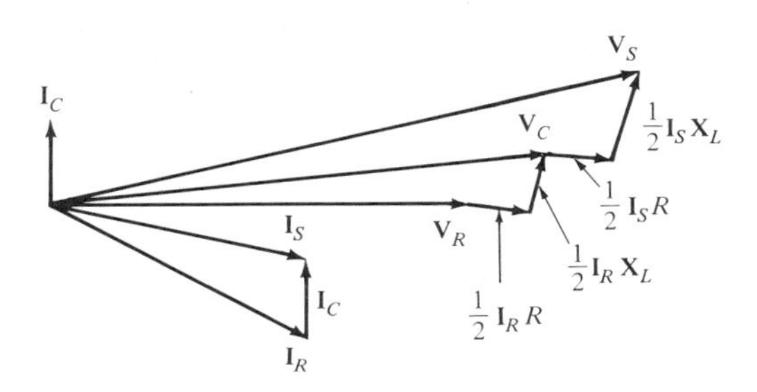

FIGURE 10-6. (a) A nominal-T circuit; (b) corresponding phasor diagram.

EXAMPLE 10-4

Consider the transmission line of Example 10-3, but let the line be 200 km long. Determine the voltage regulation of the line for the operating conditions given in Example 10-3 using (a) the nominal-Π circuit and (b) the nominal-T circuit.

From the data of Examples 10-2 and 10-3 we already know that

$$R = 2.07\ \Omega$$

$$L = 310.8\ \text{mH}$$

$$C = 1.4774\ \mu\text{F}$$

$$V_R = 124.13\ \text{kV}$$

$$I_R = 298.37\ \underline{/-25.8^\circ}\ \text{A}$$

(a) Using the nomenclature of Fig. 10-5, we have

$$\mathbf{I}_{CR} = \frac{\mathbf{V}_R}{\mathbf{X}_{c/2}} = \frac{124.13 \times 10^3\ \underline{/0^\circ}}{1/(377 \times 0.5 \times 1.4774 \times 10^{-6})\ \underline{/90^\circ}}$$

$$= 34.57\ \underline{/90^\circ}$$

$$\mathbf{I} = \mathbf{I}_R + \mathbf{I}_{CR} = 298.37 \angle{-25.8^\circ} + 34.57 \angle{90^\circ}$$

$$= 285 \angle{-19.5^\circ} \text{ A}$$

$$R + jX_L = 2.07 + j377 \times 0.3108 \simeq 117.19 \angle{88.98^\circ} \ \Omega$$

$$\mathbf{I}(R + jX_L) = 285 \angle{-19.5^\circ} \times 117.19 \angle{88.98^\circ}$$

$$= 33.34 \angle{69.48^\circ} \text{ kV}$$

$$\mathbf{V}_s = \mathbf{V}_R + \mathbf{I}(R + jX_L) = 124.13 \angle{0^\circ} + 33.4 \angle{69.48^\circ}$$

$$= 139.39 \text{ kV}$$

$$\text{per cent regulation} = \frac{139.39 - 124.13}{124.13} \times 100 = 12.3\%$$

(b) Using the nomenclature of Fig. 10-6, we have

$$\mathbf{V}_C = \mathbf{V}_R + \tfrac{1}{2} \mathbf{I}_R(R + jX_L)$$

$$= 124.13 \angle{0^\circ} + \frac{10^{-3}}{2} \times 298.37 \angle{-25.8^\circ} \times 117.19 \angle{88.98^\circ}$$

$$= 132.92 \angle{67.4^\circ} = (132 + j15.6) \text{ kV}$$

$$\mathbf{I}_C = \frac{\mathbf{V}_c}{\mathbf{X}_c} = \frac{132.92 \times 10^3 \angle{67.4^\circ}}{1/(377 \times 1.4774 \times 10^{-6}) \angle{90^\circ}} = 74 \angle{157.4^\circ} \text{ A}$$

$$\mathbf{I}_S = \mathbf{I}_R + \mathbf{I}_C = 298.37 \angle{-25.8^\circ} + 74 \angle{157.4^\circ} = 287.14 \angle{-20.68^\circ} \text{ A}$$

$$\mathbf{V}_S = \mathbf{V}_C + \tfrac{1}{2} \mathbf{I}_S(R + jX_L)$$

$$= 132.92 \angle{67.4^\circ} + \frac{10^{-3}}{2} \times 287.14 \angle{-20.68^\circ} \times 117.19 \angle{88.98^\circ}$$

$$= 140.0 \text{ kV}$$

$$\text{per cent regulation} = \frac{140 - 124.13}{124.13} \times 100 = 12.78\%$$

■

Finally, we consider transmission lines over 240 km long. Such a line is known as a long line. Parameters of long lines are distributed over the entire length of the line. On a per phase basis, a long line is shown in Fig. 10-7. Let z be the series impedance of the line per unit length, and let y be the shunt admittance per unit length. With the currents and voltages shown in Fig. 10-7, for an elemental length dx, we have

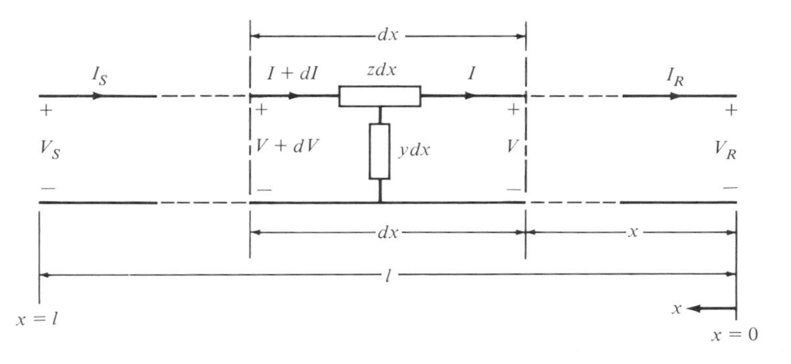

FIGURE 10-7. Distributed parameter long line.

$$d\mathbf{V} = \mathbf{I}\mathbf{z}\,dx \quad \text{or} \quad \frac{d\mathbf{V}}{dx} = \mathbf{z}\mathbf{I} \tag{10-10}$$

$$d\mathbf{I} = \mathbf{V}\mathbf{y}\,dx \quad \text{or} \quad \frac{d\mathbf{I}}{dx} = \mathbf{y}\mathbf{V} \tag{10-11}$$

Eliminating I from (10-10) and (10-11) yields

$$\frac{d^2\mathbf{V}}{dx^2} = \boldsymbol{\gamma}^2 V \tag{10-12}$$

where $\boldsymbol{\gamma} = \sqrt{\mathbf{yz}}$, and is known as the *propagation constant*. Solution to (10-12) may be written as

$$\mathbf{V} = \mathbf{C}_1 e^{\gamma x} + \mathbf{C}_2 e^{-\gamma x} \tag{10-13}$$

where $\mathbf{C}_1$ and $\mathbf{C}_2$ are arbitrary constants. Differentiating (10-13) and substituting (10-10) gives

$$\mathbf{C}_1 \boldsymbol{\gamma} e^{\gamma x} - \mathbf{C}_2 \boldsymbol{\gamma} e^{-\gamma x} = \mathbf{z}\mathbf{I} \tag{10-14}$$

Let $\boldsymbol{\gamma}/\mathbf{z} = \mathbf{Z}_c = \sqrt{\mathbf{z}/\mathbf{y}}$. In terms of $\mathbf{Z}_c$, we obtain

$$\mathbf{I} = \frac{1}{\mathbf{Z}_c}(\mathbf{C}_1 e^{\gamma x} - \mathbf{C}_2 e^{-\gamma x}) \tag{10-15}$$

The quantity $\mathbf{Z}_c$ is called the *characteristic impedance* of the line.

To evaluate $\mathbf{C}_1$ and $\mathbf{C}_2$, we refer to Fig. 10-7 from which $\mathbf{V} = \mathbf{V}_R$ and $\mathbf{I} =$

$\mathbf{I}_R$ at $x = 0$. Substituting these conditions in (10-13) and (10-15) and solving for $\mathbf{C}_1$ and $\mathbf{C}_2$ yields

$$\mathbf{C}_1 = \tfrac{1}{2}(\mathbf{V}_R + \mathbf{Z}_c\mathbf{I}_R)$$

$$\mathbf{C}_2 = \tfrac{1}{2}(\mathbf{V}_R - \mathbf{Z}_c\mathbf{I}_R)$$

In terms of these constants, the voltage and current distributions are given by

$$\mathbf{V} = \tfrac{1}{2}\mathbf{V}_R(e^{\gamma x} + e^{-\gamma x}) + \tfrac{1}{2}\mathbf{I}_R\mathbf{Z}_c(e^{\gamma x} - e^{-\gamma x}) \tag{10-16}$$

$$\mathbf{I} = \tfrac{1}{2}\frac{\mathbf{V}_R}{\mathbf{Z}_c}(e^{\gamma x} - e^{-\gamma x}) + \tfrac{1}{2}\mathbf{I}_R(e^{\gamma x} + e^{-\gamma x}) \tag{10-17}$$

In terms of hyperbolic functions, (10-16) and (10-17) are finally expressed as

$$\mathbf{V} = \mathbf{V}_R \cosh \boldsymbol{\gamma} x + \mathbf{I}_R\mathbf{Z}_c \sinh \boldsymbol{\gamma} x \tag{10-18}$$

$$\mathbf{I} = \mathbf{I}_R \cosh \boldsymbol{\gamma} x + \frac{\mathbf{V}_R}{\mathbf{Z}_c} \sinh \boldsymbol{\gamma} x \tag{10-19}$$

Since $\mathbf{V} = \mathbf{V}_s$ and $\mathbf{I} = \mathbf{I}_s$ at $x = l$, (10-18) and (10-19) become

$$\mathbf{V}_s = \mathbf{V}_R \cosh \boldsymbol{\gamma} l + \mathbf{I}_R\mathbf{Z}_c \sinh \boldsymbol{\gamma} l \tag{10-20}$$

$$\mathbf{I}_s = \mathbf{I}_R \cosh \boldsymbol{\gamma} l + \frac{\mathbf{V}_R}{\mathbf{Z}_c} \sinh \boldsymbol{\gamma} l \tag{10-21}$$

These equations are used in evaluating the performance of long lines. In carrying out numerical computations, the following relationships are often useful:

$$\boldsymbol{\gamma} = \alpha + j\beta$$

$$\cosh \boldsymbol{\gamma} l = \cosh(\alpha l + j\beta l) = \cosh \alpha l \cos \beta l + j \sin \alpha l \sin \beta l$$

$$\sinh \boldsymbol{\gamma} l = \sinh(\alpha l + j\beta l) = \sinh \alpha l \cos \beta l + j \cosh \alpha l \sin \beta l$$

$$\cosh \boldsymbol{\gamma} l = 1 + \frac{(\gamma l)^2}{2!} + \frac{(\gamma l)^4}{4!} + \cdots \simeq 1 + \tfrac{1}{2}\mathbf{YZ}$$

$$\sinh \boldsymbol{\gamma} l = \gamma l + \frac{(\gamma l)}{3!} + \frac{(\gamma l)^5}{5!} + \cdots \simeq \sqrt{\mathbf{YZ}}\,(1 + \tfrac{1}{6}\mathbf{YZ})$$

EXAMPLE 10-5

The parameters of a 215-kV 400-km 60-Hz three-phase transmission line are as follows:

$$\mathbf{y} = j3.2 \times 10^{-6}\ \Omega/\text{km} \qquad \mathbf{z} = (0.1 + j0.5)\ \Omega/\text{km}$$

The line supplies a 150-MW load at unity power factor. Determine (a) the voltage regulation, (b) the sending-end power, and (c) the efficiency of transmission.

$$\mathbf{z} = 0.1 + j0.5 = 0.51\ \underline{/78.7^\circ}$$

$$\mathbf{y} = j3.2 \times 10^{-6} = 3.2 \times 10^{-6}\ \underline{/90^\circ}$$

$$\boldsymbol{\gamma} l = l\sqrt{\mathbf{zy}} = 400\sqrt{0.51 \times 3.2 \times 10^{-6}}\ \underline{/\tfrac{1}{2}}\,(90 + 78.7)^\circ$$

$$= 0.51\ \underline{/84.35^\circ} = 0.05 + j0.5 = \alpha + j\beta \text{ rad}$$

$$\mathbf{z}_c = \sqrt{\frac{\mathbf{z}}{\mathbf{y}}} = \sqrt{\frac{0.51}{3.2 \times 10^{-6}}}\ \underline{/\tfrac{1}{2}}\,(78.7 - 90)^\circ = 399.2\ \underline{/-5.65^\circ}\ \Omega$$

$$\mathbf{V}_R = \frac{215 \times 10^3}{\sqrt{3}} = 124.13\ \underline{/0^\circ}\text{ kV}$$

$$\mathbf{I}_R = \frac{150 \times 10^6}{\sqrt{3} \times 215 \times 10^3} = 402.8\ \underline{/0^\circ}\text{ A}$$

$$\cosh \boldsymbol{\gamma} l = \cosh 0.05 \cos 0.5 + j \sinh 0.05 \sin 0.5$$

$$= 0.877 + j0.024 = 0.877\ \underline{/1.57^\circ}$$

$$\sinh \boldsymbol{\gamma} l = \sinh 0.05 \cos 0.5 + j \cosh 0.05 \sin 0.5$$

$$= 0.044 + j0.479 = 0.48\ \underline{/84.75^\circ}$$

From (10-20) and (10-21) we respectively obtain

$$\mathbf{V}_s = (124.13\ \underline{/0^\circ} \times 0.877\ \underline{/1.57^\circ} + 402.8\ \underline{/0^\circ} \times 10^{-3}$$

$$\times\ 399.2\ \underline{/-5.65^\circ} \times 0.48\ \underline{/84.75^\circ})\text{ kV}$$

$$= 146.4\ \underline{/32.55^\circ}\text{ kV}$$

and

$$\mathbf{I}_s = \left(402.8\ \underline{/0^\circ} \times 0.877\ \underline{/1.57^\circ} + \frac{124.13\ \underline{/0^\circ} \times 10^3}{399.2\ \underline{/-5.65^\circ}} \times 0.48\ \underline{/84.75^\circ}\right)\text{ A}$$

$$= 386.52\ \underline{/24.28^\circ}\text{ A}$$

Receiving-end voltage on no-load becomes, from (10-20) (with $\mathbf{I}_R = 0$),

$$(V_R)_{\text{no-load}} = \frac{V_s}{\cosh \boldsymbol{\gamma} l} = \frac{146.4}{0.877} = 166.93 \text{ kV}$$

(a) $$\text{percent voltage regulation} = \frac{166.93 - 124.13}{124.13} \times 100 = 34.48\%$$

(b) $$\begin{aligned} \text{sending-end power} &= 3V_s I_s \cos \varphi \\ &= 3 \times 146.4 \times 10^3 \times 386.5 \cos (32.55 - 24.28) \\ &= 167.98 \text{ MW} \end{aligned}$$

(c) $$\begin{aligned} \text{efficiency of transmission} &= \frac{\text{power received}}{\text{power sent}} \times 100 \\ &= \frac{150}{167.98} = 89.3\% \end{aligned}$$

■

10-4

TRANSMISSION LINE AS A TWO-PORT NETWORK AND POWER FLOW

In representing transmission lines by their equivalent circuits, we notice from Section 10-3 that the sending-end voltage and current can be expressed in terms of the receiving-end voltage and current and the line parameters as in (10-20) and (10-21). In general, a transmission line may be viewed as a four-terminal network, as shown in Fig. 10-8, such that the terminal voltages and currents are mutually related by

$$\mathbf{V}_s = \mathbf{A}\mathbf{V}_r + \mathbf{B}\mathbf{I}_r \tag{10-22}$$

$$\mathbf{I}_s = \mathbf{C}\mathbf{V}_r + \mathbf{D}\mathbf{I}_r \tag{10-23}$$

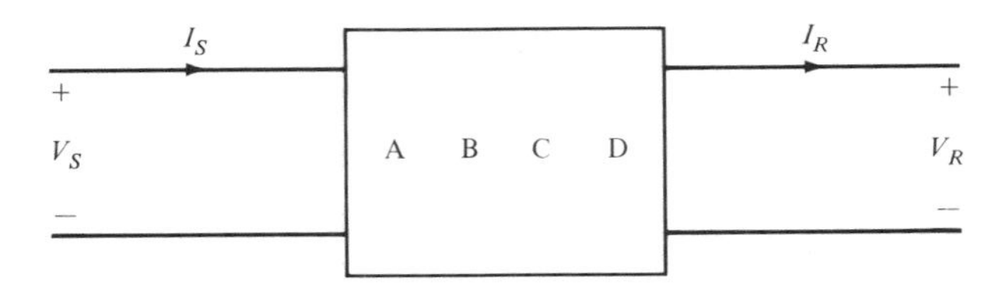

FIGURE 10-8. A transmission line as a four-terminal network.

TABLE 10-2 ***ABCD*** **Constants for Transmission Lines**

Line	Equivalent Circuit	A	B	C	D
Short	Series impedance, Fig. 10-3	1	$\mathbf{Z}$	0	1
Medium length	Nominal-Π, Fig. 10-5	$1 + \frac{1}{2}\mathbf{YZ}$	$\mathbf{Z}$	$\mathbf{Y}(1 + \frac{1}{4}\mathbf{YZ})$	$1 + \frac{1}{2}\mathbf{YZ}$
	Nominal-T, Fig. 10-6	$1 + \frac{1}{2}\mathbf{YZ}$	$\mathbf{Z}(1 + \frac{1}{4}\mathbf{YZ})$	$\mathbf{Y}$	$1 + \frac{1}{2}\mathbf{YZ}$
Long	Distributed parameter, Fig. 10-7	$\cosh \boldsymbol{\gamma} l$	$\mathbf{Z}_c \sinh \boldsymbol{\gamma} l$	$\sinh \boldsymbol{\gamma} l / \mathbf{Z}_c$	$\cosh \boldsymbol{\gamma} l$

where **ABCD** are constants dependent on line parameters, are called *generalized circuit constants,* and in general are complex. The validity of (10-22) and (10-23) is based on the fact that a transmission line can be represented by a linear, passive, and bilateral network. By virtue of reciprocity, the generalized constants are related to each other by the following equation:

$$\mathbf{AD} - \mathbf{BC} = 1 \tag{10-24}$$

Transmission lines of various lengths can be represented by equivalent four-terminal networks, as depicted by Fig. 10-8. The **ABCD** constants for various lines are summarized in Table 10-2.

Power flow calculations at a point on a transmission line can be conveniently carried out in terms of **ABCD** constants. To show the procedure, let us evaluate the power at the receiving end of a transmission line. Since the **ABCD** constants, in general, are complex, we let

$$\mathbf{A} = |\mathbf{A}| \underline{/\alpha} \qquad \text{and} \qquad \mathbf{B} = |\mathbf{B}| \underline{/\beta}$$

Choosing $\mathbf{V}_R$ as the reference phasor, we assume that

$$\mathbf{V}_R = |\mathbf{V}_R| \underline{/0^\circ} \qquad \text{and} \qquad \mathbf{V}_S = |\mathbf{V}_s| \underline{/\delta}$$

Consequently, from (10-22) we obtain

$$\mathbf{I}_R = \frac{|\mathbf{V}_s|}{|\mathbf{B}|} \underline{/(\delta - \beta)} - \frac{|\mathbf{A}||\mathbf{V}_R|}{|\mathbf{B}|} \underline{/(\alpha - \beta)}$$

The complex power $\mathbf{V}_R \mathbf{I}_R^*$ at the receiving end is given by

$$P_R + jQ_R = \frac{|\mathbf{V}_R||\mathbf{V}_s|}{|\mathbf{B}|} \underline{/\beta - \delta} - \frac{|\mathbf{A}||\mathbf{V}_R|^2}{|\mathbf{B}|} \underline{/\beta - \alpha} \tag{10-25}$$

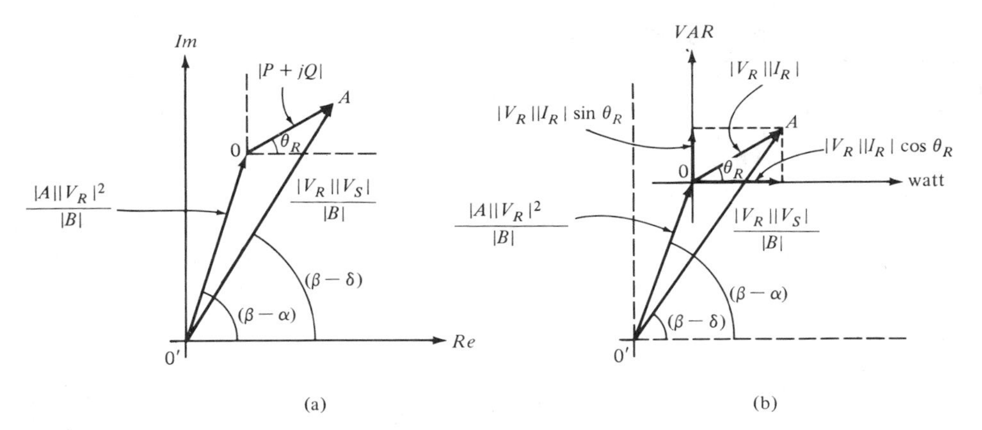

FIGURE 10-9. (a) Phasor diagram representing (10-25); (b) corresponding power diagram.

which yields

$$P_R = \frac{|\mathbf{V}_R||\mathbf{V}_s|}{|\mathbf{B}|}\cos(\beta - \delta) - \frac{|\mathbf{A}||\mathbf{V}_R|^2}{|\mathbf{B}|}\cos(\beta - \alpha) \tag{10-26}$$

$$Q_R = \frac{|\mathbf{V}_R||\mathbf{V}_s|}{|\mathbf{B}|}\sin(\beta - \delta) - \frac{|\mathbf{A}||\mathbf{V}_R|^2}{|\mathbf{B}|}\sin(\beta - \alpha) \tag{10-27}$$

Figure 10-9(a) depicts (10-25) phasorially. By shifting the origin from O' to O, Fig. 10-9(a) becomes a power diagram, as shown in Fig. 10-9(b). Because $O'A = |\mathbf{V}_R||\mathbf{V}_s|/|\mathbf{B}|$ for a given line and a given value of $|\mathbf{V}_R|$, the locus of A will be a set of circles (of radii $O'A$) for a set of values of $|\mathbf{V}_s|$. Portions of two such loci are given in Fig. 10-10. These circles are sometimes called *receiving-end circles*. A number of interesting results may be obtained from the figure. For instance, for the line to transmit maximum power, $O'A$, must become parallel to the real axis; that is, $\beta - \delta = 0$. For this condition, we have, from (10-26),

$$(P_R)_{\max} = \frac{|\mathbf{V}_R||\mathbf{V}_s|}{|\mathbf{B}|} - \frac{|\mathbf{A}||\mathbf{V}_R|^2}{|\mathbf{B}|}\cos(\beta - \alpha) \tag{10-28}$$

Line OA in Fig. 10-10 is the load line the intersection of which with the power circle determines the operating point. Thus, for a load having a lagging power factor angle θ_R, **A** and **C** are the respective operating points for sending-end voltages $|\mathbf{V}_{s1}|$ and $|\mathbf{V}_{s2}|$. These operating points determine the real and reactive powers received at the two sending-end voltages.

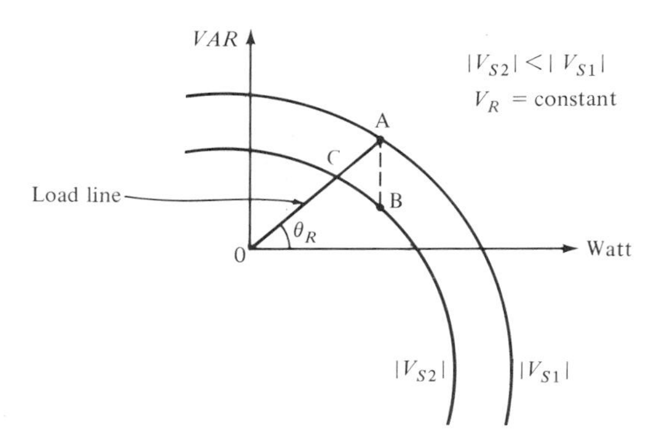

FIGURE 10-10. Receiving end power circles for a constant $|V_R|$ and two sending-end voltages $|V_{S1}|$ and $|V_{S2}|$.

From transmission-line models, we know that the receiving-end voltage will tend to go down with the sending-end voltage. Suppose that the sending-end voltage goes down from $|\mathbf{V}_{s1}|$ to $|\mathbf{V}_{s2}|$. In order to maintain a constant $|\mathbf{V}_R|$, the amount of reactive power to be supplied by capacitors at the receiving end (in parallel with the load) is simply given by the length *AB*, which is parallel to the reactive-power axis. Such information can be obtained very quickly from the circle diagram. Although circle diagrams yield quick estimates of certain results pertinent to transmission lines, digital computers are generally used in practice. Circle diagrams illustrate certain concepts very clearly.

10-5

HIGH-VOLTAGE DC TRANSMISSION [3–5]

Historically, DC power transmission has been in existence since 1882. However, commercially feasible high-voltage dc (HVDC) transmission lines began to emerge after 1954. Since then HVDC systems have been steadily growing. From 1954 to 1970, dc power transmission capability increased from 20 MW to 4000 MW, and between 1970 to 1978, from 4000 MW to 14,750 MW. By 1985, it is estimated that the total capability of HVDC transmission lines throughout the world will reach 25,000 MW. The HVDC transmission voltage has increased from 100 kV to ± 533 kV between 1950 and 1979. From the current activities relating to HVDC systems it is evident that the overhead transmission voltage will range from ± 600 to ± 750 kV, with transmitting power up to 6000 MW.

A single-circuit dc transmission line, also known as a *dc link,* is shown in Fig. 10-11. The line usually has two conductors, although in some cases the line may have only one conductor, the return path being in the earth. Notice

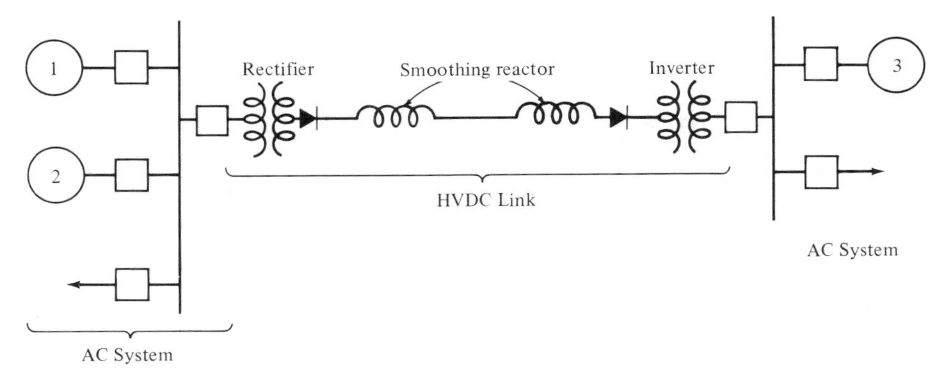

FIGURE 10-11. A single-current HVDC transmission line (one-line representation).

from the figure that the line has *converters* at both ends. At the sending end, the converter takes the form of a *rectifier,* whereas at the receiving end the converter operates as an inverter. (See Chapter 12 for a discussion of converters.) However, at both ends the converter is capable of operating in either mode, thus permitting a bidirectional power flow. From Fig. 10-11 it is clear that the major component in a HVDC system, as distinct from an ac system, is the converter. The two major elements of the converter are *valves* and *converter transformer*. Over the years the valve has undergone a radical change—from mercury pool valve to solid-state thyristors. Thyristors of 3000 A and 5 kV ratings are commercially available and are now being used in HVDC systems. Converter transformers are very much similar to conventional ac transformers. However, operating conditions of the two are not similar. The primary of the converter transformer is connected to the ac system, but the secondary is connected to the valves carrying dc. This condition requires that the oil and paper insulation used in the transformer must withstand ac as well as dc voltage stresses.

As mentioned earlier, a HVDC link may have only one conductor, usually of negative polarity, and a ground return. Such a link is called a *monopolar link*. On the other hand, a link with two conductors—one positive and the other negative—is known as a *bipolar link*. The voltage rating of a bipolar link is expressed as $\pm$kV. The third type of the link is the *homopolar link,* in which all conductors have the same polarity and which always has a ground return.

Some of the advantages of HVDC systems are as follows:

1. Because bipolar HVDC transmission lines require only two-pole conductors compared to three-phase conductors for ac systems, the HVDC system is economical when savings in conductor costs, losses, towers, and right-of-way costs offset the additional costs of converting equipment.

2. The HVDC system has a greater flexibility of operation in that the transmission line may be operated in the monopolar mode with ground return when one of the poles requires opening under a permanent faulted condition.
3. Staging of facilities—connecting the valves in series–parallel combinations—to accommodate desired voltage and current levels is another advantage of the dc system.
4. Addition of a dc transmission line does not increase the short-circuit duty of the existing ac system.
5. Control of converter valves permits rapid changes in magnitude and direction of power flow and enhances system stability.
6. An HVDC link may be used to interconnect two ac systems of different frequencies.
7. The dc system has a relatively less reactive power flow problem.
8. For a given power, its transmission at higher voltages is economical because of the reduced I^2R losses.
9. Radio interference for a dc system is less than that for an ac system.

Among the major problems associated with HVDC systems are ion drift during corona, corrosion during monopolar operation with earth return, and complexity of the dc circuit breaker design. A HVDC system is considered economical only for transmission over distances exceeding 800 km. For overhead transmission along a distance of 600 to 1000 km, ± 400 kV is considered an optimum voltage for a power level of 800 to 1000 MW. As the power level increases, the optimum voltage also increases, such that ± 1200 kV is the optimum for power levels ranging from 4000 to 10,000 MW.

The major application of HVDC systems in the United States is from mine-mouth or hydroelectric plants and nuclear plants to urban areas. Developments in related technology to produce compact terminals, HVDC circuit breakers and controls, and thyristors of increased current and voltage ratings will certainly enhance the prospects of HVDC transmission.

REFERENCES

1. **W. D. Stevenson, Jr.,** *Elements of Power System Analysis,* 4th ed., McGraw-Hill, New York, 1982.
2. **B. M. Weedy,** *Electric Power Systems,* 3rd ed., Wiley, New York, 1979.
3. **F. J. Ellert** and **N. G. Hingorani,** "HVDC for the Long Run," *IEEE Spectrum,* August 1976, pp. 37–42.
4. **K. R. Shah, W. F. Griffard,** and **F. A. Denbrock,** "HVDC Transmission Assessment," *Proceedings, American Power Conference,* 1977, Vol. 39, pp. 1183–1190.
5. **E. W. Kimbark,** *Direct Current Transmission,* Vol. 1, Wiley-Interscience, New York, 1971.

PROBLEMS

10-1 A three-phase 60-Hz 25-km transmission line is made of hard-drawn copper conductor 500 mils in diameter. What is the per phase resistance of the line at 60°C? Refer to Table 10-1 for the values of constants.

10-2 The conductors of the line of Problem 10-1 are arranged in the form of an equilateral triangle, the sides of which are 6 m. Evaluate the per phase inductive and capacitive reactances of the line.

10-3 A three-phase wye-connected 15-MW 0.866 lagging power factor load is supplied by the line of Problem 10-1 at 115 kV line-to-line voltage. The operating temperature of the line is 60°C. Calculate the line losses and the percent voltage regulation.

10-4 A 138-kV three-phase short transmission line has a per phase impedance of $(2 + j4)$ ohms. If the line supplies a 25-MW 0.8 lagging power factor load, calculate (a) the efficiency of transmission and (b) the sending-end voltage and power factor.

10-5 Consider the data of Problem 10-4. If the sending-end voltage is 152 kV (line to line), determine (a) the receiving-end voltage, (b) the efficiency of transmission, and (c) the angle between the sending-end and receiving-end voltages.

10-6 A three-phase short transmission having a per phase impedance of $(2 + j4)$ ohms has equal line-to-line receiving-end and sending-end voltages of 115 kV (at both ends) while supplying a load at a 0.8 leading power factor. Calculate the power supplied by the line.

10-7 A three-phase wye-connected 20-MW 0.866 power factor load is to be supplied by a transmission line at 138 kV. It is desired that the line losses not exceed 5 percent of the load. If the per-phase resistance of the line is 0.7 Ω, what is the maximum length of the line?

10-8 The impedance of a single-phase transmission line is $Z = R + jX$. The sending-end and receiving-end voltages are V_s and V_R, respectively. Show that the maximum power that can be transmitted by the line is given by

$$P_{\max} = \frac{V_s V_R}{Z} - \frac{R V_R^2}{Z^2}$$

10-9 The per phase constants of a 345-kV three-phase 150-km-long transmission are

$$\text{resistance} = 0.1\ \Omega/\text{km}$$

$$\text{inductance} = 1.1\ \text{mH/km}$$

$$\text{capacitance} = 0.02\ \mu\text{F/km}$$

The line supplies a 180-MW 0.9 lagging power factor load. Using the nominal-Π circuit, determine the sending-end voltage.

10-10 Repeat Problem 10-9 using the nominal-T current.

10-11 Determine the efficiency of transmission of the line of Problem 10-9 using (a) the nominal-Π and (b) the nominal-T circuit.

10-12 The per phase parameters of a 345-kV 500-km 60-Hz three-phase transmission line are as follows:

$$y = j\,4 \times 10^{-6}\ \text{S/km} \qquad z = (0.08 + j0.6)\ \Omega/\text{km}$$

If the line supplies a 200-MW 0.866 lagging power factor load, calculate the sending-end voltage and power.

10-13 Determine the **ABCD** constants of the line of Problem 10-9. Hence determine the sending-end voltage.

10-14 The generalized constants of a 345-kV three-phase transmission line are as follows:

$$\mathbf{A} = \mathbf{D} = 0.932\,\underline{/1.04^\circ}$$

$$\mathbf{B} = 92.77\,\underline{/75.96^\circ}$$

$$\mathbf{C} = 0.0014745\,\underline{/90.24^\circ}$$

The line supplies a 140-MW load at 0.8 lagging current. Using the nominal-T circuit, determine the charging current and the voltage regulation.

10-15 Repeat Problem 10-14 using the nominal-Π current.

CHAPTER 11

Electric Power Systems

In this chapter we study the electric power system as a whole and discuss various phases of power system engineering. This discussion will give us an insight into the diverse activities of power system engineers. The subject matter of electric power systems has a very broad scope, and we shall restrict ourselves to three major topics: fault calculations, power-flow studies, and power system stability.

11-1 PHASES OF POWER SYSTEM ENGINEERING

The various phases of power system engineering include:

1. System planning.
2. System design.
3. Construction.
4. System operation.

These are discussed in some detail in the following sections.

System Planning

Power systems cannot now be planned "from scratch" since the populated areas of the United States are already served by central-station supply, and we must start where we are now. The task of *system* planning is that of defining the optimum expansion program that will provide for orderly system growth, achieving a balance between technical adequacy and long-term investment costs. Therefore, the problem is not just to find a solution to expansion needs, but to find the solution that is adequate and costs the least. Alternative plans that achieve the same ultimate goals must be compared in terms of the present worth of the capital required to implement them. The optimum plan will emerge only after both technical and financial aspects of each plan are evaluated.

System planning is done by specialists using a formidable array of computers, system data, and computational techniques. A mathematical model of the system as it exists becomes the starting point. Such a model is constructed from a system diagram which represents all the basic elements of the system with their interconnections and electrical characteristics. Material purchasing and construction time required are such that transmission lines must be planned 3 to 4 years in advance, and generating stations must be specified on paper 5 to 8 years before they are required to be in service. *Load forecasting* is an important part of planning, because it determines the timing and magnitude of system additions needed. Having established the projected peak load for each of several years into the future, it is necessary to specify additional facilities each year which will serve the growing load. Using various computer programs to analyze voltage, power flows, losses, and system stability, the system model is modified according to several proposed expansion plans in turn until the optimum plan is found. This procedure is followed for each successive year, always building on the previous years' results. System planning is a continual process. Each year the previous 8- to 10-year set of plans are refined to compensate for actual developments, recognizing that some long-range commitments have been made which cannot be altered.

System Design

The second phase of power system engineering is *system design*. Whereas the timing, scale, and location of system additions are determined in the planning phase, the system elements must be specified in great detail in the design phase. Only the very largest electric utilities find it feasible to have power stations designed by their own employees. Consulting firms are usually employed to design them within the constraints of size, type of fuel, water supply, and numerous preferences which the purchaser may have accumulated through past experience. Reliability and protection of the environment are features that must be designed into every system, with no increase in productivity but at considerable increase in cost. Company management must define the philosophy which ultimately sets the limit on optional features and costs. Transmission and distribution systems are quite massive, but their individual components are less complex. Therefore, most companies have the necessary staff to assemble their own system using some off-the-shelf materials and specifying part of the equipment made to special order. The final results of the design phase is a set of drawings and specifications from which contractors can submit bids and construction can be completed.

Construction

After contracts are awarded, the *construction* phase begins. In the case of power stations the power company usually purchases the major components, such as boiler, turbines, generators, pumps, transformers, auxiliary equipment, and building materials, from separate specialized manufacturers. Two predominant methods are employed to coordinate the work of contractors as they erect the building and install the equipment. The simplest method is to give the design engineering firm the responsibility of both design and construction, which results in a finished product ready to place in service. The other is to have a sufficient in-house engineering staff to enable the company to act as its own prime contractor, coordinating and inspecting the work of all subcontractors and retaining ultimate responsibility for successful and timely completion.

Transmission-line construction may be done either by company personnel or by contractors. Engineering personnel and inspectors from the company usually work closely with the contractor. After decision by the system planners to build a line of certain specifications from one point to another, typical time lags for a 100-mile line would require a year for right-of-way acquisition, another year for material ordering and delivery, and a year or two for actual construction.

System Operation

Once a power station, substation, or transmission line has been built and thoroughly checked out for correct operation, it is turned over to *systems operation* personnel for *commercial operation*. From that time on the facility is an integral part of the system and its operation is coordinated from the *system control center*. Activity there is devoted to two broad categories: *power control* and transmission system *switching supervision*. A one-line diagram is used to represent the entire network in its simplest form (see Fig. 3-9).

A *complex communications* system is essential to these functions. Some principal modes of communication are two-way radio, privately owned microwave system, leased microwave, leased telephone circuits, and power-line carrier. The intelligence transmitted may be voice, telemetering, remote control signals, contact closures indicating specific messages or alarms, digital data to or between computers, teletype, facsimile, or video pictures. By these various media, the system operator is able to keep informed of virtually every kind of activity on the system, and exert the necessary control actions for safe, reliable, and economic operation. *Telemetering* is the process whereby signals representing generator outputs and power flow through tie lines with other systems are transmitted to the control center.

The power control function starts by analyzing power flows to determine total system generation, system load, *net interchange,* and immediate generation requirements. A *system operations computer* or other automatic controller, using telemetered quantities, will compute the need and send out control signals to appropriate generators, keeping the system in dynamic balance. Many functions are performed by the on-line control computer. Its primary function is *automatic generation control,* but its myriad responsibilites include data gathering, economic allocation of generation, logging, alarming, system security, and operations planning. Unlike a general-purpose data-processing computer which works by job batches on input from punched cards and presents its output on a printed page, the *process control* computer has as its main inputs telemetered quantities on a continuous basis from the system and occasional data entries by the operator. Outputs are in the form of control signals, alarm messages, and periodic data logs relative to the controlled process, which includes the entire generation and transmission facilities of the system. It can be seen that the computer, telemetry, and process (power system) constitute a *closed-loop* control circuit.

On a typical power system, *switching supervision and coordination* occupies one or more people each shift around the clock. On a widely spread system many work crews may need to do routine or emergency maintenance at the same time. If the integrity of the system is to be protected, all such work must be coordinated centrally, giving consideration to overall system loads and equipment outages. The switching coordinator serves as a clearing house for all such work, sometimes running computer studies to determine the adequacy of the system under the proposed conditions. Strict switching procedures are followed which assure the personal safety of field personnel. A *state estimation* and *security assessment* program running in real time at intervals of a few minutes can warn the coordinator of the approach of dangerous conditions in time to make necessary adjustments.

Equipment operating at high voltages must be protected from damage in the event of short circuits or other abnormal conditions. Millions of dollars in damage can occur in a fraction of a second. It is difficult to imagine the destructive potential that can be unleashed in a few microseconds if an arc is started. The art of *protective relaying* seeks to detect faults in any component of the system and isolate that component before it or other equipment suffers damage, or to hold damage to a minimum. Relays are composed of highly sophisticated circuitry which can be specialized to detect the many kinds of faults that can develop in the broad range of components. First, a short-circuit study is made on a computer, using a mathematical model of the system to simulate the more common faults which can be expected to occur, and recording the fault currents and voltages that would be present if a fault should occur. Such data are tabulated for each point in the system. Then the protective relays can be calibrated to detect and distinguish between the many types of fault at a given location, and supply a tripping signal to the appropriate circuit breakers, which will open and isolate the faulted component until it can be repaired.

11-2

INTERCONNECTED SYSTEMS

At first, central-station supply consisted of one central generating station with distribution feeder circuits branching out radially over a town. Before the invention of labor saving devices in such profusion as we know today, the electric load was concentrated in the early evening hours and during one or two mornings a week, catering to the needs of the housewife. Because of the lack of load, the generating station was shut down late at night and during most of the daytime hours. It was natural that generators would tend to fail, if at all, during peak-load periods, leaving all of the customers without power. Full duplication of facilities was too expensive to be feasible, so loss of a generator caused interruption of all electric service to the area until repairs could be made. A *power system* was formed when the generating stations in several isolated towns or areas of a city were tied together with transmission lines to form a network of facilities which could continue to serve all customer loads even after the loss of one or more generators. A combination of residential, commercial, and industrial loads created a load *diversity* which resulted in total demand on a system being somewhat more uniform. As customers became more dependent on electricity, they demanded a higher degree of reliability in their service. Excess generating capacity had to be installed and held in reserve to guarantee reliability of service any time the largest unit on the system was unavailable.

Engineers reasoned that the concept of diversity could be extended to include two or more systems, resulting in more reliable supply and greatly reduced generation reserve requirements. Thus the concept of *interconnected system operation* was born. The first documented case of two large power systems operating in *parallel* occurred on September 7, 1926, when a 132-kV tie was closed between the Duquesne Light Company of Pittsburgh and the West Penn Power Company. On January 17, 1927, several systems in the eastern United States stretching from Boston to Chicago to Florida achieved parallel (synchronous) operation in the modern sense for a brief time. The ultimate in parallel operation on a continental scope was attained on Feburary 7, 1967, when the 230-kV East-West tie in the Northwest was closed for the first time, connecting in parallel virtually all major power systems from coast to coast.

What are the advantages of parallel operation? We mentioned the ability to share reserve capacity so that each company could reduce its ratio of reserve capacity to load. Another inherent advantage of such operation arises from the synchronous nature of the tie. If a generator is tripped off-line suddenly, phase-angle relationships develop which will automatically cause power to flow to the deficient system before manual intervention by an operator would be possible. Some systems have enough ties with adjacent systems on all sides that they could continue to serve all their load even after the loss of all internal generation. Voluntary system operations committees have been formed to study on a continuing basis the problems associated with accurate control and accounting procedures among systems. Engineers from the North American Power Systems Interconnection Committee (NAPSIC) meet regularly in regional and national groups to improve interconnected operation.

One further attempt to improve the quality of service while reducing the cost of power has been the establishment of *power pools*. Two or more systems may enter into a contractual agreement to pool their facilities and operate them as one larger system. Larger, more efficient generators can be afforded by power pools than can be justified by one system alone. System planning and system operation are both coordinated such that the maximum benefits are realized. To facilitate the exchange of large amounts of power on a regular basis, a more substantial transmission *grid* is required by the pool systems. An extra-high-voltage (EHV) grid is useful to the extent that load diversity can be exploited in providing more economical, reliable operation.

From the preceding remarks and from Chapter 3 it is clear that even a small portion of an electric power system corresponds to a very large and complex electrical network. The pertinent problems are numerous and complex. In this chapter we will restrict ourselves to the following three types of power system problems:

1. Fault analysis.
2. Power flow.
3. System stability.

To handle problems relating to the above-mentioned topics, it is important that we represent the system by an electrical circuit. Such a representation of power systems has been discussed in Chapter 3.

11-3 FAULT ANALYSIS

Under normal conditions, a power system operates as a balanced three-phase ac system. A significant departure from this condition is often caused by a fault. A fault may occur on a power system for a number of reasons, some of the common ones being lightning, high winds, snow, ice, and frost. Faults give rise to abnormal operating conditions, usually excessive currents and voltages

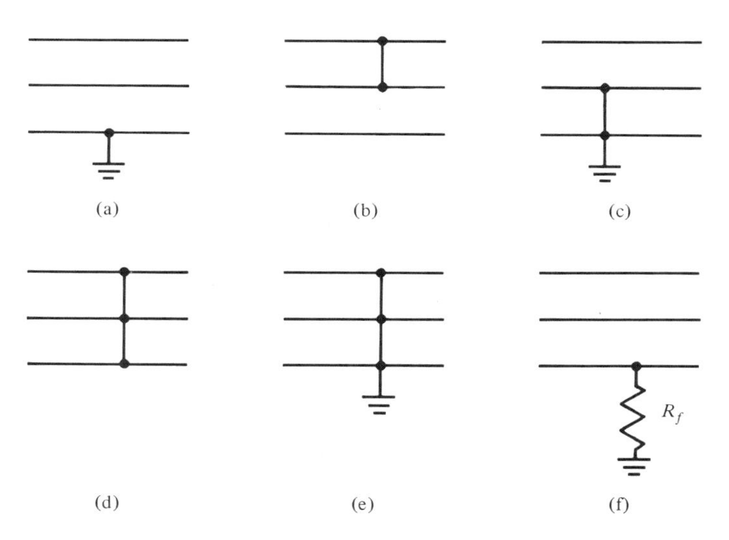

FIGURE 11-1. Various types of faults: (a) line-to-ground; (b) line-to-line; (c) line-to-line-to-ground, or double-line-to-ground; (d) balanced three-phase; (e) three-phase-to-ground; (f) line-to-ground through a fault resistance R_f. *Note:* Other types of faults may occur through fault resistances.

at certain points on the system. Protective equipment is used on the system to guard against abnormal conditions. For example, the magnitudes of fault currents determine the interrupting capacity of the circuit breakers and settings of protective relays. Faults may occur within a generator or at the terminals of a transformer. However, here we will be mostly concerned with faults on transmission lines.

Various types of faults that occur on a transmission line are depicted in Fig. 11-1. In the order of frequency of occurrence are the types of faults shown in Fig. 11-1(a) through (d). The balanced three-phase short circuit is the least common but most severe fault, and therefore determines the rating of the circuit breaker. Consequently, it is almost invariably included in fault studies. In summary, a fault study includes the following:

1. Determination of maximum and minimim three-phase short-circuit currents.
2. Determination of unsymmetrical fault currents, as in single line-to-ground, double line-to-ground, line-to-line, and open-circuit faults.
3. Determination of ratings of circuit breakers.
4. Investigating schemes of protective relaying.
5. Determination of voltage levels at strategic points during the fault.

Of the items listed above, we consider only the first two here. Per unit values and system representation, discussed in Chapter 3, should be reviewed at this point.

Balanced Three-Phase Short Circuit

Balanced three-phase fault calculations can be carried out on a per phase basis so that only single-phase equivalent circuits are used. Invariably, the circuit constants are expressed in per unit, and all calculations are made on a per unit basis. In short-circuit calculations, we often evaluate the short-circuit MVA, which is equal to $\sqrt{3}\, V_l I_f 10^6$, where V_l is the nominal line voltage and I_f is the fault current. We illustrate the procedure by the following examples. Also, at this point it will be useful to refer back to Chapter 3 and review Section 3-13 on per unit representation. The next example further illustrates the application of per unit representation.

EXAMPLE 11-1

A one-line diagram of a generator supplying a load through a step-up transformer, a transmission line, and a step-down transformer is shown in Fig. 11-2 on a per phase basis. The pertinent numerical values are as labeled in the figure. Taking the generator voltage and kVA as base values, express the line and load impedances in per unit. Also calculate the per unit current. The transformers are ideal.

Because the voltage (and current) levels change due to the transformers, different base voltages prevail at different locations on the system.

For the generator:

$$(V_{\text{base}})_{\text{gen}} = 480\ \text{V} = 1\ \text{pu}$$

$$(\text{kVA}_{\text{base}})_{\text{gen}} = 20\ \text{kVA} = 1\ \text{pu}$$

$$(I_{\text{base}})_{\text{gen}} = \frac{20{,}000}{480} = 41.67\ \text{A} = 1\ \text{pu}$$

$$(Z_{\text{base}})_{\text{gen}} = \frac{480}{41.67} = 11.52\ \Omega = 1\ \text{pu}$$

For the transmission line:

$$(\text{V}_{\text{base}})_{\text{line}} = \frac{480}{0.5} = 960\ \text{V} = 1\ \text{pu}$$

$$(\text{kVA}_{\text{base}})_{\text{line}} = 20\ \text{kVA} = 1\ \text{pu}$$

$$(I_{\text{base}})_{\text{line}} = \frac{20{,}000}{960} = 20.83 = 1\ \text{pu}$$

$$(Z_{\text{base}})_{\text{line}} = \frac{960}{20.83} = 46.08\ \Omega = 1\ \text{pu}$$

$$\mathbf{Z}_{\text{line}} = \frac{1 + j3}{46.08} = 0.022 + j0.065\ \text{pu}$$

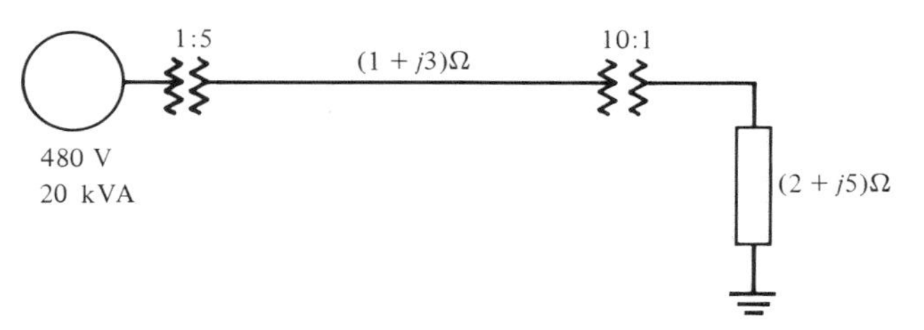

FIGURE 11-2. Example 11-1.

For the load:

$$(V_{\text{base}})_{\text{load}} = \frac{960}{10} = 96 \text{ V} = 1 \text{ pu}$$

$$(\text{kVA}_{\text{base}})_{\text{load}} = 20 \text{ kVA} = 1 \text{ pu}$$

$$(I_{\text{base}})_{\text{load}} = \frac{20{,}000}{96} = 208.3 = 1 \text{ pu}$$

$$(Z_{\text{base}})_{\text{load}} = \frac{96}{208.3} = 0.4608\ \Omega = 1 \text{ pu}$$

$$\mathbf{Z}_{\text{load}} = \frac{2 + j5}{0.4608} = 4.34 + j10.85 \text{ pu}$$

Per unit current:

$$\begin{aligned}\mathbf{Z}_{\text{total}} &= \mathbf{Z}_{\text{line}} + \mathbf{Z}_{\text{load}} \\ &= (0.022 + j0.065) + (4.34 + j10.85) \\ &= 4.362 + j10.915 = 11.75\angle 68^\circ \text{ pu}\end{aligned}$$

$$\mathbf{I} = \frac{1\angle 0^\circ}{\mathbf{Z}_{\text{total}}} = \frac{1\angle 0^\circ}{11.75\angle 68^\circ} = 0.085\angle -68^\circ \text{ pu}$$

Notice that the two ideal transformers are eliminated as circuit components from the calculations. This is one of the major advantages of representing a system in per unit. ■

EXAMPLE 11-2

An interconnected generator–reactor system is shown in Fig. 11-3(a). Generator and reactor ratings are as shown. The values of the corresponding reactances in percent, with the ratings of the equipments as base values, are also shown. A

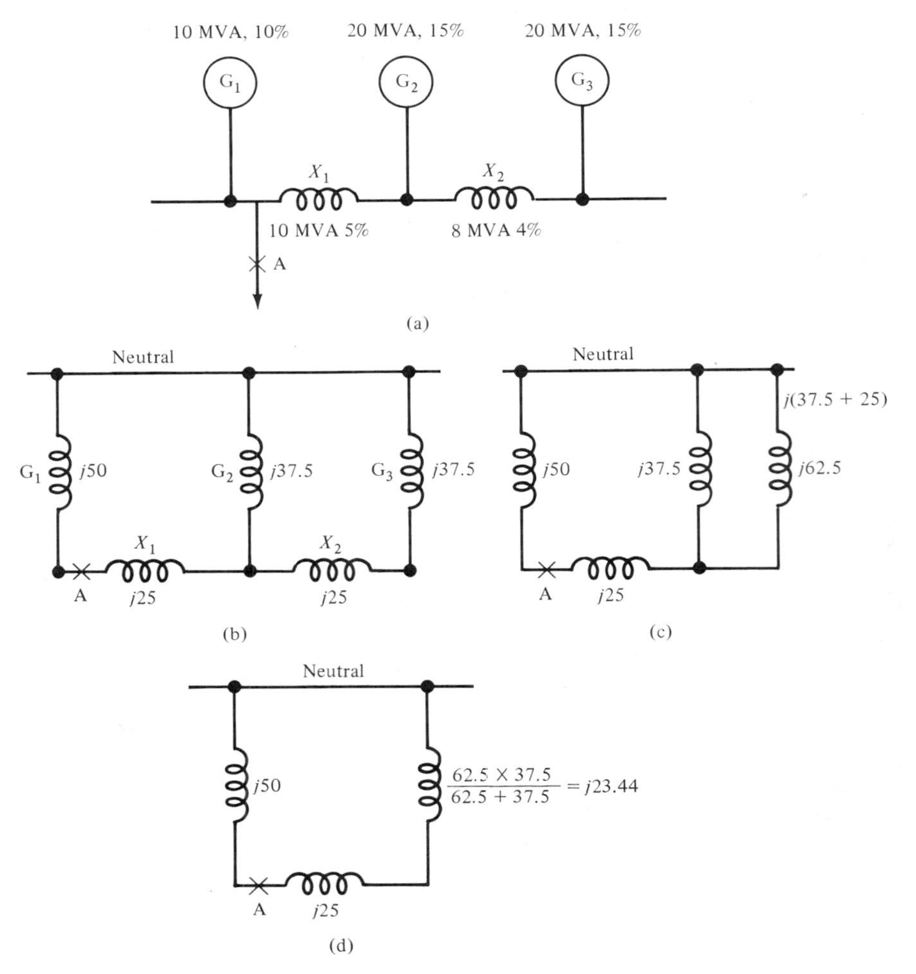

FIGURE 11-3. Example 11-2 (reactance values in percent on 50 MVA base).

three-phase short-circuit occurs at A. Determine the fault current and the fault kVA if the busbar line-to-line voltage is 11 kV.

First, we choose a base MVA for the system, and express the reactances in percent for this base value. Let 50 MVA be the base MVA. On this base

$$\text{reactance of generator } G_1 = \frac{50}{10} \times 10 = 50\%$$

$$\text{reactance of generator } G_2 = \frac{50}{20} \times 15 = 37.5\%$$

$$\text{reactance of generator } G_3 = \frac{50}{20} \times 15 = 37.5\%$$

$$\text{reactance } X_1 = \frac{50}{10} \times 5 = 25\%$$

$$\text{reactance } X_2 = \frac{50}{8} \times 4 = 25\%$$

With these values of reactances, a reactance diagram for the system is drawn, as given in Fig. 11-3(b). This diagram is drawn per phase and does not contain any sources. The reduction of the reactance diagram is illustrated in Fig. 11-3(c) and (d). Finally, the total reactance from the neutral to the fault (at A) is

$$\text{percent } X = j\,\frac{50 \times (23.44 + 25)}{50 + (23.44 + 25)} = j24.6\%$$

$$\text{fault MVA} = \frac{50}{24.6} \times 100 = 203.25 \text{ MVA}$$

$$\text{fault current} = \frac{203.25 \times 10^6}{\sqrt{3} \times 11 \times 10^3} = 10{,}668 \text{ A}$$

Recall from the discussions in the beginning of this section that a three-phase fault is the most severe fault and often determines the ratings of circuit breakers. ■

EXAMPLE 11-3

A three-phase short-circuit fault occurs at point A on the system shown in Fig. 11-4(a). The ratings, reactances, and impedance values are as shown. Calculate the fault current. This example is a further illustration of a three-phase short-circuit calculation.

Let the base MVA be 30 MVA and 33 kV the base voltage. Then on this base we have the following values of reactances and impedance:

$$\text{reactance of generator } G_1 = \frac{30}{20} \times 15 = 22.5\%$$

$$\text{reactance of generator } G_2 = \frac{30}{10} \times 10 = 30\%$$

$$\text{reactance of transformer} = \frac{30}{30} \times 5 = 5\%$$

$$\text{impedance of line} = (3 + j15)\,\frac{30}{(33)^2} \times 100 = (8.26 + j41.32)\%$$

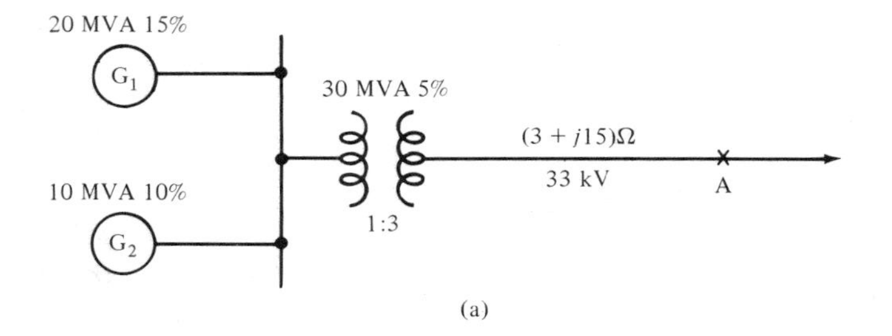

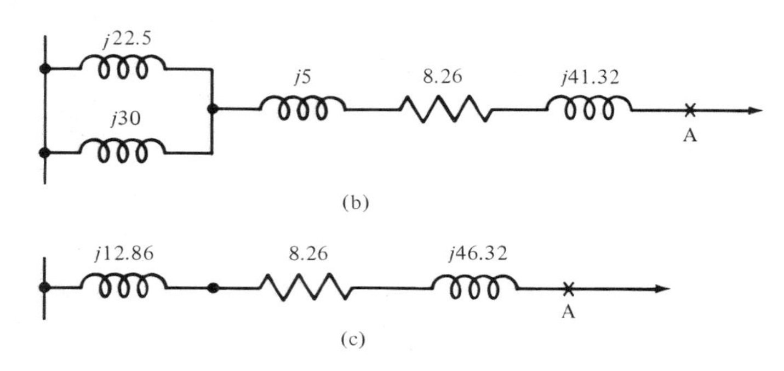

FIGURE 11-4. Example 11-3.

These values in percent are shown on the diagram of Fig. 11-4(b), which is then reduced to Fig. 11-4(c). Finally, the total impedance from the generator neutral to the fault is

$$\text{percent } Z = 8.26 + j59.18 = 59.75\%$$

$$\text{short-circuit MVA} = \frac{30}{59.75} \times 100 = 50.21 \text{ MVA}$$

$$\text{short-circuit current} = \frac{50.21 \times 10^6}{\sqrt{3} \times 33 \times 10^3} = 878.5 \text{ A}$$

■

Unbalanced Faults: Method of Symmetrical Components

The preceding method of fault calculations is valid only for balanced three-phase short circuits. However, for unsymmetrical faults such as line-to-line and line-to-ground faults (which occur more frequently than three-phase short circuits), the method of symmetrical components is used. The method is based on the fact that a set of three-phase unbalanced phasors (voltages or currents) can be resolved into three sets of symmetrical components, which are termed the *positive-sequence*, *negative-sequence*, and *zero-sequence components*. The pha-

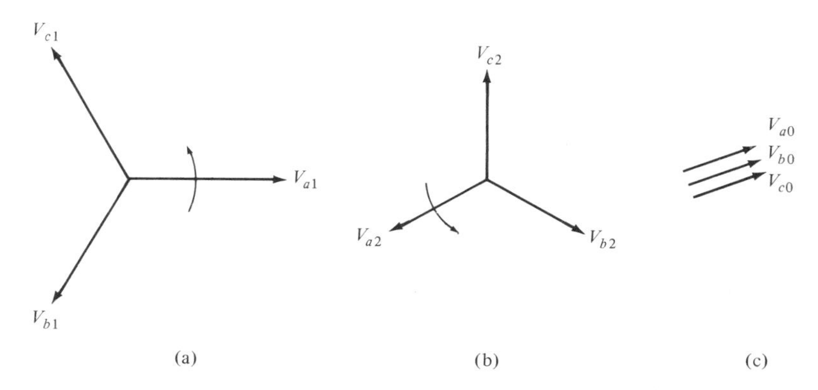

FIGURE 11-5. (a) Positive-sequence components; (b) negative-sequence components; (c) zero-sequence components.

sors of a set of positive-sequence components have a counterclockwise phase rotation (or phase sequence), *abc*; the negative-sequence components have the reverse phase sequence, *acb*; and the zero-sequence components are all in phase with each other. These sequence components are represented geometrically in Fig. 11-5, and can be used to form the unbalanced system of Fig. 11-6. In other words, the unbalanced system shown Fig. 11-6 can be resolved into its sym-

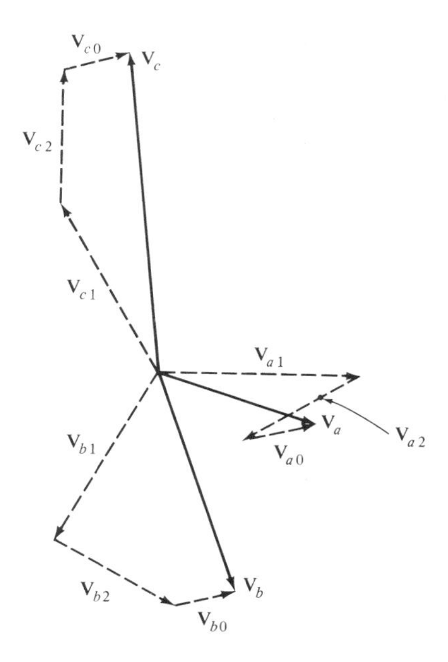

FIGURE 11-6. A three-phase unbalanced system and its symmetrical components.

metrical components shown in Fig. 11-5. The positive-sequence component is designated with a subscript 1. The subscripts 2 and 0 are used for the negative- and zero-sequence components, respectively.

According to the method of symmetrical components, as as illustrated in Fig. 11-6, we have

$$\mathbf{V}_a = \mathbf{V}_{a0} + \mathbf{V}_{a1} + \mathbf{V}_{a2} \tag{11-1}$$

$$\mathbf{V}_b = \mathbf{V}_{b0} + \mathbf{V}_{b1} + \mathbf{V}_{b2} \tag{11-2}$$

$$\mathbf{V}_c = \mathbf{V}_{c0} + \mathbf{V}_{c1} + \mathbf{V}_{c2} \tag{11-3}$$

We now introduce an operator **a** such that it causes a rotation of 120° in the counterclockwise direction (just as the j-operator produces a 90° rotation), such that

$$\mathbf{a} = 1\underline{/120^\circ} = 1 \cdot e^{j120} = -0.5 + j0.866 \tag{11-4a}$$

$$\mathbf{a}^2 = 1\underline{/240^\circ} = -0.5 - j0.866 = \mathbf{a}^* \tag{11-4b}$$

$$\mathbf{a}^3 = 1\underline{/360^\circ} = 1\underline{/0^\circ} \tag{11-4c}$$

$$1 + \mathbf{a} + \mathbf{a}^2 = 0 \tag{11-4d}$$

Using the above-mentioned properties of the **a**-operator, we may write the components of a given sequence in terms of any chosen component. Expressed mathematically, we have, from Fig. 11-5,

$$\mathbf{V}_{b1} = \mathbf{a}^2\mathbf{V}_{a1}$$

$$\mathbf{V}_{c1} = \mathbf{a}\mathbf{V}_{a1}$$

$$\mathbf{V}_{b2} = \mathbf{a}\mathbf{V}_{a2}$$

$$\mathbf{V}_{c2} = \mathbf{a}^2\mathbf{V}_{a2}$$

$$\mathbf{V}_{a0} = \mathbf{V}_{b0} = \mathbf{V}_{c0}$$

Consequently, (11-1)–(11-3) become (in terms of components of phase a)

$$\mathbf{V}_a = \mathbf{V}_{a0} + \mathbf{V}_{a1} + \mathbf{V}_{a2} \tag{11-5}$$

$$\mathbf{V}_b = \mathbf{V}_{a0} + \mathbf{a}^2\mathbf{V}_{a1} + \mathbf{a}\mathbf{V}_{a2} \tag{11-6}$$

$$\mathbf{V}_c = \mathbf{V}_{a0} + \mathbf{a}\mathbf{V}_{a1} + \mathbf{a}^2\mathbf{V}_{a2} \tag{11-7}$$

Solving for the sequence components from (11-5)–(11-7) yields

$$\mathbf{V}_{a0} = \tfrac{1}{3}(\mathbf{V}_a + \mathbf{V}_b + \mathbf{V}_c) \tag{11-8}$$

$$\mathbf{V}_{a1} = \tfrac{1}{3}(\mathbf{V}_a + \mathbf{a}\mathbf{V}_b + \mathbf{a}^2\mathbf{V}_c) \tag{11-9}$$

$$\mathbf{V}_{a2} = \tfrac{1}{3}(\mathbf{V}_a + \mathbf{a}^2\mathbf{V}_b + \mathbf{a}\mathbf{V}_c) \tag{11-10}$$

In deriving (11-8)–(11-10), properties of **a** such as those given in (11-4a)–(11-4d) have been used.

Relationships similar to those of (11-5)–(11-10) are valid for phase and sequence currents also; that is, for current relationships, we simply replace the **V**'s in (11-5)–(11-10) by **I**'s.

Corresponding to sequence currents, we may define sequence impedances. Thus the *positive-sequence impedance* corresponds to an impedance through which only positive-sequence currents flow. Similarly, when only negative-sequence currents flow, the impedance is known as the *negative-sequence impedance,* and when zero-sequence currents alone are present, the impedance is called the *zero-sequence impedance*.

Unsymmetrical (or unbalanced) fault calculations are facilitated by the use of the concepts of sequence voltages, currents, and impedances.

EXAMPLE 11-4

A three-phase wye-connected load is connected across a three-phase balanced supply system. Obtain a set of equations giving the relationships between the symmetrical components of line and phase voltages.

The symmetrical system, the assumed directions of voltages, and the nomenclature are shown in Fig. 11-7, from which we have

$$\mathbf{V}_{ab} = \mathbf{V}_a - \mathbf{V}_b$$

$$\mathbf{V}_{bc} = \mathbf{V}_b - \mathbf{V}_c$$

$$\mathbf{V}_{ca} = \mathbf{V}_c - \mathbf{V}_a$$

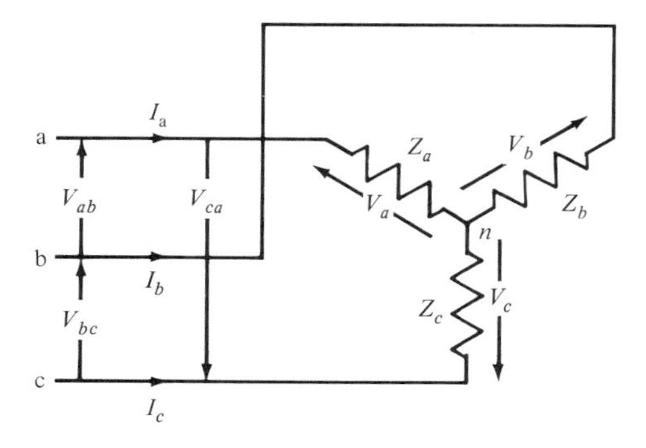

FIGURE 11-7. Example 11-4.

Because $\mathbf{V}_{ab} + \mathbf{V}_{bc} + \mathbf{V}_{ca} = 0$, we get

$$\mathbf{V}_{ab0} = \mathbf{V}_{bc0} = \mathbf{V}_{ca0} = 0$$

We choose $\mathbf{V}_{ab}$ as the reference phasor. For the positive-sequence component, we have

$$\begin{aligned}
\mathbf{V}_{ab1} &= \tfrac{1}{3}(\mathbf{V}_{ab} + \mathbf{a}\mathbf{V}_{bc} + \mathbf{a}^2\mathbf{V}_{ca}) \\
&\ \tfrac{1}{3}[(\mathbf{V}_a - \mathbf{V}_b) + \mathbf{a}(\mathbf{V}_b - \mathbf{V}_c) + \mathbf{a}^2(\mathbf{V}_c - \mathbf{V}_a)] \\
&= \tfrac{1}{3}[(\mathbf{V}_a + \mathbf{a}\mathbf{V}_b + \mathbf{a}^2\mathbf{V}_c) - (\mathbf{a}^2\mathbf{V}_a + \mathbf{V}_b + \mathbf{a}\mathbf{V}_c)] \\
&= \tfrac{1}{3}[(\mathbf{V}_a + \mathbf{a}\mathbf{V}_b + \mathbf{a}^2\mathbf{V}_c) - \mathbf{a}^2(\mathbf{V}_a + \mathbf{a}\mathbf{V}_b + \mathbf{a}^2\mathbf{V}_c)] \\
&= \tfrac{1}{3}[(1 - \mathbf{a}^2)(\mathbf{V}_a + \mathbf{a}\mathbf{V}_b + \mathbf{a}^2\mathbf{V}_c)] = (1 - \mathbf{a}^2)\mathbf{V}_{a1} \\
&= \sqrt{3}\, V_{a1}e^{j30^\circ}
\end{aligned} \tag{11-11}$$

Similarly, for the negative-sequence component, we obtain

$$\begin{aligned}
\mathbf{V}_{ab2} &= \tfrac{1}{3}(\mathbf{V}_{ab} + \mathbf{a}^2\mathbf{V}_{bc} + \mathbf{a}\mathbf{V}_{ca}) \\
&= \tfrac{1}{3}[(\mathbf{V}_a - \mathbf{V}_b) + \mathbf{a}^2(\mathbf{V}_b - \mathbf{V}_c) + \mathbf{a}(\mathbf{V}_c - \mathbf{V}_a)] \\
&= \tfrac{1}{3}[(\mathbf{V}_a + \mathbf{a}^2\mathbf{V}_b + \mathbf{a}\mathbf{V}_c) - (\mathbf{a}\mathbf{V}_a + \mathbf{V}_b + \mathbf{a}^2\mathbf{V}_c)] \\
&= \tfrac{1}{3}[(\mathbf{V}_a + \mathbf{a}^2\mathbf{V}_b + \mathbf{a}\mathbf{V}_c) - \mathbf{a}(\mathbf{V}_a + \mathbf{a}^2\mathbf{V}_b + \mathbf{a}\mathbf{V}_c)] \\
&= \tfrac{1}{3}(1 - \mathbf{a})(\mathbf{V}_a + \mathbf{a}^2\mathbf{V}_b + \mathbf{a}\mathbf{V}_c) = (1 - \mathbf{a})\mathbf{V}_{a2} \\
&= \sqrt{3}\, V_{a2}e^{-j30^\circ}
\end{aligned} \tag{11-12}$$

In (11-11) and (11-12), $\mathbf{V}_{a1}$ and $\mathbf{V}_{a2}$ are, respectively, the positive- and negative-sequence components of the phase voltage $\mathbf{V}_a$.

Proceeding as in (11-11) and (11-12), but choosing $\mathbf{V}_{bc}$ as the reference phasor, it may be shown that

$$\mathbf{V}_{bc1} = -j\sqrt{3}\, V_{a1} \tag{11-13}$$

$$\mathbf{V}_{bc2} = j\sqrt{3}\, V_{a2} \tag{11-14}$$

■

EXAMPLE 11-5

The line voltages across a three-phase wye-connected load, consisting of a 10-Ω resistance in each phase, are unbalanced such that $\mathbf{V}_{ab} = 220\angle 131.7^\circ$ V, $\mathbf{V}_{bc} = 252\angle 0^\circ$ V, and $\mathbf{V}_{ca} = 195\angle -122.6^\circ$ V. Determine the sequence phase voltages. Hence find the value of voltages across the 10-Ω resistances, and calculate the line currents.

Since line voltages are given, we determine the sequence components of line voltages. Thus, from (11-9) and (11-10), we obtain

$$\mathbf{V}_{bc1} = \tfrac{1}{3}(\mathbf{V}_{bc} + \mathbf{a}\mathbf{V}_{ca} + \mathbf{a}^2\mathbf{V}_{ab})$$
$$= \tfrac{1}{3}(252\angle 0^\circ + 1\angle 120^\circ \times 195\angle -122.6^\circ + 1\angle -120^\circ \times 220\angle 131.7^\circ)$$
$$= 221 + j12 \text{ V}$$
$$\mathbf{V}_{bc2} = \tfrac{1}{3}(\mathbf{V}_{bc} + \mathbf{a}^2\mathbf{V}_{ca} + \mathbf{a}\mathbf{V}_{ab})$$
$$= \tfrac{1}{3}(252\angle 0^\circ + 1\angle -120^\circ \times 195\angle -122.6^\circ + 1\angle 120^\circ \times 220\angle 131.7^\circ)$$
$$= 31 - j11.9$$

From (11-8) we have

$$\mathbf{V}_{bc0} = \tfrac{1}{3}(\mathbf{V}_{bc} + \mathbf{V}_{ca} + \mathbf{V}_{ab})$$
$$= \tfrac{1}{3}(252\angle 0^\circ + 195\angle -122.6^\circ + 220\angle 131.7^\circ) = 0 \text{ V}$$

Sequence components of phase voltages are

$$\mathbf{V}_{a0} = 0$$

From (11-13) and (11-14), we get

$$\mathbf{V}_{a1} = \frac{\mathbf{V}_{bc1}}{\sqrt{3}(-j)} = \frac{221 + j12}{\sqrt{3}(-j)} = -6.9 + j127.5 \text{ V}$$
$$\mathbf{V}_{a2} = \frac{\mathbf{V}_{bc2}}{\sqrt{3}(j)} = \frac{31 - j11.9}{\sqrt{3}(j)} = -6.9 - j17.9 \text{ V}$$

Hence, from (11-5)–(11-7),

$$\mathbf{V}_a = -6.9 + j127.5 - 6.9 - j17.9 = -13.8 + j10.96 \text{ V}$$
$$\mathbf{V}_b = \mathbf{a}^2\mathbf{V}_{a1} + \mathbf{a}\mathbf{V}_{a2} = 132.8 - j54.8 \text{ V}$$

(upon substitution and simplification). The line currents are given by

$$\mathbf{I}_a = \frac{\mathbf{V}_a}{R} = \tfrac{1}{10}(-13.8 + j109.6) = -1.38 + j10.96 = 11.05 \text{ A} \angle 97.2^\circ$$
$$\mathbf{I}_b = \frac{\mathbf{V}_b}{R} = \tfrac{1}{10}(132.8 - j54.8) = 13.28 - j5.48 = 14.37 \text{ A} \angle -22.4^\circ$$

Since $\mathbf{I}_a + \mathbf{I}_b + \mathbf{I}_c = 0$, we finally get

$$\mathbf{I}_c = -\mathbf{I}_a - \mathbf{I}_b = -1.38 + j10.96 + 13.28 - j5.48$$
$$= -11.9 - j5.48 = 13.1\ \text{A}\ \underline{/-155.3^\circ}$$ ■

Sequence Networks of a Generator on No-Load

A three-phase synchronous generator, grounded through an impedance $\mathbf{Z}_n$, is shown in Fig. 11-8. The generator is not supplying any load, but because of a fault at the generator terminals, currents $\mathbf{I}_a$, $\mathbf{I}_b$, and $\mathbf{I}_c$ flow through the phases a, b, and c, respectively. Let the generator-induced voltages be $\mathbf{E}_a$, $\mathbf{E}_b$, and $\mathbf{E}_c$ in the three phases (Fig. 11-8). The induced voltages in the generator are balanced. Therefore, these voltages are of positive sequence only. For the positive-sequence (phase) voltage, we have

$$\mathbf{V}_{a1} = \mathbf{E}_a - \mathbf{I}_{a1}\mathbf{Z}_1 \tag{11-15}$$

where $\mathbf{I}_{a1}\mathbf{Z}_1$ is the positive-sequence voltage drop in the positive-sequence impedance (of the generator) $\mathbf{Z}_1$. If $\mathbf{Z}_2$ is the negative-sequence impedance of the generator, the negative-sequence voltage at the terminal of a phase is simply

$$\mathbf{V}_{a2} = -\mathbf{I}_{a2}\mathbf{Z}_2 \tag{11-16}$$

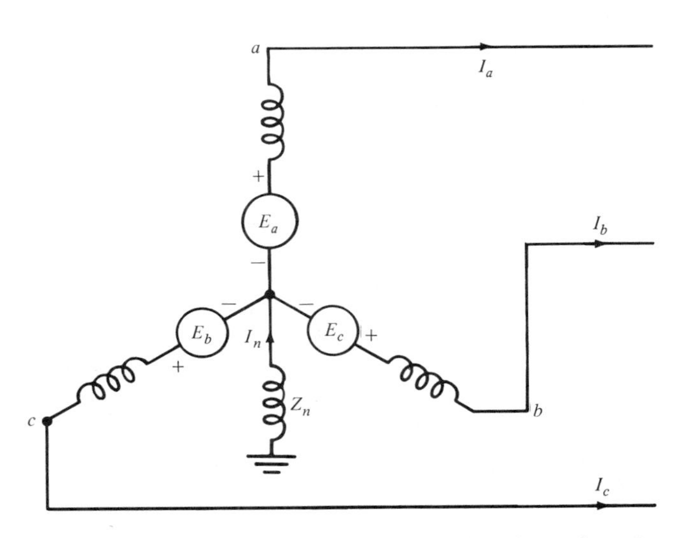

FIGURE 11-8. An unloaded generator grounded through an impedance Z_n.

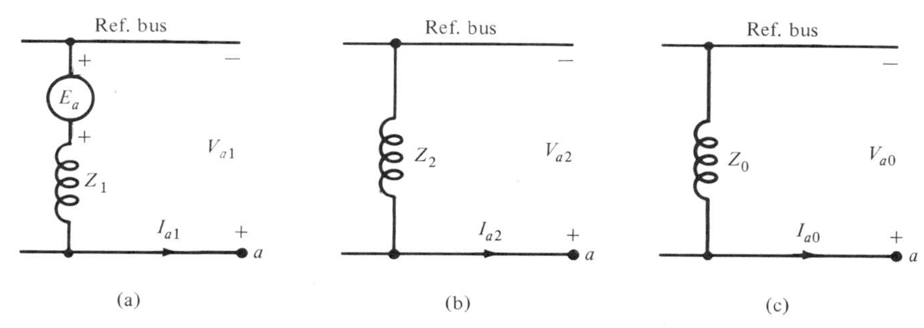

FIGURE 11-9. Networks of a grounded generator: (a) positive-sequence; (b) negative-sequence; (c) zero-sequence.

since there is no negative-sequence generated voltage. The generator zero-sequence currents flow through $\mathbf{Z}_n$ as well as through $\mathbf{Z}_{g0}$, the generator zero-sequence impedance. The total zero-sequence current through $\mathbf{Z}_n$ is $\mathbf{I}_{a0} + \mathbf{I}_{b0} + \mathbf{I}_{c0} = 3\mathbf{I}_{a0}$, but the current through $\mathbf{Z}_{g0}$ is $\mathbf{I}_{a0}$. Hence

$$\mathbf{V}_{a0} = -\mathbf{I}_{a0}\mathbf{Z}_{g0} - 3\mathbf{I}_{a0}\mathbf{Z}_n$$

which is also written as

$$\mathbf{V}_{a0} = -\mathbf{I}_{a0}\mathbf{Z}_0 \tag{11-17}$$

where

$$\mathbf{Z}_0 = \mathbf{Z}_{g0} + 3\mathbf{Z}_n \tag{11-18}$$

Sequence networks corresponding to (11-15), (11-16), and (11-17) are shown in Fig. 11-9(a), (b), and (c), respectively. These sequence networks are interconnected to represent certain fault conditions, as shown by the next example.

EXAMPLE 11-6

A line-to-ground fault occurs on phase a of the generator of Fig. 11-8. Derive a sequence network representation of this condition and determine the current in phase a.

The constraints corresponding to the fault are:

$$\mathbf{I}_b = \mathbf{I}_c = 0 \qquad \text{(lines being open-circuited)}$$

$$\mathbf{V}_a = 0 \qquad \text{(line-to-ground short-circuit)}$$

Consequently, the symmetrical components of the current in phase a are given by

$$\mathbf{I}_{a0} = \tfrac{1}{3}(\mathbf{I}_a + \mathbf{I}_b + \mathbf{I}_c) = \tfrac{1}{3}\mathbf{I}_a$$

$$\mathbf{I}_{a1} = \tfrac{1}{3}(\mathbf{I}_a + \mathbf{a}\mathbf{I}_b + \mathbf{a}^2\mathbf{I}_c) = \tfrac{1}{3}\mathbf{I}_a$$

$$\mathbf{I}_{a2} = \tfrac{1}{3}(\mathbf{I}_a + \mathbf{a}^2\mathbf{I}_b + \mathbf{a}\mathbf{I}_c) = \tfrac{1}{3}\mathbf{I}_a$$

Hence

$$\mathbf{I}_{a0} = \mathbf{I}_{a1} = \mathbf{I}_{a2} = \tfrac{1}{3}\mathbf{I}_a$$

Consequently, the sequence networks must be connected in series, as shown in Fig. 11-10. The sequence voltages appear across the respective sequence networks.

To determine the current, we have from Fig. 11-10,

$$\mathbf{V}_{a0} + \mathbf{V}_{a1} + \mathbf{V}_{a2} = \mathbf{E}_a - \mathbf{I}_{a1}\mathbf{Z}_1 - \mathbf{I}_{a1}\mathbf{Z}_2 - \mathbf{I}_{a1}\mathbf{Z}_0$$

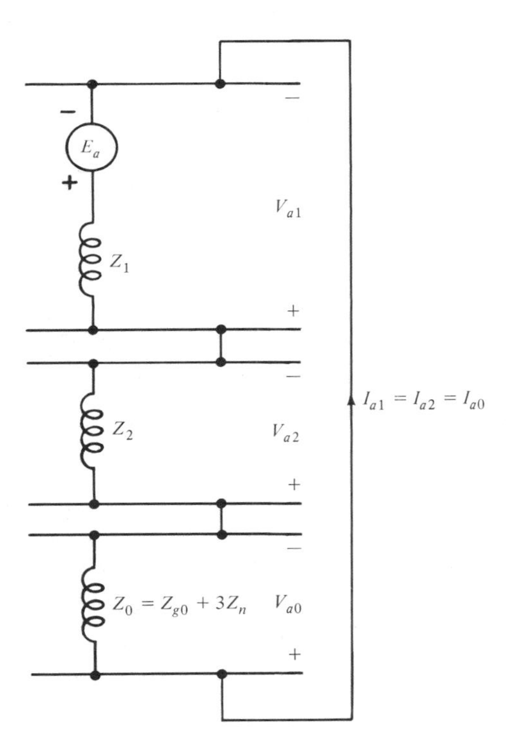

FIGURE 11-10. Sequence network representation of a line-to-ground fault of an unloaded generator.

But

$$\mathbf{V}_a = \mathbf{V}_{a0} + \mathbf{V}_{a1} + \mathbf{V}_{a2} = 0$$

Consequently,

$$\mathbf{I}_{a1} = \frac{\mathbf{E}_a}{\mathbf{Z}_1 + \mathbf{Z}_2 + \mathbf{Z}_0} = \frac{1}{3}\mathbf{I}_a$$

and

$$\mathbf{I}_a = \frac{3\mathbf{E}_a}{\mathbf{Z}_1 + \mathbf{Z}_2 + \mathbf{Z}_0}$$ ■

Example 11-6 illustrates the development of a sequence network representation of an unbalanced line-to-ground fault. Sequence networks can be derived for other cases of faults on power systems. These networks are likely to be more complex than that of Fig. 11-10, especially when three-phase transformers are present in the system, and are beyond the scope of this text. References 1–3 may be consulted for more detailed analyses of unbalanced faults on power systems.

11-4 POWER-FLOW STUDY

Power-flow studies, commonly known as load-flow studies, are extremely important in evaluating the operations of power systems, controlling them, and planning for future expansions [4-6]. Basically, a power-flow study yields the real and reactive power and phasor voltage at each bus on the system, although a wealth of information is available from the printout of a digital computer solution of a typical power-flow study conducted by a power company. As a consequence of a power-flow study, we can optimize the system operation with regard to system losses and load distribution. The effect of temporary loss of generation capacity or transmission circuits can also be investigated via a power-flow study.

Whereas the principles of a power-flow study are straightforward, a realistic study relating to a power system can be carried out only with the digital computer. In such a case, numerical computations are carried out in a systematic manner by an iterative procedure. Two of the commonly used numerical methods are the Gauss–Seidel method and the Newton–Raphson method. For details pertaining to these methods, any of the references cited at the end of this chapter may be consulted. In the following, we consider only simple examples to illustrate the principles and certain procedures of power-flow calculations.

In order to develop a feeling for the parameters that control power flow, let us first consider the power relationships for a transmission line.

EXAMPLE 11-7

A short transmission line, shown in Fig. 11-11(a), has negligible resistance and a series of reactance jX ohms per phase. If the per phase sending- and receiving-end voltages are $\mathbf{V}_s$ and $\mathbf{V}_R$, respectively, determine the real and reactive powers at the sending end and at the receiving end. Assume that $\mathbf{V}_s$ leads $\mathbf{V}_r$ by an angle δ.

The complex power, **S**, in voltamperes, in general, is given by

$$\mathbf{S} = P + jQ = \mathbf{V}\mathbf{I}^* \qquad \text{VA} \tag{11-19}$$

where $\mathbf{I}^*$ is the complex conjugate of **I**. Thus, on a per phase basis, at the sending end we have

$$\mathbf{S}_s = P_s + jQ_s = \mathbf{V}_s\mathbf{I}^* \qquad \text{VA} \tag{11-20}$$

From Fig. 11-11(a), **I** is given by

$$\mathbf{I} = \frac{1}{jX}(\mathbf{V}_s - \mathbf{V}_R)$$

and (11-21)

$$\mathbf{I}^* = \frac{1}{-jX}(\mathbf{V}_s^* - \mathbf{V}_R^*)$$

Substituting (11-21) in (11-20) yields

$$\mathbf{S}_s = \frac{\mathbf{V}_s}{-jX}(\mathbf{V}_s^* - \mathbf{V}_R^*) \tag{11-22}$$

From the phasor diagram of Fig. 11-11(b), we have

$$\mathbf{V}_R = |\mathbf{V}_R| \angle 0^\circ \qquad \mathbf{V}_R = \mathbf{V}_R^*$$

and

$$\mathbf{V}_S = |\mathbf{V}_S| \angle \delta$$

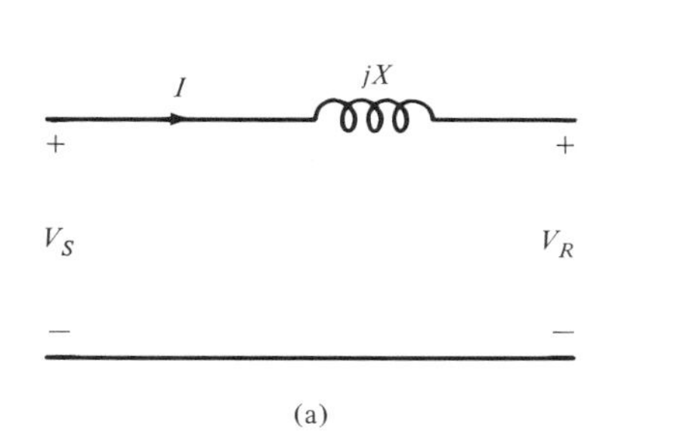

(a)

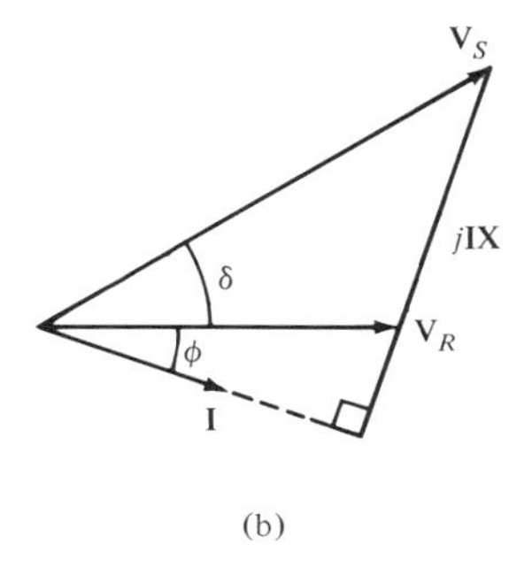

(b)

FIGURE 11-11. Example 11-7.

Hence (11-22) becomes

$$\mathbf{S}_s = \frac{|\mathbf{V}_s|^2 - |\mathbf{V}_R||\mathbf{V}_s|e^{j\delta}}{-jX} = P_s + jQ_s$$

$$= \frac{|\mathbf{V}_s||\mathbf{V}_R|}{X}\sin\delta + j\frac{1}{X}(|\mathbf{V}_s|^2 - |\mathbf{V}_s||\mathbf{V}_R|\cos\delta)$$

Finally,

$$P_s = \frac{1}{X}(|\mathbf{V}_s||\mathbf{V}_R|\sin\delta) \qquad \text{W} \tag{11-23}$$

and

$$Q_s = \frac{1}{X}(|\mathbf{V}_s|^2 - |\mathbf{V}_s||\mathbf{V}_R|\cos\delta) \qquad \text{VA} \tag{11-24}$$

Similarly, for the receiving end, we have

$$\mathbf{S}_R = P_R + jQ_R = \mathbf{V}_R\mathbf{I}^*$$

Proceeding as above finally yields

$$P_R = \frac{1}{X}(|\mathbf{V}_s||\mathbf{V}_R|\sin\delta) \qquad \text{W} \tag{11-25}$$

$$Q_R = \frac{1}{X}(|\mathbf{V}_s||\mathbf{V}_R|\cos\delta - |V_R|^2 \qquad \text{VA} \tag{11-26}$$

■

From this simple example, a number of significant conclusions may be derived. First, the angle δ is known as the *power angle* (see Chapter 6). The transfer of real power depends on δ alone, not on the relative magnitudes of the sending- and receiving-end voltages (unlike in a dc system). The transmitted power varies approximately as the square of the voltage level. The maximum power transfer occurs when $\delta = 90°$ and

$$(P_R)_{\max} = (P_s)_{\max} = \frac{|\mathbf{V}_s||\mathbf{V}_R|}{X} \tag{11-27}$$

Finally, from (11-24) and (11-26), it is clear that reactive power flow will be in the direction of the lower voltage. If the system operates with $\delta \simeq 0$, the average reactive-power flow over the line is given by

$$Q_{\text{av}} = \tfrac{1}{2}(Q_s + Q_R) = \frac{1}{2X}(|\mathbf{V}_s|^2 - |\mathbf{V}_R|^2) \qquad \text{var} \tag{11-28}$$

Equation (11-28) shows the strong dependence of the reactive-power flow on the difference of the voltages.

Up to this point we have neglected the I^2R loss in the line. If R is the resistance of the line per phase, line loss is given by

$$P_{\text{line}} = |\mathbf{I}|^2 R \qquad \text{W} \tag{11-29}$$

From (11-19),

$$\mathbf{I}^* = \frac{P + jQ}{\mathbf{V}}$$

and

$$\mathbf{I} = \frac{P - jQ}{\mathbf{V}^*}$$

Thus

$$\mathbf{II}^* = |\mathbf{I}|^2 = \frac{P^2 + Q^2}{|\mathbf{V}|^2}$$

and (11-29) becomes

$$P_{\text{line}} = \frac{(P^2 + Q^2)R}{|\mathbf{V}|^2} \qquad \text{W} \tag{11-30}$$

indicating that real and reactive powers both contribute to the line losses. Thus it is important to reduce reactive power flow to reduce the line losses. Later we will see that reactive power is required for the voltage control on a bus. This is accomplished by injecting reactive power locally by installing shunt capacitors at the buses.

Preceding remarks give us some idea of real and reactive power flow over transmission lines. In particular, we observe that to raise the voltage level at a given bus, we must supply reactive power at the bus. Furthermore, reactive power flows from higher to lower voltages, whereas the direction of flow of real power depends not on the relative magnitudes of the voltages, but on their phase displacements. In essence, a power-flow analysis yields the voltage levels and the flow of real and reactive powers at various buses of the system. However, explicit analytical solutions are not forthcoming because of load fluctuations on the buses and the receiving-end voltage may not be known. As mentioned in the beginning of this section, in such cases numerical methods are used to solve the problem. An iterative procedure may be used to handle simple problems such as the following. To solve the problem by a noniterative procedure is very cumbersome.

EXAMPLE 11-8

A two-bus system is shown in Fig. 11-12. The load on bus 2 requires 1.0 pu real power and 0.6 pu reactive power per phase. The line impedance per phase is $0.05 + j0.02$ pu. The voltage on bus 1 is $1\underline{/0^\circ}$ pu. Determine on a per phase basis the voltage on bus 2 and the real and reactive powers on bus 1.

In Fig. 11-12, we show the real power by solid arrows and reactive power by dashed arrows. The governing equations for the system are (on a per phase basis)

$$\mathbf{S}_2 = \mathbf{V}_2\mathbf{I}^*$$

$$\mathbf{V}_1 = \mathbf{V}_2 + \mathbf{Z}_l\mathbf{I}$$

where the symbols are defined in Fig. 11-12. Solving for $\mathbf{V}_2$ and eliminating $\mathbf{I}$ from these equations yields

$$\mathbf{V}_2 = \mathbf{V}_1 - \mathbf{Z}_l\mathbf{I} = \mathbf{V}_1 - \mathbf{Z}_l\frac{\mathbf{S}_2^*}{\mathbf{V}_2^*} \tag{11-31}$$

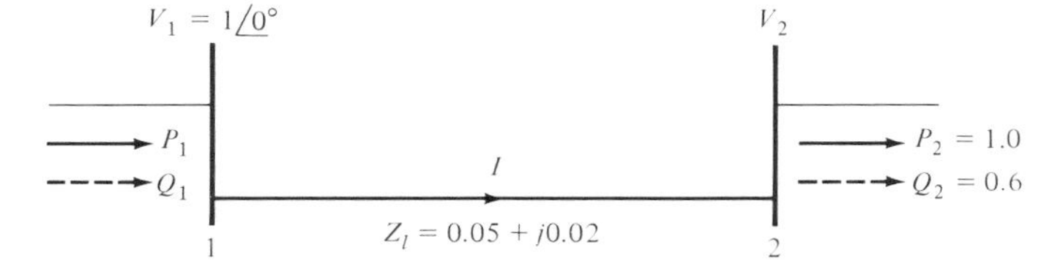

FIGURE 11-12. Example 11-8.

To solve (11-31) iteratively, we assume a value for $\mathbf{V}_2$ and call it $\mathbf{V}_2^{(0)}$. Substitute this in the right-hand side of (11-31) and solve for $\mathbf{V}_2$, calling this new value of $\mathbf{V}_2$ after the first iteration, $\mathbf{V}_2^{(1)}$. We substitute $\mathbf{V}_2^{(1)}$ in the right-hand side of (11-32) and obtain $\mathbf{V}_2^{(2)}$. This procedure is continued until convergence is achieved. The iterative procedure is thus given by the general equation, or algorithm, as

$$\mathbf{V}_2^{(k)} = \mathbf{V}_1 - \frac{\mathbf{Z}_l S_2^*}{\mathbf{V}_2^{(k-1)*}} \tag{11-32}$$

For the given numerical values, we first assume that $\mathbf{V}_2 = 1 \angle 0°$ and use (11-32) in succession to obtain the following table.

Iteration	$\mathbf{V}_2 = 1$ **pu**
0	$1.0 + j0$
1	$0.962 - j0.05$
2	$0.9630 - j0.054$
3	$0.9635 - j0.054$
4	$0.9635 - j0.054$

Notice that convergence is achieved in just four iterations. For other data, such as a greater load, it may take more iterations to converge to the solution. Of course, in some cases convergence may not be achieved because a solution may not exist or the starting point of the iteration process may not be appropriate.

To pursue the problem further, we have

$$\mathbf{V}_2 = 0.9635 - j0.054 \text{ pu}$$

Hence

$$\mathbf{I} = \frac{\mathbf{S}_2^*}{\mathbf{V}_2^*} = \frac{1.0 - j0.6}{0.9635 + j0.054} = 1.208 \angle -27.75° \text{ pu}$$

Or

$$\mathbf{I}^* = 1.208 \angle 27.75°$$

and

$$\mathbf{V}_1 = 1 \angle 0° \quad \text{(given)}$$

Thus

$$P_1 + jQ_1 = \mathbf{S}_1 = \mathbf{V}_1\mathbf{I}^* = (1.208 \angle 27.75^\circ)(1 \angle 0^\circ)$$

$$= 1.069 + j0.5625$$

The real and reactive powers on bus 1 are, therefore,

$$P_1 = 1.069 \text{ pu}$$

$$Q_1 = 0.5625 \text{ pu}$$

■

EXAMPLE 11-9

For the system presented in Example 11-8 it is desired to have $|\mathbf{V}_1| = |\mathbf{V}_2| = 1.0$ pu by supplying reactive power at bus 2. Determine the value of the reactive power.

From (11-19) we have

$$\mathbf{I} = \frac{\mathbf{S}^* + jQ_2}{\mathbf{V}_2^*}$$

which, when substituted in (11-31), yields

$$\mathbf{V}_1 = \mathbf{V}_2 + \frac{\mathbf{Z}_l}{\mathbf{V}_2^*}(\mathbf{S}_2^* + jQ_2) \tag{11-33}$$

where Q_2 represents the added var at bus 2. We now substitute the following numerical values in (11-33):

$$|\mathbf{V}_1| = 1, \quad \mathbf{V}_2 = 1 \angle 0^\circ, \quad \mathbf{Z}_l = 0.05 + j0.02, \quad \mathbf{S}_2^* = 1 - j0.6$$

and obtain

$$1 = |1 + (0.05 + j0.02)[1 + j(Q_2 - 0.6)]|$$

Hence

$$Q_2 = 4.02 \text{ pu}$$

■

As already mentioned, these examples merely illustrate the iteration process of a power-flow study. Detailed calculations for an actual power system must be carried out on a digital computer.

11-5

POWER SYSTEM STABILITY

In Chapter 6 we referred to the maximum power capability of a synchronous machine. We indicated that if a load requires a greater power than the maximum power of the machine, it will pull out of synchronism. This concept of loss of synchronism can be extended to power system stability considerations. By stability of a power system we mean that the system will remain in operating equilibrium, or synchronism, while disturbances occur on the system. The three types of stability are:

1. Steady-state stability.
2. Dynamic stability.
3. Transient stability.

Steady-state stability essentially relates to the maximum power capability of a synchronous machine when the load on the machine is gradually increasing, until the machine pulls out of synchronism. The power-angle characteristics of cylindrical-rotor and salient-rotor machines are, respectively, given by (6-17) and (6-21), from which the steady-state stability limit can be obtained.

Dynamic stability relates to small disturbances occurring on the system, thereby producing oscillations. If these oscillations are of successively smaller amplitudes, the system is considered dynamically stable. If the oscillations grow in amplitude, the system is dynamically unstable. The source of this type of instability is usually an interaction between control systems, and may be slow in becoming apparent. Times of the order of 10 to 30 s are considered sufficient to assess the dynamic stability of the system.

Transient stability relates to a sudden change of load, introducing a large disturbance on the system. A large disturbance on the system causes rather large changes in rotor speeds, power angles, and power transfers. The stability of this transient response is usually evident in less than 1 s for a generator close to the disturbance.

Invariably, stability studies of multimachine power systems are carried out on a digital computer. Therefore, in the following we will only consider special cases to illustrate certain principles and basic concepts. For details, the References at the end of this chapter may be consulted.

Inertia Constant and Swing Equation

The angular momentum, or the inertia constant, plays an important role in determining the stability of a synchronous machine. The per unit inertia constant, H, is defined as the kinetic energy stored in the rotating parts of the machine at synchronous speed per unit MVA (megavoltampere) rating of the machine. Thus if G is the MVA rating of the machine, then

$$GH = \tfrac{1}{2} J\omega_s^2 \qquad (11\text{-}34)$$

where J is the polar moment of inertia of all rotating parts in kg-m^2 and ω_s is the angular synchronous velocity in electrical rad/s. If M is the corresponding angular momentum, then

$$M = J\omega_s \qquad (11\text{-}35)$$

Since $\omega_s = 360f$ electrical degrees per second, (11-34) and (11-35) yield

$$GH = \tfrac{1}{2}M\omega_s = \tfrac{1}{2}M(360)f$$

or (11-36)

$$M = \frac{GH}{180f} \qquad \text{MJ-s/electrical degree}$$

Consider a synchronous generator developing an electromagnetic torque T_e while running at the synchronous speed, ω_s. If the input torque provided by the prime mover at the generator shaft is T_i, then under steady-state conditions (with no disturbance) we have

$$T_e = T_i$$

or

$$T_e\omega_s = T_i\omega_s$$

and

$$T_i\omega_s - T_e\omega_s = P_i - P_e = 0 \qquad (11\text{-}37)$$

If a departure from steady state occurs, such as a change in load or a fault, the "power in," P_i, does not equal the "power out," P_o, and (11-37) does not equal zero. An accelerating torque comes into play. If P_a is the corresponding accelerating (or decelerating) power, then

$$P_a = P_i - P_e = M\frac{d^2\theta}{dt^2} \qquad (11\text{-}38)$$

where M has been defined in (11-36), P_a is the accelerating power in megawatts, and θ is the angular position of the rotor such that

$$\frac{d\theta}{dt} = \omega_s$$

and

$$\theta = \omega_s t + \delta \tag{11-39}$$

Recall from Chapter 6 that δ is the power angle of the synchronous machine. Substituting (11-39) in (11-38) yields

$$M \frac{d^2\delta}{dt^2} = P_i - P_e = P_a \tag{11-40}$$

which is known as the *swing equation*. If we combine (11-36) and (11-40), we obtain the per unit swing equation as

$$\frac{H}{180f} \frac{d^2\delta}{dt^2} = P_i \text{ (pu)} - P_e \text{ (pu)} = P_a \text{ (pu)} \tag{11-41}$$

The swing equation contains information regarding the machine dynamics and stability. Two basic assumptions used in deriving the swing equation are: (1) M of (11-35) has been taken as a constant, although strictly speaking, this is not so; and (2) the damping term proportional to $d\delta/dt$ has been neglected. The swing equation may be used to determine the transient stability of the machine. The approach used is as follows.

EXAMPLE 11-10

The inertia constan, H, of a 60-Hz 100-MVA hydroelectric generator is 4.0 MJ/MVA. What is kinetic energy stored in the rotor at synchronous speed? If the input to the generator is suddenly increased by 20 MW, determine the rotor acceleration.

The energy stored is given by (11-34), so that

$$GH = 100 \times 4 = 400 \text{ MJ}$$

The acceleration is given by (11-40) where P_a = 20 MW-accelerating power and M is found from (11-36). Thus

$$M = \frac{GH}{180f} = \frac{400}{180 \times 60} = \frac{1}{27}$$

Hence

$$\frac{1}{27}\frac{d^2\delta}{dt^2} = 20$$

and

$$\frac{d^2\delta}{dt^2} = \text{acceleration} = 20 \times 27 = 540 \text{ electrical degrees/s}^2$$

■

Equal-Area Criterion

Consider the swing equation, (11-40). In an unstable system, δ increases indefinitely with time and the machine loses synchronism. For a stable system δ undergoes oscillations which eventually die out, as shown in Fig. 11-13. From the figure it is clear that for a stable system, $d\delta/dt = 0$ at some instant. This criterion can be simply obtained from (11-40), and is derived for the simplified case of a two-machine system. Furthermore, it is assumed that H is constant, damping is negligible, and the control system is ignored.

$$2\frac{d\delta}{dt}\frac{d^2\delta}{dt^2} = \frac{2P_a}{M}\frac{d\delta}{dt}$$

which, upon integration, gives

$$\left(\frac{d\delta}{dt}\right)^2 = \frac{2}{M}\int_{\delta_0}^{\delta} P_a \, d\delta$$

or (11-42)

$$\frac{d\delta}{dt} = \sqrt{\frac{2}{M}\int_{\delta_0}^{\delta} P_a \, d\delta}$$

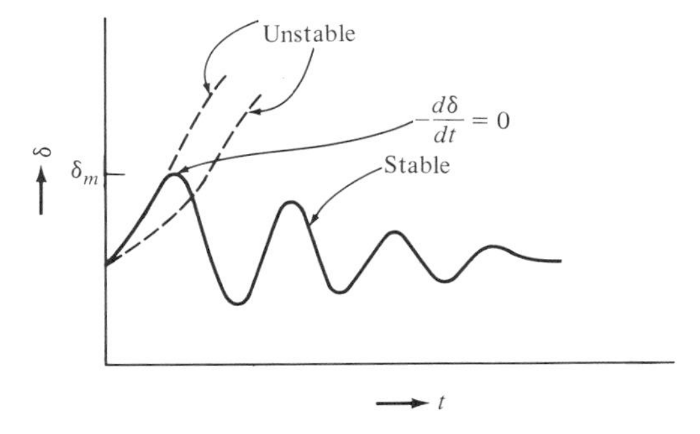

FIGURE 11-13. Stable and unstable responses due to a disturbance.

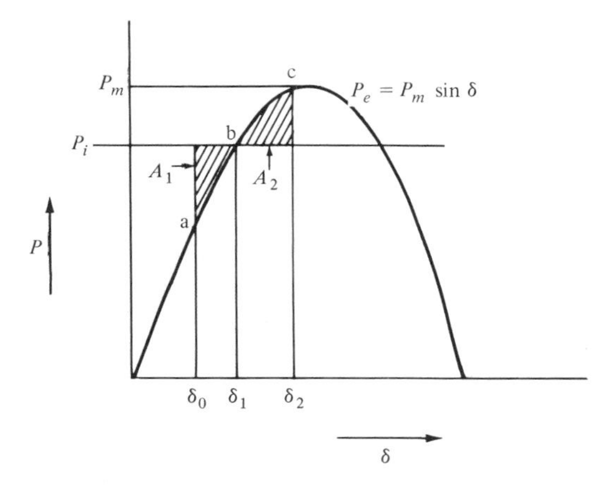

FIGURE 11-14. Equal-area criterion for stability.

where δ_0 is the initial power angle before the rotor begins to swing because of the disturbance. The stability criterion $d\delta/dt = 0$ implies that

$$\int_{\delta_0}^{\delta} P_a \, d\delta = 0 \tag{11-43}$$

This condition requires that for stability the area under the accelerating power P_a versus δ curve must be zero at some value of δ; that is, the positive (or accelerating) area under the P_a–δ curve must be equal to the negative (or deceleration) area. This criterion is therefore known as the *equal-area criterion* of stability. From (11-40), the accelerating power is the difference between the input power and the electromagnetic power of the generator. Hence the criterion for stability is that the area between the P_e versus δ curve and the input power line must be zero, as shown in Fig. 11-14. Physically, the criterion implies that the rotor must be able to store electrically in the system the entire kinetic energy gained by the motors during the accelerating period.

Referring to Fig. 11-14, let a be the initial operating point corresponding to the power angle δ_0. Let the mechanical input power suddenly increase to P_i, which causes the rotor to accelerate, and thereby δ begins to increase. At the point b, P_a is zero and beyond b, P_a is negative or decelerating. Although $P_a = 0$ at b, δ continues to increase because of the rotor inertia. At point c, where area A_1 = area A_2, the rotor stops. The power angle swings between δ_0 and δ_2 until oscillations are damped and equilibrium is reached at the point b. The equal-area criterion requires that, for stability,

$$\text{area } A_1 = \text{area } A_2$$

or

$$\int_{\delta_0}^{\delta_1} (P_i - P_m \sin \delta)\, d\delta = \int_{\delta_1}^{\delta_2} (P_m \sin \delta - P_i)\, d\delta$$

or

$$P_i(\delta_1 - \delta_0) + P_m(\cos \delta_1 - \cos \delta_0) = P_i(\delta_1 - \delta_2) + P_m(\cos \delta_1 - \cos \delta_2) \tag{11-44}$$

But

$$P_i = P_m \sin \delta_1$$

so (11-44) becomes

$$(\delta_2 - \delta_0) \sin \delta_1 + \cos \delta_2 - \cos \delta_0 = 0 \tag{11-45}$$

Knowing δ_0 and δ_1 (11-45) can be solved for δ_2.

EXAMPLE 11-11

A synchronous generator, capable of developing 500 MW of power, operates at a power angle of 8°. Determine by how much the input shaft power can be increased suddenly without loss of stability.

Initially at $\delta_0 = 8°$, the electromagnetic power developed is

$$P_{e0} = P_m \sin \delta_0 = 500 \sin 8° = 69.6 \text{ MW}$$

Let δ_m (Fig. 11-15) be the power angle up to which the rotor can swing before losing synchronism. Then the equal-area criterion requires that (11-45) be satisfied, except that instead of δ_2 we must have δ_m. From Fig. 11-15, $\delta_m = \pi - \delta_1$, which when substituted in (11-45) yields

$$(\pi - \delta_1 - \delta_0) \sin \delta_1 + \cos (\pi - \delta_1) - \cos \delta_0 = 0$$

or (11-46)

$$(\pi - \delta_1 - \delta_0) \sin \delta_1 - \cos \delta_1 - \cos \delta_0 = 0$$

Substituting $\delta_0 = 8° = 0.13885$ rad in (11-46) gives

$$(3 - \delta_1) \sin \delta_1 - \cos \delta_1 - 0.99 = 0$$

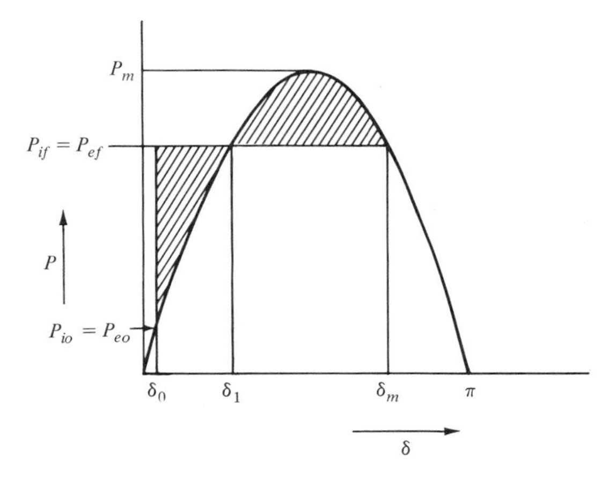

FIGURE 11-15. Example 11-11.

Solving for δ_1, we obtain

$$\delta_1 = 50°$$

Corresponding to this δ_1,

$$P_{ef} = P_m \sin \delta_1 = 500 \sin 50° = 383.02 \text{ MW}$$

The initial power developed by the machine was 69.6 MW. Hence, without loss of stability,

$$\text{sudden increase in load} = P_{ef} - P_{e0} = 383.02 - 69.6 = 313.42 \text{ MW} \blacksquare$$

Other types of transient stability problems, such as the sudden loss of one parallel line, or a sudden short circuit of one parallel line, can also be handled by the equal-area criterion. In the majority of practical cases, solutions are obtained through use of a digital computer.

REFERENCES

1. **W. D. Stevenson, Jr.,** *Elements of Power System Analysis,* 4th ed., McGraw-Hill, New York, 1982.
2. **B. M. Weedy,** *Electric Power System,* 3rd ed., Wiley, New York, 1979.
3. **C. A. Gross,** *Power System Analysis,* Wiley, New York, 1979.
4. **G. W. Stagg,** *Computer Methods in Power System Analyis,* McGraw-Hill, New York, 1968.
5. **M. A. Pai,** *Computer Techniques in Power System Analysis,* Tata McGraw-Hill, New Delhi, 1980.
6. **I. J. Nagrath** and **D. P. Kothari,** *Modern Power System Analysis,* Tata McGraw-Hill, New Delhi, 1980.

11-1 A portion of a power system is shown in Fig. 11P-1, which also shows the ratings of the generators and the transformer and their respective percent reactances. A symmetrical short circuit appears on a feeder at F. Find the value of the reactance X (in percent) such that the short-circuit MVA does not exceed 300.

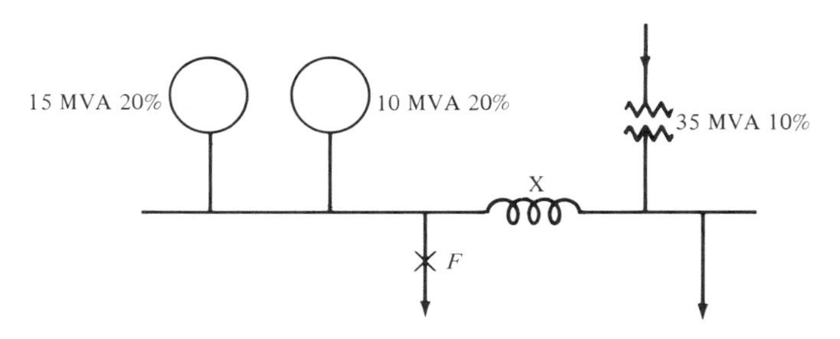

FIGURE 11-P1. PROBLEM 11-1.

11-2 Three generators, each rated at 10 MVA and having a reactance of 10%, are connected to common busbars, and supply the load through two 15-kVA step-up transformers. Each transformer has 7% reactance. Determine the maximum fault MVA on (a) the high-voltage side and (b) the low-voltage side.

11-3 For the system shown in Fig. 11P-3, calculate the short-circuit MVA at A and at B.

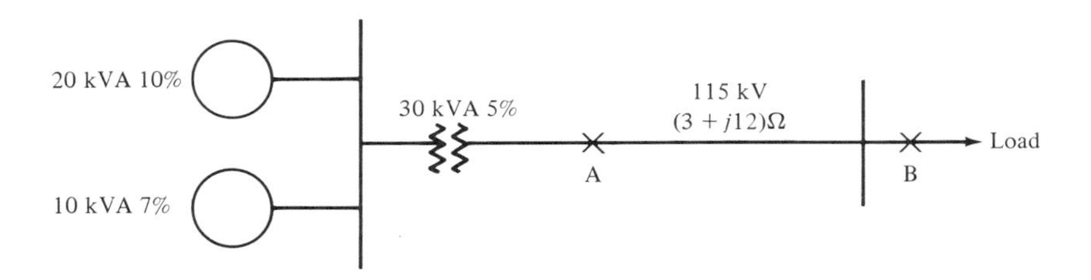

FIGURE 11-P3. PROBLEM 11-3.

11-4 The line currents in a three-phase four-wire system are $\mathbf{I}_a = (300 + j400)$ A, $\mathbf{I}_b = (200 + j200)$ A, and $\mathbf{I}_c = (-400 - j200)$ A. Determine the positive-, negative-, and zero-sequence components.

11-5 The line currents in a delta-connected load are $\mathbf{I}_a = 5\angle 0^\circ$, $\mathbf{I}_b = 7\angle 200^\circ$, and $\mathbf{I}_c = 5\angle 90^\circ$. Calculate the positive-, negative-, and zero-sequence components of currents for phase a. Also determine the positive- and negative-sequence components of the current $\mathbf{I}_{ab}$ and hence calculate $\mathbf{I}_{ab}$.

11-6 A three-phase unbalanced delta load draws 100 A of line current from a balanced three-phase supply. An open-circuit fault occurs on one of the lines. Determine the sequence components of the currents in the unfaulted lines.

11-7 The positive-, negative-, and zero-sequence reactances of a 15-MVA 11-kV three-phase wye-connected generator are 11%, 8%, and 3% respectively. The neutral of the generator is grounded, and the generator is excited to the rated voltage on open circuit. A line-to-ground fault occurs on phase a of the generator. Calculate the phase voltages and currents.

11-8 A line-to-line fault occurs between phases b and c of the generator of Problem 11-7 while phase a remains open-circuited. Determine the phase voltages and currents.

11-9 For the system shown in Fig. 11P-9, it is desired that $|\mathbf{V}_1| = |\mathbf{V}_2| = 1$ pu. The loads are as shown in Fig. 11-18 and are

$$\mathbf{S}_1 = 6 + j10 \text{ pu}$$

$$\mathbf{S}_2 = 14 + j8 \text{ pu}$$

The line impedance is $j0.05$ pu. If each generator supplies 10 pu of real power, calculate the power and the power factors at the two ends.

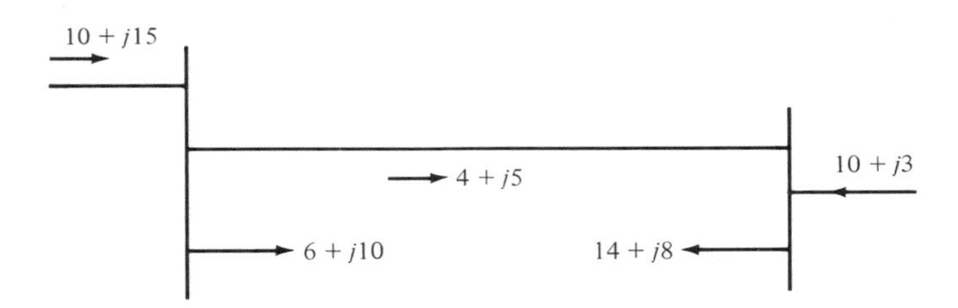

FIGURE 11-P9. PROBLEM 11-9.

11-10 Two buses are interconnected by a transmission line of impedance $(0.3 + j1.2)$ pu. On one bus the voltage is $1 \angle 0°$, and the load on the other bus is $(2 + j0.8)$ pu. Determine the pu voltage on this bus. Also calculate the pu real and reactive power on the first bus.

11-11 It is desired that the voltages on the two buses of Problem 11-10 be equal in magnitude, by supplying reactive power at the second bus. What is the value of this reactive power?

11-12 At a certain leading power factor load the sending- and receiving-end voltages of a short transmission line of impedance $(R + jX)$ are equal. Determine the ratio X/R so that maximum power is transmitted over the line.

11-13 The sending- and receiving-end voltages of a transmission line at a 100-MW load are equal at 115 kV. The per phase line impedance is $(4 + j7)$ ohms. Calculate (a) the steady-state power limit and (b) the maximum additional load that could suddenly be taken on by the line without the loss of stability.

11-14 A synchronous generator is operating at an infinite bus while supplying 0.4 pu of the maximum capacity. Because of a fault, the reactance between the generator and the line becomes twice the value before the fault. The maximum power that can be delivered after fault clearance is 0.7 of the original maximum value. Determine the critical clearing angle to maintain stability.

11-15 A two-pole 60-Hz generator has a moment of inertia of 50×10^3 kg-m^2. What is the energy stored in the rotor at the rated speed? What is the corresponding angular momentum? Determine the inertia constant H.

11-16 The input to the generator of Problem 11-15 is suddenly increased by 25 MW. Determine the rotor acceleration.

11-17 If the acceleration calculated in Problem 11-16 remains constant for 12 cycles, calculate the change in the power angle and the speed at the end of 12 cycles.

11-18 A 60-Hz generator, connected directly to an infinite bus at $1 \angle 0°$ V pu, has a synchronous reactance of 1.35 pu. The generator no-load voltage is 1.1 pu, and its inertia constant H is 4 MJ/MVA. If the generator is suddenly loaded to 60 percent of its maximum power limit, determine the frequency of natural oscillations.

CHAPTER 12

Power Electronics

12-1 INTRODUCTION

In Chapters 5 through 9 we have discussed the three basic types of dc and ac machines. In Chapter 11 we presented an overview of electric power systems, including a discussion of faults in power systems. Most methods of motor control and protection of power systems against various types of faults involve switching operations—switches may be required to be opened or closed to achieve the desired goal. Modulation of power by turning switches on or off can be accomplished by mechanical switches, such as contactors, or by solid-state electronic switches, such as transistors and thyristors. Because power levels in electric motors and power systems are high compared to those in conventional electronics circuits (such as amplifiers, oscillators, etc.), the study of electronic circuits pertinent to electric machines and power systems is known as *power electronics*. Thus the scope of power electronics includes the applications of solid-state switches* to the control and modulation of power in electric motors and electric power systems.

*We will restrict ourselves to solid-state devices, as gas-filled and vacuum tubes are gradually becoming obsolete.

There is a great variety of solid-state components and systems used to control electric motors. In terms of analysis and applications, no other aspect of electric machines has undergone such dramatic changes in recent years or holds greater potential for improving machine characteristics in the future than does the solid-state control of electric machines. Similarly, in application to the protection of electric power systems against faults and in high-voltage dc transmission solid-state switches hold a great promise.

In this chapter we discuss the various solid-state devices used in power electronics. This discussion will be from a circuit viewpoint, and the physics of semiconductors will not be included. Next, we review waveform analysis because invariably the output waveforms from solid-state switching devices are nonsinusoidal. This is followed by several dc and ac motor control schemes, including a brief review of thyristor commutation (or turn-off) techniques. Finally, we present certain applications of solid-state devices to electric power systems.

12-2 POWER SOLID-STATE DEVICES

Many types of solid-state devices exist which are suitable for power electronics applications. [1] For a specific application, the choice depends on the power, voltage, and current requirements; environmental considerations such as ambient temperature; circuit considerations; and overall system cost. Some of the devices commonly used in power electronics circuits are listed in Table 12-1, which also lists the symbols and maximum ratings of the devices. Obviously, there are numerous other power solid-state devices not included in the table, and some of these may be found in the References at the end of the chapter.

In Table 12-1, the state-of-the-art voltage, current, and time of response (or speed) are given. However, these capabilities are seldom achievable simultaneously in a single device. In practice, a device is chosen primarily either for its voltage or current rating. The devices listed here are *pn*-junction devices, having two layers as in a silicon rectifier; three layers, as in a power transistor; or four layers, as in a thyristor. Let us now consider these devices in some detail.

Silicon Rectifier

Silicon rectifiers are high-power diodes capable of operating at high junction temperatures. The principal parameters of a silicon rectifier are the repetitive peak reverse voltage (PRV) or blocking voltage, average forward current, and maximum operating junction temperature ($\simeq$125°C). The terminal (V–I) characteristic of a typical silicon rectifier is shown in Fig. 12-1, which also shows the switching characteristic of an ideal diode. The silicon rectifier has a forward voltage drop of about 1 V at all current levels within its rating.

After the forward current in a silicon rectifier has ceased, a reverse current flows for a very short time. The rectifier assumes its full reverse blocking after the reverse current goes to zero. This characteristic of a diode is known as the reverse *recovery performance* and the time interval during which the reverse current flows is known as the *recovery time*, which is of the order of a few microseconds. The recovery time determines the rate at which blocking voltage can be reapplied to the diode and thus governs its frequency of operation. For applications requiring very short recovery times (of the order of several hundred nanoseconds), fast recovery devices have been developed. Figure 12-2 compares the recovery characteristics of conventional and fast-recovery silicon rectifiers.

Two of the major applications of silicon rectifiers in power electronics are as freewheeling diodes (providing a bypass for the flow of current) in motor controllers and, in general, as rectifiers.

Silicon-Controlled Rectifiers or Thyristors

A silicon-controlled rectifier (or *SCR*), also known as a *thyristor*, is a four-layer *p-n-p-n* semiconductor switch. Unlike the diode, which has only two terminals—anode and cathode—the thyristor has three terminals—anode, cathode, and gate. The reverse characteristic of the thyristor is similar to that of silicon rectifier just discussed. However, the forward conduction of a thyristor can be controlled by utilizing the gate. Normally, a thyristor will not conduct in the forward direction unless it is "turned on" by applying a triggering signal to the gate. However, the full conduction in a thyristor is not instantaneous. We define the turn-on time, t_{on}, when the anode current reaches 90 percent of its final value. Once the thyristor starts to conduct, it continues to do so until turned off by external means. The turn-off of the thyristor is known as *commutation*.

Figure 12-3 shows a copy of the data sheet of a commercial thyristor. The principal parameters characterizing this thyristor are as follows:

1. Repetitive peak reverse voltage (PRV) (1800 V).
2. Maximum value of average on-state current; this parameter is related to the heating within the semiconductor; ($2/\pi \times 850 = 540$ A).
3. Maximum value of rms on-state current; this is the current rating of metal conductor portions of the device, such as the anode pigtail in stud devices (850 A).
4. Peak one-cycle on-state current; this is the surge current limit (6500 A).
5. Critical rate of rise of forward blocking voltage; there are usually two ratings: initial (when the device is turned on) and reapplied (following commutation) (200 V/μs).

TABLE 12-1 Power Solid-State Devices

Device	Abbreviation	Symbol[a]	Maximum Ratings		
			Volts	RMS Amperes	Speed (μs)
Silicon rectifier	—	A, K	5000	7500	—
Silicon-controlled rectifier (thyristor)	SCR	G, A, K	5000	3000	1
Bidirectional switch	Triac	A, K, G	1000	2000	1

Gate turn-off SCR Gate-controlled switch Gate-assisted thyristor Light-activated thyristor	GTO GCS GAT LAT or LASCR	A K G	400 1000	200 200	0.2 2
Power transistor	—	E B C	3000	500	0.2
Power Darlington	—	C B E	1000	200	1

[a] A, anode; G, gate; K, cathode; B, base; C, collector; E, emitter.

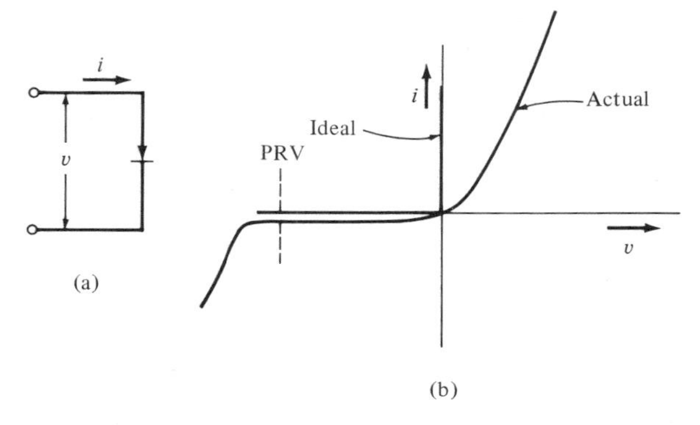

FIGURE 12-1. Silicon rectifier characteristics: (a) symbols; (b) *v-i* characteristics.

6. Turn-off time; the off-time required following commutation before forward voltage can be reapplied (see Section 12-4) (40 μs).
7. Maximum rate of rise of anode current during turn-on; too high di/dt may result in local hot-spot heating, a main cause of device failure (300 A/μs).
8. Maximum operating junction temperature (125°C).

Thermal management of the thyristor is extremely critical in all applications. Most of the parameters just mentioned vary considerably as a function of device temperature. Therefore, much of the engineering required in the application of thyristors is in the design of its heat sink, mounting method, and auxiliary cooling (if required). The use of thyristors in series or parallel electrical connections generally assists in meeting thermal requirements.

Triacs

The Triac, often called a bidirectional switch, is approximately equivalent to a pair of back-to-back or antiparallel thyristors fabricated on a single chip of

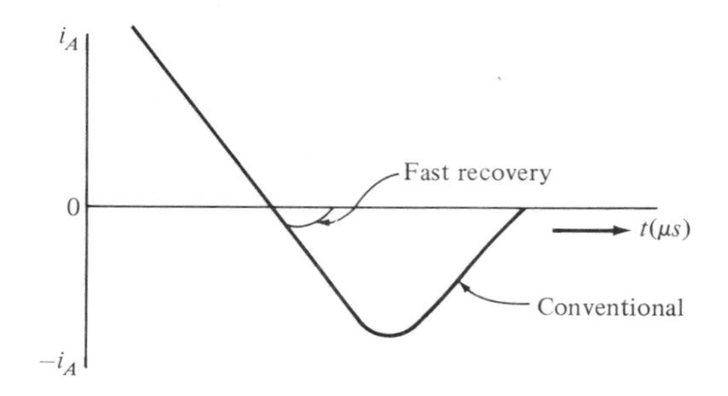

FIGURE 12-2. Comparison of conventional and "fast recovery" silicon rectifiers in the reverse-recovery region.

Maximum Allowable Ratings

Types	Repetititve peak off-state voltage, V_{DRM}[a] $T_j = 40$ to $\leq 125°C$	Repetitive peak reverse Voltage, V_{RRM}[a] $T_j = 0.40$ to $+125°C$	Nonrepetitive peak reverse voltage, V_{RSM}[a] $T_j = +125°C$
C449PN	1800 volts	1800 volts	2040 volts
C449PS	1700	1700	1920
C449PM	1600	1600	1790
C449PE	1500	1500	1700

[a] Half sinewave waveform, 10 ms max pulse width. Consult factory for lower-rated voltage devices.

Peak one cycle surge (nonrepetitive) on-state current, I_{TSM}	6500 amperes
Critical rate-of-rise of on-state current, nonrepetitive	500 A/μs
Critical rate-of-rise of on-state current, repetitive	300 A/μs
Average gate power dissipation, $P_{G(av)}$	5 watts
Storage temperature, T_{stg}	−40 to +150°C
Operating temperature, T_j	−40 to +125°C
Mounting force required	3.3000 lb + 500 lb − 0 lb 13.3 KN + 2.2 KN − 0 KN

FIGURE 12-3. Portion of SCR data sheet. (Courtesy General Electric Company)

semiconductor material. Triggered conduction may occur in both directions, that is, the Triac is a quasi-bilateral device. Triac applications include light dimming, heater control, and ac motor speed control. The parameters listed earlier as important in the application of thyristors generally apply also to the Triac. However, it should be noted that the Triac is a three-terminal device with only one gate, which has an effect on its time response compared with that of two distinct thyristors connected in antiparallel position. The turn-off time of a Triac is in the same order of magnitude as that of a thyristor. This implies that a time period approximately equal to the turn-off time must be observed before applying reverse voltage to a Triac. In an antiparallel pair, however, reverse voltage can be immediately applied after cessation of forward current in one thyristor. Triacs are not available in as high voltage and current ratings as thyristors at the present time and therefore are used in control of motors of relatively low power ratings.

Diverse Thyristors

In addition to the Triac, other forms of thyristors include the following:

1. *Gate turn-off thyristor* (GTO). This thyristor can be turned off at a high temperature, and normal commutation circuit is therefore not required. This type of a thyristor has a high blocking voltage rating and is capable of handling large currents.
2. *Gate-assisted thyristor* (GAT). This thyristor requires large power for triggering. It has a small turn-off time and is specially suited for series-type inverters.
3. *Light-activated thyristor* (LAT or LASCR). This thyristor is turned on by photon or light. Such thyristors find application in high-voltage dc transmission.

Power Transistor

When used in motor control circuits, power transistors are almost always operated in a switching mode. The transistor is driven into saturation and the linear gain characteristics are not used. The common-emitter configuration is the most common, because of the high power gain in this connection. The collector–emitter saturation voltage, $V_{CE(\mathrm{SAT})}$ for typical power transistors is from 0.2 to 0.8 V. This range is considerably lower than the on-state anode-to-cathode voltage drop of a thyristor. Therefore, the average power loss in a power transistor is lower than that in a thyristor of equivalent power rating. The switching times of power transistors are also generally faster than those of thyristors, and the problems associated with turning off or commutating a thyristor are almost nonexistent in transistors. However, a power transistor is more expensive than a thyristor of equivalent power capability. In addition, the voltage and current ratings of available power transistors are much lower than those of existing thyristors. It has already been stated that the maximum ratings listed in Table 12-1 are generally unobtainable concurrently in a single device. This is particularly true of power transistors. Devices with voltage ratings of 1000 V or above have limited current ratings of 10 A or less. Similarly, the devices with higher current ratings, 50 A and above, have voltage ratings of 200 V or less. For handling motor control requiring large current ratings at 200 V or below, it has been common to parallel transistors of lower current rating. This requires great care to assure equal sharing of collector currents and proper synchronization of base currents among the paralleled devices.

Ratings of significance for motor control application include:

1. Breakdown voltage, specified by the symbols BV_{CEO}, collector-to-emitter breakdown voltage with base open, and BV_{CBO}, collector-to-base breakdown voltage with emitter open.
2. Collector saturation voltage, $V_{CE(\mathrm{SAT})}$.
3. Emitter–base voltage rating, V_{EBO}.
4. Maximum collector current, I_C, average and peak.
5. Forward current transfer ratio, H_{FE}, the ratio of collector to base current in the linear region.
6. Power dissipation.
7. Maximum junction temperature, typically 150° to 180°C.
8. Switching times: rise time, t_r; storage time, t_s; and fall time, t_f. Sometimes these switching times are related to a maximum frequency of switching.

The thermal impedances and temperature coefficients are also important parameters. In paralleling power transistors, the variation of device characteristics with temperature becomes especially significant. The I_C–V_{BE} characteristic is extremely temperature sensitive.

Power Darlington

This designation generally refers to the well-known Darlington-connected transistor pair fabricated on a single chip. The same characteristics are, of course, achievable through the use of two discrete transistors, albeit usually in a larger, more complex, and more costly package. The principal merit of the Darlington device is its high current gain. The operating parameters and failure modes discussed earlier for transistors are also applicable to the Darlington.

Darlington amplifiers are a recent entry into the area of motor controls but have met with considerable acceptance due to their potential for reducing the size, cost, and weight of motor controllers. These devices are used both in choppers for dc commutator motor control and in inverters for ac motor control, generally for lower-power applications. Recently, larger devices have been developed with ratings as high as 200 A and 100 V or 100 A and 450 V and have been applied to the control of traction motors used in lift trucks and industrial electric vehicles. Current gains as high as 1600 A have been achieved at these high current levels.

12-3

RMS AND AVERAGE VALUES OF WAVEFORMS

A characteristic of electronic control systems in motor controls, power system protection and high-voltage dc transmission is that pertinent voltage and current waveforms are nonsinusoidal and often discontinuous. Furthermore, these waveforms change as a function of the level of operation. Some of the consequences of the above-mentioned nonlinearities in power electronic system are as follows.

1. The measurement of voltages and currents must be performed with instruments capable of accurately indicating the type of waveforms being measured. Thermocouple instruments are adequate for measuring power components in most electronic motor control systems. Oscilloscopes are usually essential for waveform analysis of both the power and control signal parameters.

2. Loss measurements should be performed, if possible, with the motor excited as it is to be used, rather than with standard sinusoidal or dc excitation. Magnetic material loss data are obtained with sinusoidal excitation and are often incorrect for other types of excitation. The measurement of core losses is difficult when the waveforms are like those described above. In that case, special wattmeters, such as electronic-multipler, Hall-effect, or thermal-type instruments, should be used.

3. Standard circuit theory based on single-frequency sine-wave parameters is inadequate in the analysis of electronic motor control circuits. It is frequently necessary to evaluate the instantaneous time variation of both power and control signal currents and voltages. Fourier methods are also useful, as noted above.

4. The standard numerical values for the relationships between average, rms, and maximum values of current and voltages are seldom applicable.

5. The range of frequencies of the voltage and current components in an electronic motor control system is always much greater than the fundamental frequency applied to the motor. This is readily apparent if one considers the Fourier components of a nonsinusoidal periodic function. The fundamental frequency results from the switching action of power semiconductors in the control system and is usually in the power or low audio range of frequencies, seldom more than 3000 Hz. The range of frequencies in various currents and voltages may easily be 100,000 Hz or higher, however. This fact must be recognized when the choice of instrumentation used in the laboratory is made, when considering audible and electromagnetic noise interference that results from the control system, when designing filters, and when protecting the control logic circuitry used to switch the power devices.

The calculation of average and rms values of voltage and current is quite important in electronic control systems for calculating motor power and torque, for heating of wires and other components, and for sizing components and instrumentation. To make these calculations, it is often necessary to return to the definitions of average and rms values, which are respectively defined as

$$A_{\text{ave}} = \frac{1}{T_0} \int_0^{T_0} a \, dt \tag{12-1}$$

$$A_{\text{rms}} \equiv A = \left(\frac{1}{T_0} \int_0^{T_0} a^2 \, dt \right)^{1/2} \tag{12-2}$$

where a represents instantaneous value of the parameter and T_0 is the period over which the average (or rms) value is evaluated. In motor control circuits involving power semiconductors, T_0 is usually the "on-time" duration. The fundamemtal frequency of the signal referred to above is defined by

$$f_p = \frac{1}{T_p} \tag{12-3}$$

where T_p is the length of a full period.

We illustrate the calculations of rms and average values of waveforms by the following examples.

EXAMPLE 12-1

An electronic motor-controller has a chopped half-wave rectified sinusoidal output voltage waveform, as shown in Fig. 12-4. Determine the average and rms value of the output voltage.

The output voltage, from Fig. 12-4 is given by

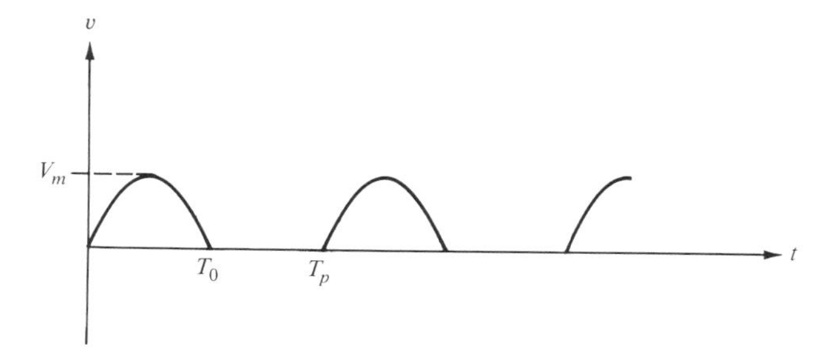

FIGURE 12-4. Chopped sine wave.

$$v(t) = V_m \sin \frac{\pi t}{T_0} \qquad 0 < t < T$$

$$= 0 \qquad T < t < T_0$$

Hence from (12-1) and (12-2) we obtain

$$V_{\text{ave}} = \frac{1}{T_p}\left(\int_0^{T_0} V_m \sin \frac{\pi t}{T_0}\, dt + 0\right) = \frac{2V_m}{\pi}\frac{T_0}{T_p}$$

and

$$V_{\text{rms}} = \left[\frac{1}{T_p}\left(\int_0^{T_0} V_m^2 \sin^2 \frac{\pi t}{T_0}\, dt + 0\right)\right]^{1/2} = V_m \sqrt{\frac{T_0}{2T_p}}$$

■

EXAMPLE 12-2

A silicon rectifier is connected to an inductive load shown in Fig. 12-5(a). With the parameters shown, determine the average value of the load current over a period $T = 2\pi/\omega$.

The instantaneous load current is given by

$$L\frac{di}{dt} + Ri = v_o \tag{12-4}$$

which has a solution of the form (see Problem 12-2)

$$i = \begin{cases} \dfrac{V_m}{Z}\left[\sin(\omega t - \varphi) + e^{-(R/L)t}\sin\varphi\right] & 0 < \omega t < \beta \\ 0 & \beta < \omega t < 2\pi \end{cases} \tag{12-5}$$

where $Z = \sqrt{R^2 + (\omega L)^2}$ and $\tan\varphi = \omega L/R$. Notice that, from Fig. 12-5(b),

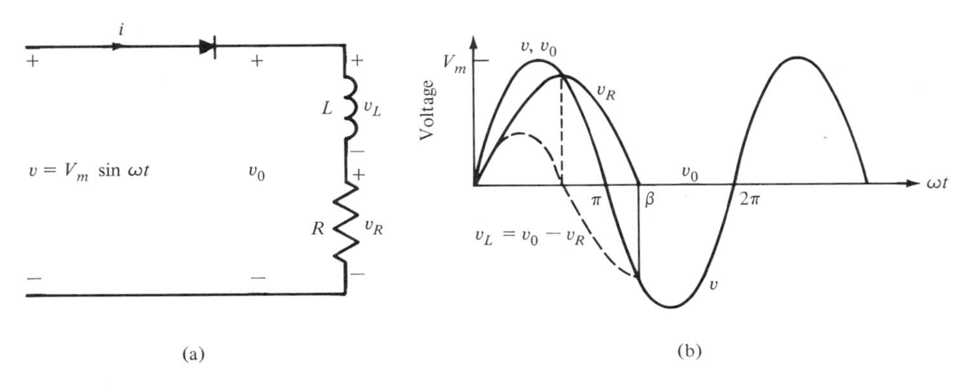

FIGURE 12-5. (a) An RL-circuit fed from a half-wave rectifier; (b) voltage waveforms.

β/ω is the time when the diode stopped conducting. To determine the average current we may use (12-5) directly or (12-4) indirectly. Choosing the latter, we get

$$\frac{L}{T}\int_0^T \frac{di}{dt}\,dt + R\left(\frac{1}{T}\int_0^T i\,dt\right) = \frac{1}{T}\int_0^T v_o\,dt \tag{12-6}$$

The first integral in (12-6) equals $[i(t) - i(0)](L/T) = 0$ since it is periodic, of period T. The second integral is simply RI_{ave}, and (12-4) becomes

$$RI_{\text{ave}} = \frac{1}{T}\int_0^{\beta/\omega} V_m \sin \omega t\,dt = \frac{\omega}{2\pi}\frac{V_m}{\omega}(1 - \cos \beta)$$

Hence

$$I_{\text{ave}} = \frac{V_m}{2\pi R}(1 - \cos \beta) \tag{12-7}$$

■

12-4 THYRISTOR COMMUTATION TECHNIQUES

We mentioned in Section 12-2 that a thyristor can be turned on by injection of energy into it through the gate connection. Once the anode current exceeds a certain minimum value, the thyristor latches on, and the gate loses control over the anode current. Subsequent anode current is determined by the external circuit between the anode and cathode, until the thyristor is brought into the blocking state or turned off. Commutation of a thyristor refers to the process of turning it off. The thyristor can be turned off when the forward anode current is reduced to zero and held at zero for a period of at least equal to the turn-off time. The three basic methods of commutation are as follows: [2,3,4]

1. *Line commutation.* In this case, the source is ac and in series with the thyristor. The anode current goes through zero in a cycle. If the current remains zero for a period greater than the turn-off time, the thyristor will be turned off, until turned on again by some external means.
2. *Load commutation.* Owing to the nature of the load, the anode current may go to zero and thereby turn off the thyristor. This type of commutation is useful mainly in dc circuits.
3. *Forced commutation.* Forced commutation is achieved in systems energized from dc sources by an arrangement of energy storage elements (capacitors and inductors) and by additional switching devices (usually thyristors). In systems energized from ac sources, forced commutation is brought about by means of the cyclic potential reversal of the power source.

Line and load commutation, items 1 and 2, are sometimes grouped as one, known as natural or starvation commutation. We now derive the conditions for commutation by the following examples.

EXAMPLE 12-3

An RL circuit is fed from an ac source in series with a thyristor, as shown in Fig. 12-6(a). The thyristor is fired at an angle α. Derive the condition for line commutation, and determine the thyristor conduction period.

The voltage equation is

$$L\frac{di}{dt} + Ri = V_m \sin \omega t \qquad (12\text{-}8)$$

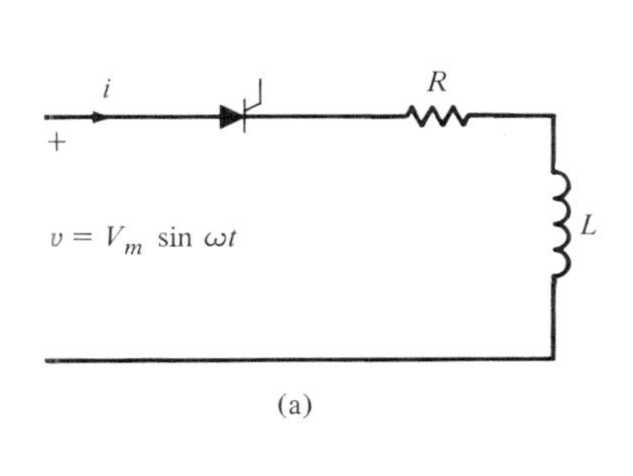

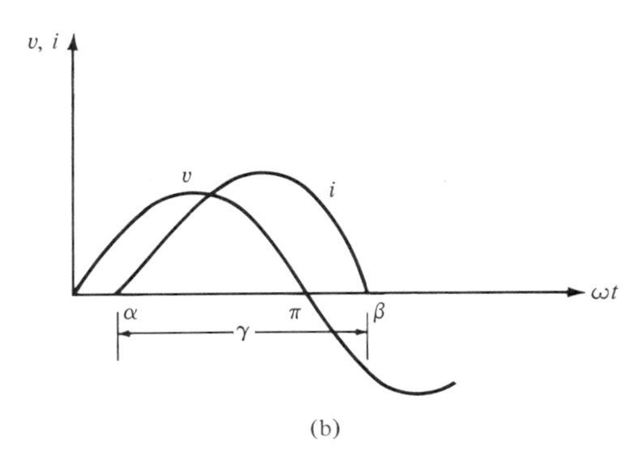

FIGURE 12-6. (a) An RL-thyristor series circuit; (b) voltage and current waveforms.

where the symbols are defined in Fig. 12-6(a). The solution to (12-8) is of the form (see also Example 12-2)

$$i = \frac{V_m}{Z} \sin(\omega t - \varphi) + ke^{-(R/L)t} \tag{12-9}$$

where $Z = \sqrt{R^2 + (\omega L)^2}$ and $\tan \varphi = \omega L/R$. To evaluate k, we use the condition $i = 0$ at $\omega t = \alpha$ in (12-12) to obtain

$$i = \frac{V_m}{Z}[\sin(\omega t - \varphi) - \sin(\alpha - \varphi)e^{R(\alpha - \beta)/\omega L}] \tag{12-10}$$

To obtain the condition for line commutation we notice that $i = 0$ at $\omega t = \beta$, as shown in Fig. 12-6(b). Thus the required condition for commutation is given by the transcendental equation

$$\sin(\beta - \varphi) = \sin(\alpha - \varphi)e^{R(\alpha - \beta)/\omega L} \tag{12-11}$$

The conduction angle γ is then given by

$$\gamma = \beta - \alpha \tag{12-12}$$

The details of solving (12-11) are available in Ref. 2. ■

EXAMPLE 12-4

From Example 12-3 it is clear that line commutation is possible only in ac systems. Similarly, in a dc system load commutation is not possible for an *RL* circuit. However, load commutation in an *RLC* circuit fed from a dc source is possible if the circuit current could be made oscillatory. For the circuit shown in Fig. 12-7, derive the equation governing the time of commutation.

The voltage equation for the given circuit is

$$L\frac{di}{dt} + Ri + \frac{1}{C}\int_0^t i\,dt + v_c(0) = V \tag{12-13}$$

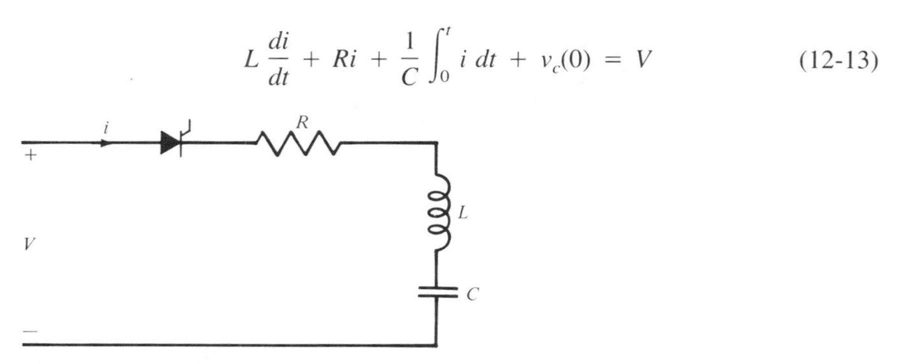

FIGURE 12-7. RLC-circuit driven by a thyristor.

where $v_c(0)$ is the charge on the capacitor at $t = 0$. Defining

$$\zeta = \frac{R}{2L} = \text{damping ratio} \tag{12-14}$$

and

$$\omega_0 = \frac{1}{\sqrt{LC}} = \text{resonant frequency} \tag{12-15}$$

for $\zeta < \omega_0$, the solution to (12-13) becomes (see Problem 12-3)

$$i(t) = e^{-\zeta t}(A \cos \omega_r t + B \sin \omega_r t) \tag{12-16}$$

where A and B are arbitrary constants and

$$\omega_r = \sqrt{\omega_0^2 - \zeta^2} = \text{ringing frequency} \tag{12-17}$$

Because of the inductance, the current in the circuit cannot change instantaneously. Hence $i(0) = 0$, which when substituted in (12-6) yields $A = 0$, and (12-16) then becomes

$$i(t) = Be^{-\zeta t} \sin \omega_r t \tag{12-18}$$

For commutation, this current must go zero, requiring that

$$\omega_r t = \pi \tag{12-19}$$

■

EXAMPLE 12-5

For the circuit shown in Fig. 12-7, $V = 96$ V, $L = 50$ mH, $C = 80$ μF, and $R = 40\ \Omega$. The initial charge on the capacitor is zero. At what time will the thyristor turn off?

From (12-14) and (12-15) we have

$$\zeta = \frac{40}{2 \times 50 \times 10^{-3}} = 400$$

and

$$\omega_0 = \frac{1}{\sqrt{50 \times 10^{-3} \times 80 \times 10^{-6}}} = 500 \text{ rad/s}$$

From (12-17)

$$\omega_r = \sqrt{500^2 - 400^2} = 300 \text{ rad/s}$$

Finally, (12-19) yields

$$300t = \pi$$

or

$$t = \frac{\pi}{300} = 10.47 \text{ ms}$$

■

Forced Commutation

Example 12-5 shows that thyristor commutation in an *RL* circuit fed from a dc source can be accomplished by including a series capacitor. The capacitor must be rated to carry full-load current. Such a scheme may not be economical in all instances, and as an alternative, forced commutation is used. Various forms of forced-commutation techniques include:

1. Series-capacitor commutation.
2. Parallel-capacitor commutation.
3. Parallel capacitor–inductor commutation.
4. External pulse commutation.

The requirements that must be fulfilled by the commutation circuit are that the thyristor current be reduced to zero, the reverse-bias voltage be applied to

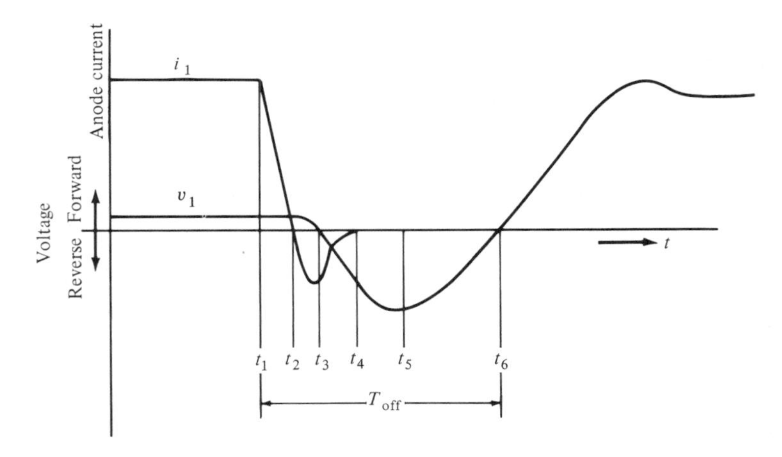

FIGURE 12-8. Thyristor voltage and current during commutation.

the thyristor for an interval greater than the turn-off time, and any stored energy be properly discharged. Numerous circuits fulfilling these requirements have been developed. In the following we illustrate the principles rather than discussing the details of various commutation circuits.

For the sake of illustration we consider Figs. 12-8 and 12-9. In Fig. 12-8 the thyristor is conducting initially. Commutation is initiated at time t_1 by introducing a negative voltage into the external anode-cathode circuit. The anode-cathode voltage drop (v_1) remains at the low on-state (1.0 to 2.0 V) until the anode current (i_1) goes to zero at time t_2. At t_2, v_1 begins to decrease and go negative at t_3, and i_1 goes to zero at t_4. At this instant the thyristor has been turned off. The voltage v_1 swings to a negative maximum at t_5 and back to zero at t_6. Until time t_6, the anode must be maintained at a negative (reverse-biased potential). Now, referring to Fig. 12-9, when T_1 is turned on, it carries only the charging current, which decays to less than the holding current when C is charged to the value V. At a later time ($>T_{\text{off}}$, the turn-off time), T_2 is turned on and C is discharged through L_c and T_2. The series circuit (RLC) is underdamped such that the voltage across C is greater than V. This reverse bias assists in turning off T_1. When T_1 is turned on, the current in the RLC series circuit is governed by

$$L\frac{di}{dt} + Ri + \frac{1}{C}\int i\,dt = V \tag{12-20}$$

In most cases of practical interest $i = 0$ at $t = 0$. Subject to this condition,

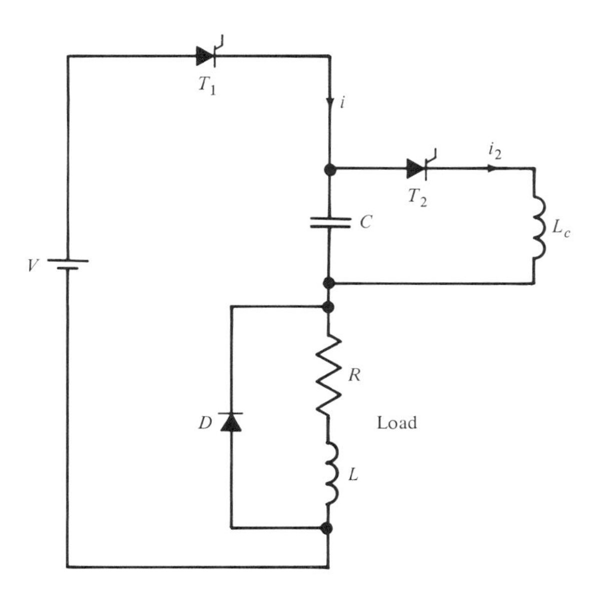

FIGURE 12-9. Series commutation circuit.

and for the underdamped case ($R^2 < 4L/C$), the solution to (12-20) becomes (see Problem 12-4)

$$i = \frac{V_0}{Z_0} e^{-\zeta t} \sin \omega_r t \qquad (12\text{-}21)$$

where $V_0 = V - V_{c0}$

V_{c0} = voltage across C at $t = 0$ (an arbitrary constant)

Z_0 = characteristic impedance = $\sqrt{(L/C - R^2/4)}$

ζ = attenuation constant = $R/2L$

ω_r = ringing frequency = Z_0/L

and other symbols are as defined in Fig. 12-9. Defining $\tan \varepsilon = \omega_r/\zeta$, the voltage across the capacitor, v_c, is given by

$$v_c = V - \frac{V - V_{c0}}{\sin \varepsilon} e^{-\zeta t} \sin (\omega_r t + \varepsilon) \qquad (12\text{-}22)$$

The pulse width, T_0, of the current given by (12-21) is

$$T_0 = \frac{\pi}{\omega_r} \qquad (12\text{-}23)$$

The time t_m after initial turn-on of T_1, when the current pulse reaches its maximum, is

$$t_m = \frac{\varepsilon}{\omega_r} \qquad (12\text{-}24)$$

From (12-21) and (12-24), the corresponding maximum current becomes

$$I_m = \frac{V_0}{Z_0} e^{-\zeta\varepsilon/\omega_r} \sin \varepsilon \qquad (12\text{-}25)$$

When the current is building up, L starts storing energy, which reaches its maximum value at $t \simeq t_m$. This energy is subsequently returned to C. At the end of the sine pulse ($t = T_0$), $i = 0$, L is fully discharged and the voltage across C is given by (with $V_{c0} = 0$),

$$v_c(T_0) = V(1 + e^{-\zeta T_0}) \qquad (12\text{-}26)$$

Because $\zeta < 1$ and $T_0 < 1$, $\zeta T_0 << 1$ for the underdamped case, (12-26) implies that at the end of the pulse the capacitance voltage becomes almost twice the source voltage. Hence the net voltage appearing between the anode-cathode terminals of T_1 is the difference between V and $v_c(T_0)$, and thereby T_1 is reverse biased and is turned off, provided the reverse bias is maintained for a period greater than the thyristor turn-off time.

The capacitance voltage polarity is next reversed by turning on T_2. As a result, a second sine pulse of current flows through the CT_2L_c circuit. At the end of this second pulse, the capacitance voltage is reversed in polarity with the upper plate of C (Fig. 12-9) now negative with respect to the lower plate. The second pulse needed to reverse capacitance voltage can be initiated if the second pulse period results in a reverse-biased condition that is maintained across T_1 for a time interval, T_Q, slightly greater than the turn-off time of T_1. This time interval (T_Q) is obtained from (for the circuit of Fig. 12-9) [1]:

$$\sin(\psi - \omega_r T_Q) = \frac{V}{V_0}\sin\psi \tag{12-27}$$

where $\tan\psi = \sin\varepsilon/(\sqrt{L/C} - \text{load impedance})$

The time T_Q is the circuit commutation time. Notice that T_2 is also turned off by a series commutation process. Finally, the steady-state voltage across C at the instant T_1 is turned on is

$$V_{c0} = V\tanh\zeta T_0 \tag{12-28}$$

As mentioned earlier, a large variety of thyristor commutation circuits are used in practice. It is beyond our scope to discuss these circuits here. The preceding analysis has been presented to illustrate the principles of operation of the series-commutation circuit. We now turn to the applications of solid-state switching devices to the control of dc and ac motors.

12-5 CONTROL OF DC MOTORS

A motor controller should be designed as an integrated part of the drive system, which includes the motor, the controller, and the source. [5,6,7] Clearly, the objective of the design is to achieve maximum efficiency at minimum cost and weight of the overall system. The controller serves as a link between the power source and the motor. Hence the controller determines the efficiency of energy transfer from the source to the motor. Consequently, the efficiency of the controller is an important factor in governing the performance of the drive system. In addition to high efficiency, some of the other desirable features of a motor controller are:

1. Quick and smooth response to the operator's signals.
2. Operation to result in minimum internal losses of the source and the motor.
3. Protection against overloads, to protect itself as well as the motor.

The torque, speed, and regeneration characteristics (under steady-state conditions) of dc motors are governed essentially by the equations derived in Chapter 5. For convenience, we now repeat these pertinent equations:

$$E = \frac{\varphi n Z}{60}\frac{p}{a} = k_a \varphi \Omega_m \tag{12-29}$$

$$T_e = k_a \varphi I_a \tag{12-30}$$

$$\Omega_m = \frac{V_t - I_a r_a}{k_a \varphi} \tag{12-31}$$

where E = voltage induced in the armature, V
I_a = armature current, A
r_a = armature resistance, Ω
φ = flux per pole, Wb
Z = number of armature conductors
a = number of parallel paths
p = number of poles
Ω_m = armature speed, rad/s
n = armature speed, r/min
V_t = armature terminal voltage, V
k_a = $Zp/2\pi a$, a constant

These equations indicate the great flexibility of controlling the dc motor. For instance, the speed of the motor may be varied by varying V_t, r_a, or φ (that is, the field current). Control of dc motors is governed by (12-29)–(12-31) and the various practical schemes are manifestations of these equations in one form or another.

From the governing equations (12-30) and (12-31) it is clear that the motor torque and speed can be controlled by controlling φ (that is, the field current), V_t, and r_a, and changes in these quantities can be accomplished as follows. In essence, the method of control involves field control, armature control, or a combination of the two. Electronically, field and/or armature control in a dc motor is achieved by modulating the voltage across the field and/or armature of the motor by means of thyristor circuits. The choice of the voltage control method depends on the nature of the available supply—whether ac or dc. For an ac source, phase-controlled rectifiers are employed; for a dc source, choppers are used. But before we discuss these, it is worthwhile to consider the analysis of a dc motor supplied by a half-wave rectifier.

Half-Wave Rectifier with DC-Motor Load

The circuit is shown in Fig. 12-10(a), where R and L are, respectively, the armature-circuit resistance and inductance, and e' is the motor back emf, assumed constant. The circuit analysis leads to the following expression for the current:

$$i = \begin{cases} 0 & 0 < \omega t < \alpha \\ \dfrac{V_m}{Z}\left[\sin(\omega t - \theta) + Be^{-(R/L)t}\right] - \dfrac{e'}{R} & \alpha < \omega t < \beta \\ 0 & \beta < \omega t < 2\pi \end{cases} \tag{12-32}$$

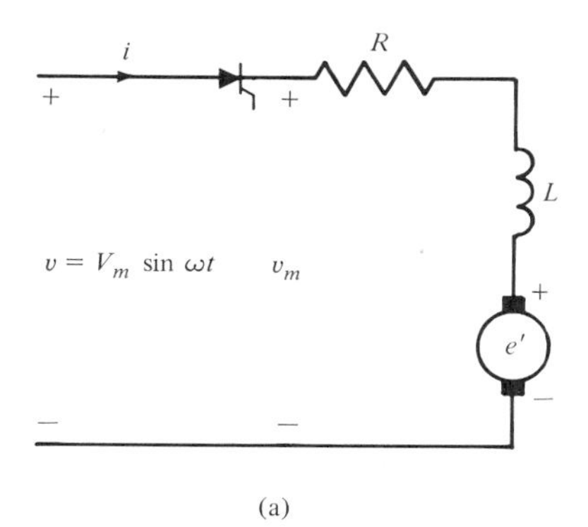

(a)

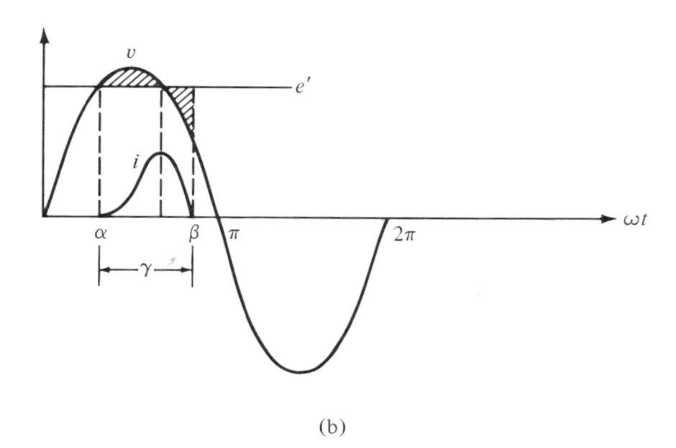

(b)

FIGURE 12-10. (a) A dc motor supplied by an ac source through a diode; (b) voltage and current waveforms.

where

$$Z = \sqrt{R^2 + (\omega L)^2} \qquad \tan\theta = \frac{\omega L}{R}$$

and where

$$B \equiv \left[\frac{e'}{V_m \cos\varphi} - \sin(\alpha - \theta)\right] e^{\alpha R/\omega L} \tag{12-33}$$

is such to make i continuous at $\omega t = \alpha$. It is seen from (12-32) that the diode starts conducting at $\omega t = \alpha$; the firing angle, α, is determined by the condition

$$\sin\alpha = \frac{e'}{V_m} \tag{12-34}$$

As shown in Fig. 12-10(b), conduction does not necessarily stop when v becomes less than e'; rather, it ends at $\omega t = \beta$, when the energy stored in the inductor during the current buildup has been completely recovered. The extinction angle, β, may be determined from the continuity of (12-32) at $\omega t = \beta$; we find that

$$\sin(\beta - \theta) + Be^{-\beta\cot\varphi} = \frac{\sin\alpha}{\cos\theta} \tag{12-35}$$

as the transendental equation for β, in which B is known from (12-33). The average value of the current over one period of the applied voltage is found to be

$$I_{\text{ave}} = \frac{1}{R} V_{R\text{ave}} = \frac{V_m}{2\pi R}(\cos\alpha - \cos\beta - \gamma\sin\alpha) \tag{12-36}$$

where $\gamma \equiv \beta - \alpha$ is the conduction angle. Figure 12-10(b) shows the waveforms.

Thyristor-Controlled DC Motor

In the example above, the dc-motor load was not controlled by the half-wave rectifier; the back emf remained constant, implying that the motor speed was unaffected by the cyclic firing and extinction of the diode. To achieve control, we use a thyristor instead of the diode, as shown in Fig. 12-11(a). The corresponding waveforms are illustrated in Fig. 12-11(b). The motor torque (or speed) may be varied by varying α. Explicitly, for the armature we integrate

$$v_m = Ri + L\frac{di}{dt} + e \tag{12-37}$$

Over the conduction period $\alpha/\omega < t < \beta/\omega$, during which v_m coincides with the line voltage v. The result is

$$V'_m = \frac{V_m(\cos\alpha - \cos\beta)}{\gamma} = RI' + E' \tag{12-38}$$

where a prime indicates an average over the conduction period. Over a full period of line voltage, the average armature current is given by

$$I_{\text{ave}} = \frac{\gamma}{2\pi} I' \tag{12-39}$$

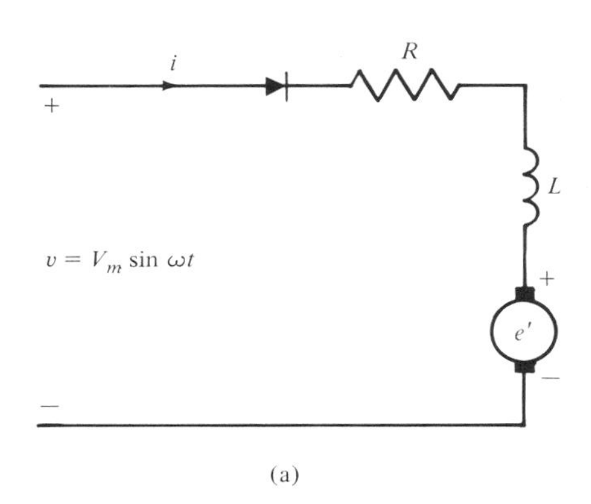

(a)

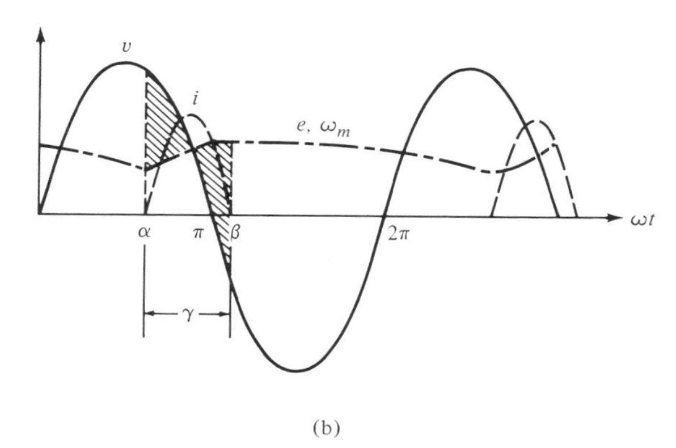

(b)

FIGURE 12-11. (a) A dc motor supplied by an ac source through a thyristor; (b) voltage, current, back emf, and speed waveforms.

and the average torque is given by

$$T_{ave} = kI_{ave} = \frac{k\gamma}{2\pi} I' \tag{12-40}$$

Equations (12-38)–(12-40) govern the steady-state behavior of the thyristor-controlled dc motor.

Chopper Control

Utilizing power semiconductors, the dc chopper is most common electronic controller for the dc motor. In principle, a chopper is an on/off switch connecting the load to and disconnecting it from the dc source, thus producing a chopped voltage across the load. Symbolically, a chopper as a switch is represented in Fig. 12-12(a), and a basic chopper circuit is shown in Fig. 12-12(b). In the circuit shown in Fig. 12-12(b), when the thyristor does not conduct, the load current flows through the freewheeling diode D. From Fig. 12-12, it is clear that the average voltage across the load, V_0, is given by

$$V_0 = \frac{t_{on}}{t_{on} + t_{off}} V_b = \frac{t_{on}}{T} V_b = \alpha V_b \tag{12-41}$$

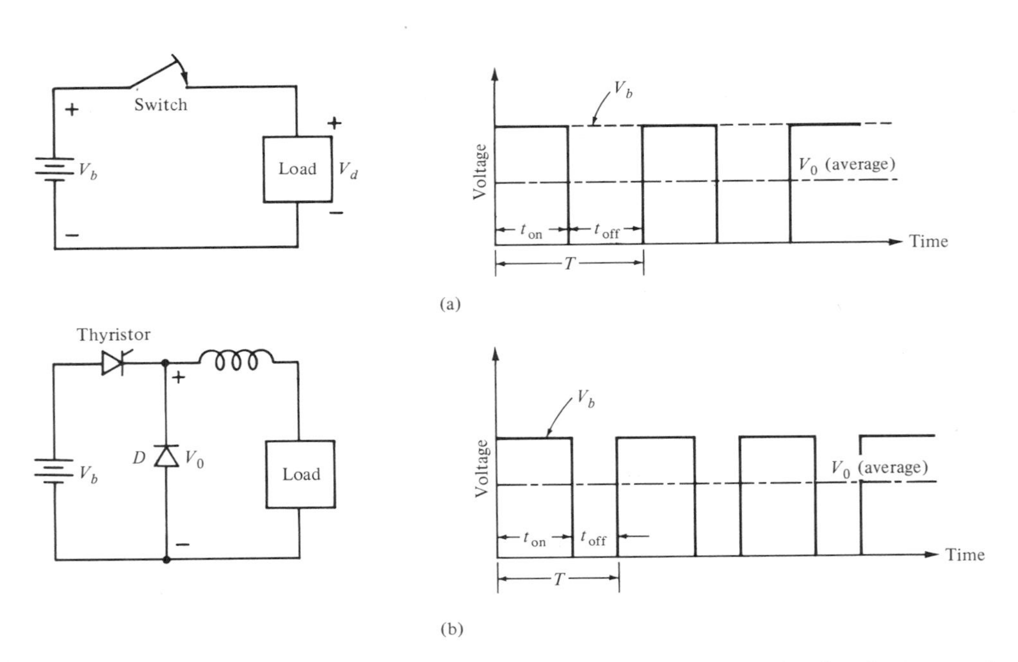

FIGURE 12-12. (a) Symbolic representation and (b) basic circuit of a chopper and output waveforms.

where the various times are shown in the figure, T is known as the chopping period, and $\alpha = t_{on}/T$ is called the duty cycle. Thus the voltage across the load varies with the duty cycle.

There are three ways in which the chopper output voltage can be varied, and these are illustrated in Fig. 12-13. In the first method, the chopping frequency is kept constant and the pulse width (or on-time, t_{on}) is varied, and the method is known as pulse-width modulation. The second method, called frequency modulation, has either t_{on} or t_{off} fixed, and a variable chopping period, as indicated in Fig. 12-13(b). The preceding two methods can be combined to obtain pulse-width and frequency modulation shown in Fig. 12-13(c), which is used in current limit control. In a method involving frequency modulation, the frequency must not be decreased to a value that may cause a pulsating effect or a discontinuous armature current, and the frequency should not be increased to such a high value

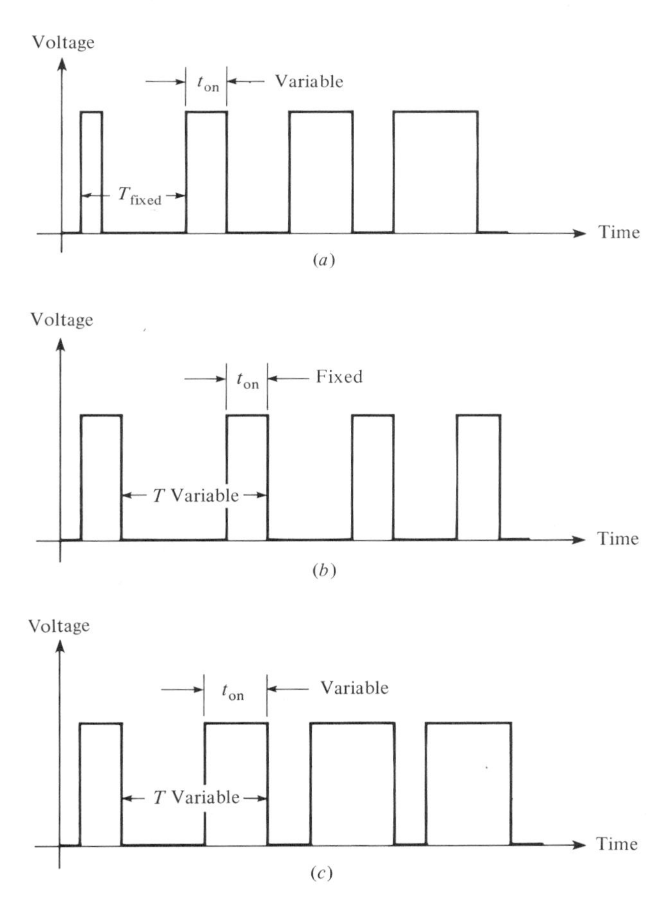

FIGURE 12-13. (a) Constant frequency, variable pulse-width; (b) variable frequency, constant pulse-width; (c) variable frequency, variable pulse-width.

as to result in excessive switching losses. The switching frequency of most chopper applications range from 50 to 500 pulses per second. The drawback of high-frequency chopper is that the current interruption generates a high-frequency noise.

We mentioned earlier that when current begins to flow through a thyristor it remains in the conductive state until turned off by external means. In a chopper we may have load commutation, in which case the load current flowing through the thyristor goes to zero and thereby turns off the thyristor. The other method of commutation is forced commutation, whereby the thyristor is turned off by forcing the current to zero. This is achieved in a dc system by an arrangement of energy storage elements (capacitors and inductors) and by additional switch-

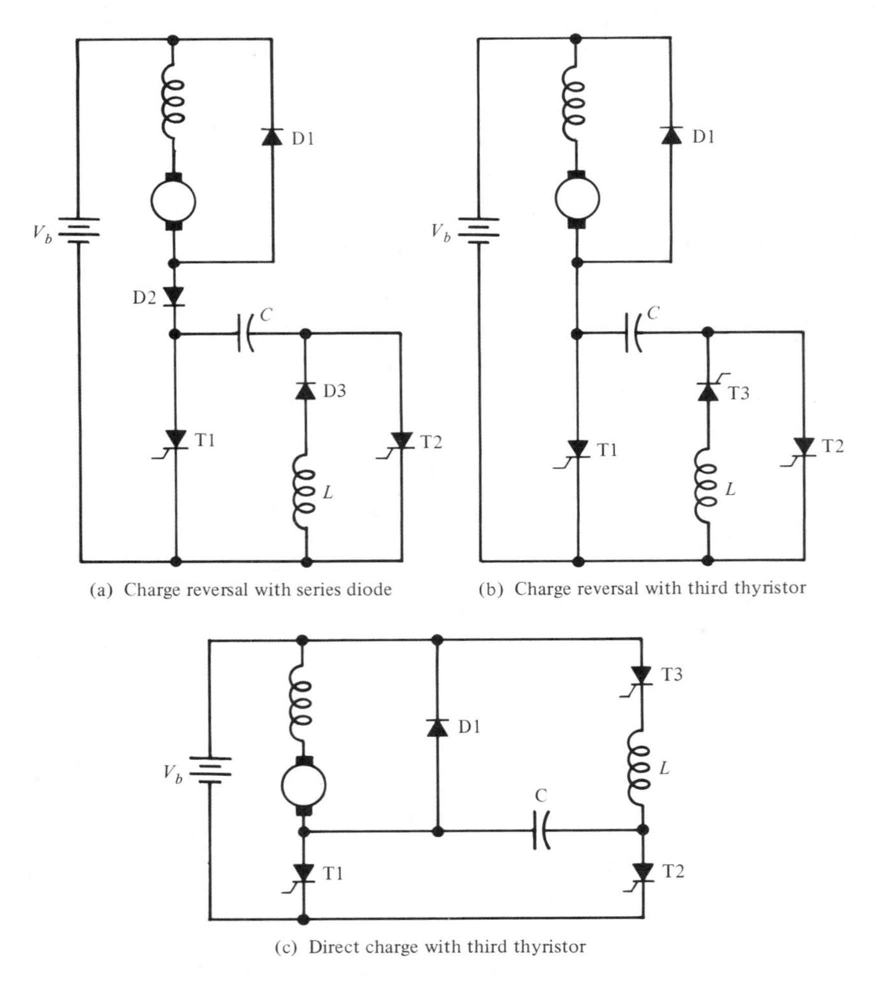

FIGURE 12-14. Commutation circuits for dc motors.

ing devices, usually thyristors. In systems energized from ac sources, forced commutation is brought about by means of the cyclic potential reversal of the power source. Several commutation circuits for a dc motor are shown in Fig. 12-14. In these circuits commutation is forced by reversing the voltage across the thyristor T_1 and diverting the load current. This is achieved by storing energy in the capacitor C during the ''on'' period and discharging C across the thyristor in the reverse polarity to turn it off.

In order to model a chopper-controlled separately excited motor, we consider a simplified circuit, and the corresponding voltage and current waveforms, as given in Fig. 12-15. Observe that when the thyristor turns off, the applied voltage v_m drops from V_1 to zero. However, armature current continues to flow through the path completed by the freewheeling diode until all the energy stored in L has been dissipated in R. Then v_m becomes equal to the motor back emf and stays at that value until the thyristor is turned on, whereupon it regains the value V_1.

If the speed pulsations are small, the motor back emf may be approximated by its average value, $k\Omega_m$, yielding

$$L\frac{di}{dt} + Ri + k\Omega_m = v_m = \begin{cases} V_1 & 0 < t < \alpha\lambda \\ 0 & \alpha\lambda < t < \gamma\lambda \\ k\Omega_m & \gamma\lambda < t < \lambda \end{cases} \tag{12-42}$$

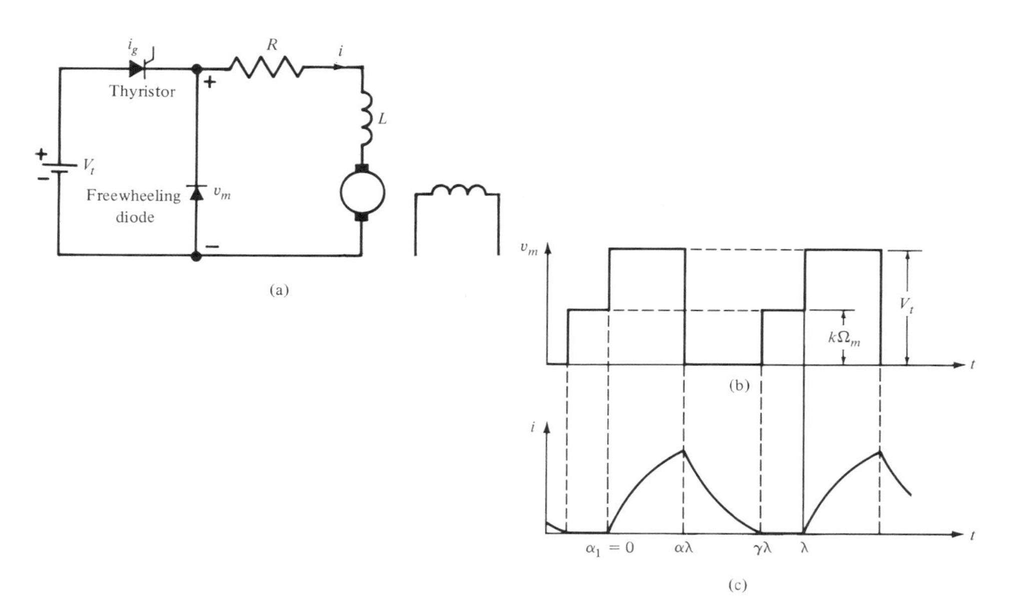

FIGURE 12-15. (a) A simplified circuit of a chopper-driven motor; (b) voltage waveform; (c) current waveform.

as the electrical equation of the system. Here λ is the period of the thyristor signal, α is the fraction of the period over which the thyristor is conductive (the duty cycle), and γ is the fraction of the period over which armature current flows.

Equation (12-42) has the solution, subject to the initial condition $i(0) = 0$,

$$i = \begin{cases} \dfrac{V_t - k\Omega_m}{R}(1 - e^{-t/\tau}) & 0 < t < \alpha\lambda \\ \dfrac{k\Omega_m}{R}[e^{(\gamma\lambda - t)/\tau} - 1] & \alpha\lambda < t < \gamma\lambda \\ 0 & \gamma\lambda < t < \lambda \end{cases} \tag{12-43}$$

where $\tau \equiv L/R$ is the armature time constant. Since $i = 0$ at $t = \alpha\lambda$ yields the following equation for γ:

$$\gamma = \frac{\tau}{\lambda} \ln \left(1 + \frac{e^{\alpha\lambda/\tau} - 1}{\Omega^*}\right) \tag{12-44}$$

in which $\Omega^* \equiv k\Omega_m/V_t$ is the normalized (dimensionless) average rotational speed of the motor. Now, it is apparent that when α is sufficiently large and Ω^* sufficiently small, (12-44) gives $\gamma > 1$, which is impossible. Thus we must distinguish between two modes of operation of the machine.

Mode I is defined by all (α, Ω^*) combinations satisfying

$$1 > \frac{\tau}{\lambda} \ln \left(1 + \frac{e^{\alpha\lambda/\tau} - 1}{\Omega^*}\right)$$

In this mode, γ is given by (12-44) and the armature current, (12-43), vanishes over a fraction $1 - \gamma$ of the basic cycle.

Mode II is defined by all (α, Ω^*) combinations satisfying

$$1 \leq \ln \left(1 + \frac{e^{\alpha\lambda/\tau} - 1}{\Omega^*}\right)$$

If the equality holds, (12-43) is valid with $\gamma = 1$; that is, the armature current becomes zero only at the period points. If the inequality holds, (12-43) is no longer valid. The governing differential equation, (12-42), and the boundary conditions must be changed to admit a strictly positive solution, for which again $\gamma = 1$.

The average torque–average speed characteristic of the motor can now be derived. Integration of (12-42) over one period of the thyristor signal gives

$$RI_{\text{ave}} + k\Omega_m = \alpha V_t + k\Omega_m(1 - \gamma) \tag{12-45}$$

On the other hand, the torque equation of the motor,

$$J\dot{\omega}_m + b\omega_m + t_0 = ki$$

where t_0 is the load torque, b is the rotational friction coefficient, and J is the moment of inertia, integrates to give

$$b\Omega_m + t_{0\text{ave}} = kI_{\text{ave}} \tag{12-46}$$

Eliminating I_{ave} between (12-45) and (12-46), we obtain the desired relation between $T^* \equiv T_{0\text{ave}}/(kV_t/R)$, the normalized (dimensionless) average torque, and $\Omega^* \equiv k\Omega_m/V_t$:

$$T^* = \alpha - \left(\frac{bR}{k^2} + \gamma\right)\Omega^*$$

that is,

$$T^* = \begin{cases} \alpha - \left[\dfrac{bR}{k^2} + \dfrac{\tau}{\lambda}\ln\left(1 + \dfrac{e^{\alpha\lambda/\tau} - 1}{\Omega^*}\right)\right] & \text{in mode I} \\ \alpha - \left(\dfrac{bR}{k^2} + 1\right)\Omega^* & \text{in mode II} \end{cases} \tag{12-47}$$

Figure 12-16 shows the torque–speed curves for several values of α. Observe the linearity of the curves in the region corresponding to mode II, which is separated from the mode I region by the dashed curve.

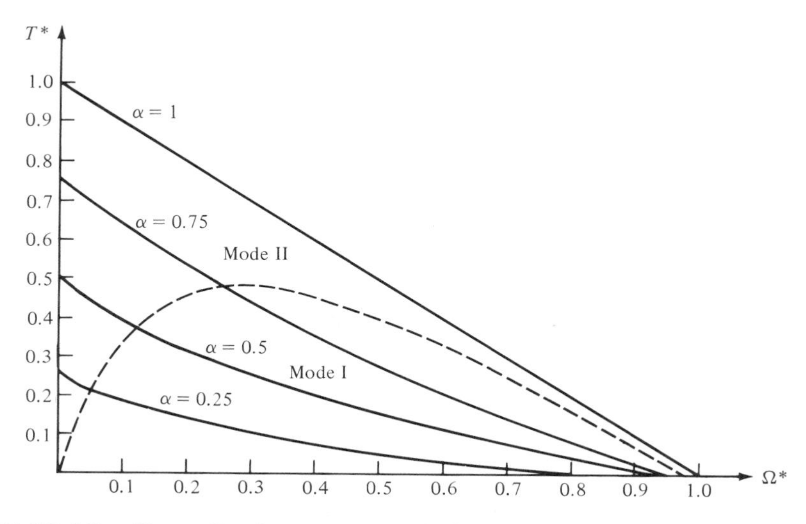

FIGURE 12-16. Normalized torque–speed characteristics.

The analysis of a chopper-driven series motor (Fig. 12-17) is slightly more involved than the analysis of the shunt motor presented earlier, because the torque of a series motor varies nonlinearly with the armature current, and magnetic saturation must be taken into account. In such a case the computations are too tedious unless a digital computer is used. The procedure is as follows:

1. Store the saturation curve φ versus I_{ave} in the computer. It suffices to use the three-segment approximation shown in Fig. 12-18.
2. Integration of the circuit equation over a period λ gives the average motor speed as

$$\Omega_m = \frac{\alpha V_t - I_{ave}R}{k_a} \tag{12-48}$$

   Setting $\Omega_m = 10$ rad/s, iteratively choose $[I_{ave}, \varphi(I_{ave})]$ pairs from step 1 until (12-48) is satisfied; the final I_{ave} is the average armature current for the speed 10 rad/s.
3. Repeat step 2 for a set of evenly space values of Ω_m and thus obtain $I_{ave}(\Omega_m)$.
4. For each Ω_m, compute the corresponding developed power from

$$P_d = \alpha V_t I_{ave}\Omega_m - R[I_{ave}\Omega_m]^2 \tag{12-49}$$

   thus obtaining the power–speed characteristic.

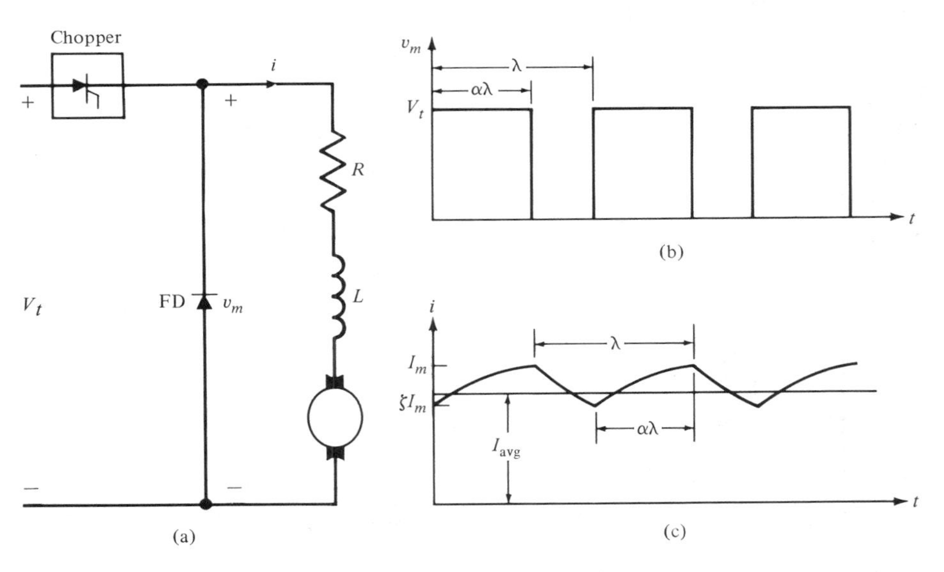

FIGURE 12-17. (a) A chopper-driven series motor; (b) motor voltage and duty cycle; (c) motor current.

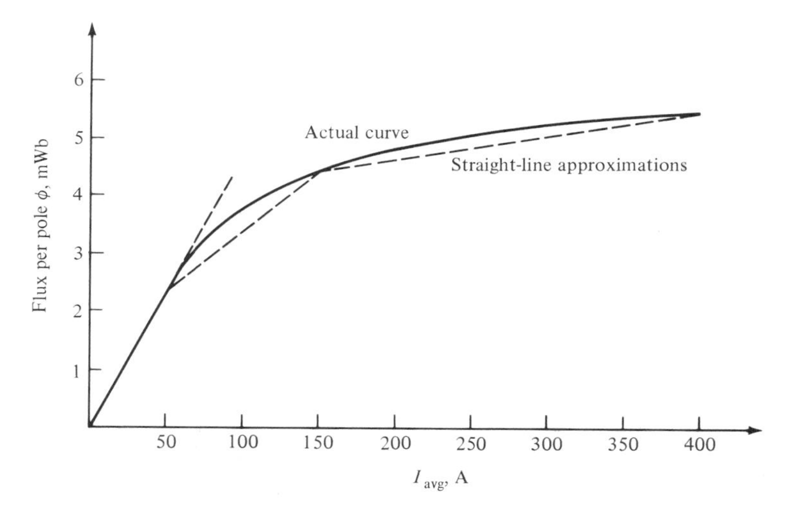

FIGURE 12-18. Straight-line approximation of the saturation curve of a series motor.

5. For each Ω_m, compute, from step 4,

$$T_{ave} = \frac{P_d}{\Omega_m} \tag{12-50}$$

generating the torque–speed characteristic.

12-6 CONTROL OF AC MOTORS

We recall from Chapters 6 and 7 that the two major categories of ac motors are induction motors and synchronous motors. The synchronous speed, n_s, of an ac motor is given by

$$n_s = \frac{120f}{p} \tag{12-51}$$

where f is the supply frequency (Hz) and p is the number of poles. The most efficient method of varying the speed of an ac motor is by varying the supply frequency. If the available source of power is dc, the variable frequency is obtained by means of an inverter. On the other hand, if the source is ac, it may be converted into dc and then inverted, or the variable frequency may be obtained using a cycloconverter. These methods of speed control of ac motors are illustrated in terms of functional block diagrams in Fig. 12-19(a) and (b). We

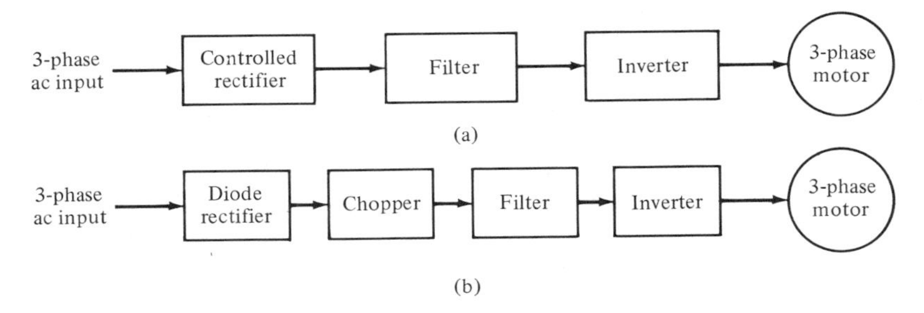

FIGURE 12-19. Two schematics of ac motor speed control by frequency variation.

will discuss the components of these block diagrams later in this chapter. With a variable-frequency controller, the torque–speed characteristics of an induction motor for several frequencies are shown in Fig. 12-20(a). The dashed-line envelope defines two distinct operating regions of constant maximum torque and constant output power. In the constant maximum torque region, the ratio of applied motor voltage to supply frequency (V/Hz) is held constant by increasing motor voltage directly with frequency, as shown in Fig. 12-20(b). The volts/hertz ratio defines the air-gap magnetic flux in the motor and can be held constant for

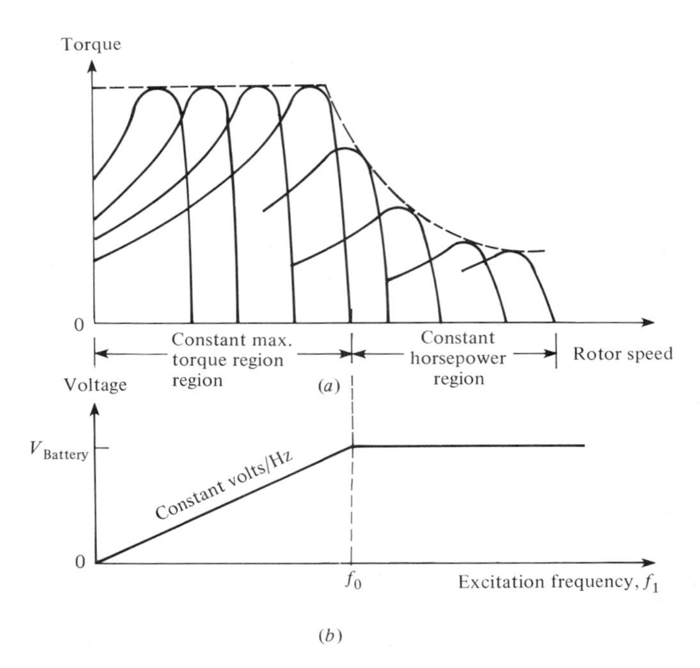

FIGURE 12-20. (a) Induction motor speed–torque characteristics; (b) voltage characteristics for variable frequency operation.

constant torque within the excitation of frequency range and corresponding rotor speeds extending to frequency f_0. Beyond frequency f_0, the motor voltage cannot be increased to maintain a constant volts/hertz ratio due to the limitations of a finite source voltage. For motor speeds in the range where f_r is greater than f_0, the supply voltage is held constant at its maximum level, and the supply frequency is increased to provide the motor speed demanded. This is the constant maximum horsepower region of operation, since the maximum developed torque decreases nonlinearly with speed.

The envelope illustrated in Fig. 12-20(a) characterizes induction motor operation in a typical electric traction system. Any speed–torque combination within the envelope can be provided by appropriate voltage and frequency control. For heavy loads, such as vehicle acceleration, the motor is operated in the constant-torque region to provide the torque demanded. For high-speed operation at vehicle cruising speeds, the motor operates in the constant-horsepower region at a frequency to satisfy the demanded speed. Voltage control is usually accomplished in the constant-torque region by pulse-width (duty-cycle) modulation. The motor operates with a fixed-voltage, variable-frequency square wave in the constant-horsepower region. Voltage and frequency control are used in both driving and braking operating modes of the vehicle. Regenerative braking is accomplished by reducing the supply frequency below the rotor frequency, that is, f_0 less than f_r. Under these conditions, the motor acts as a generator, developing a negative braking torque. Both the degree of braking and the level of power returned to the battery are controlled by voltage and frequency control.

We now turn our attention to some of the solid-state controllers for ac motors.

Inverters

There are a wide variety of inverter circuits which may be used to control the speed of an ac motor. For the present, however, we consider the full-bridge inverter circuit shown in Fig. 12-21, which also shows the voltage and current waveforms. When T1 and T3 are conducting, the battery voltage appears across the load with the polarities shown in Fig. 12-21(b). But when T2 and T4 are conducting, the polarities across the load are reversed. Thus we get a square-wave voltage across the load, and the frequency of this wave can be varied by varying the frequency of gating signals. If the load is not purely resistive, the load current will not reverse instantaneously with the voltage. The antiparallel connected diodes shown in Fig. 12-21(a) allow the load current to flow after voltage reversal.

The principle of the bridge inverter mentioned above can be extended to form a three-phase bridge inverter of Fig. 12-22(a). The gating signals and the output voltages are shown in Fig. 12-22(b). The fundamental components of the line-to-line voltages will form a balanced three-phase system. The antiparallel connected diodes are used to allow flow of currents out of phase with the voltage.

Some commonly used inverters are discussed below.

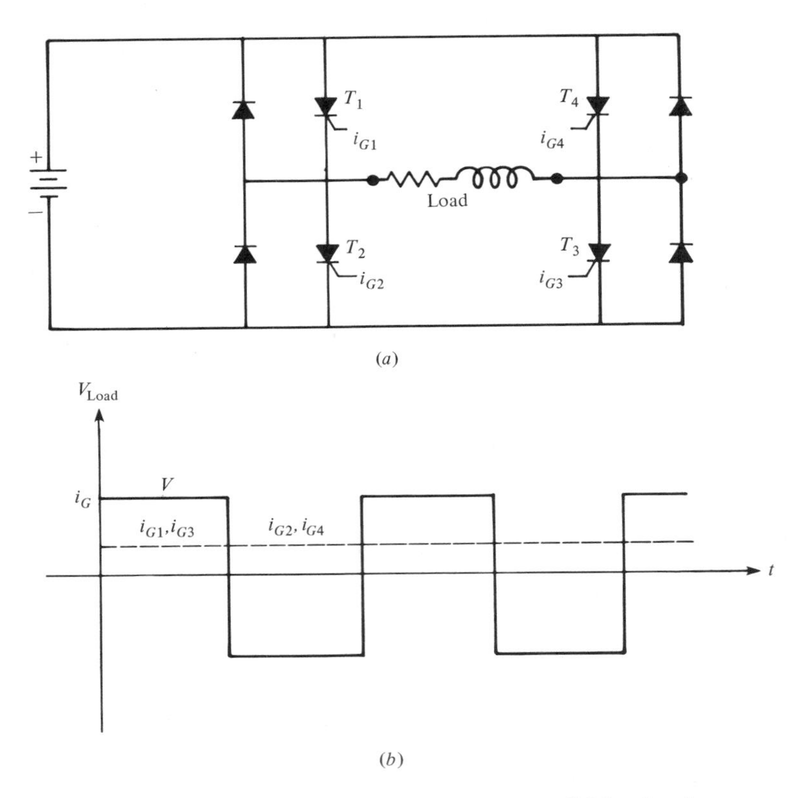

FIGURE 12-21. (a) Single-phase full bridge inverter; (b) load voltage waveform.

Adjustable Voltage Inverter. In an adjustable voltage inverter (AVI), the output voltage and frequency both can be varied. The voltage is controlled by including a chopper between the battery and the inverter, whereas the frequency is varied by the frequency of operation of the gating signals. In an AVI, the amplitude of the output decreases with the output frequency and the ratio V/f remains essentially constant over the entire operating range. The AVI output waveform does not contain 2nd, 3rd, 4th, 6th, 8th, and 10th harmonics. Other harmonic contents as a fraction of the total rms output voltage are:

$$\text{fundamental} = 0.965$$

$$\text{5th harmonic} = 0.1944$$

$$\text{7th harmonic} = 0.138$$

$$\text{11th harmonic} = 0.087$$

The losses produced by these harmonics tend to heat the motor, and its efficiency

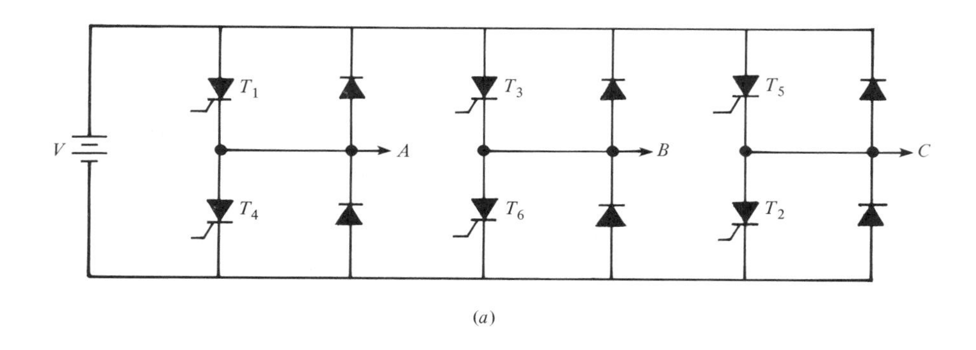

(a)

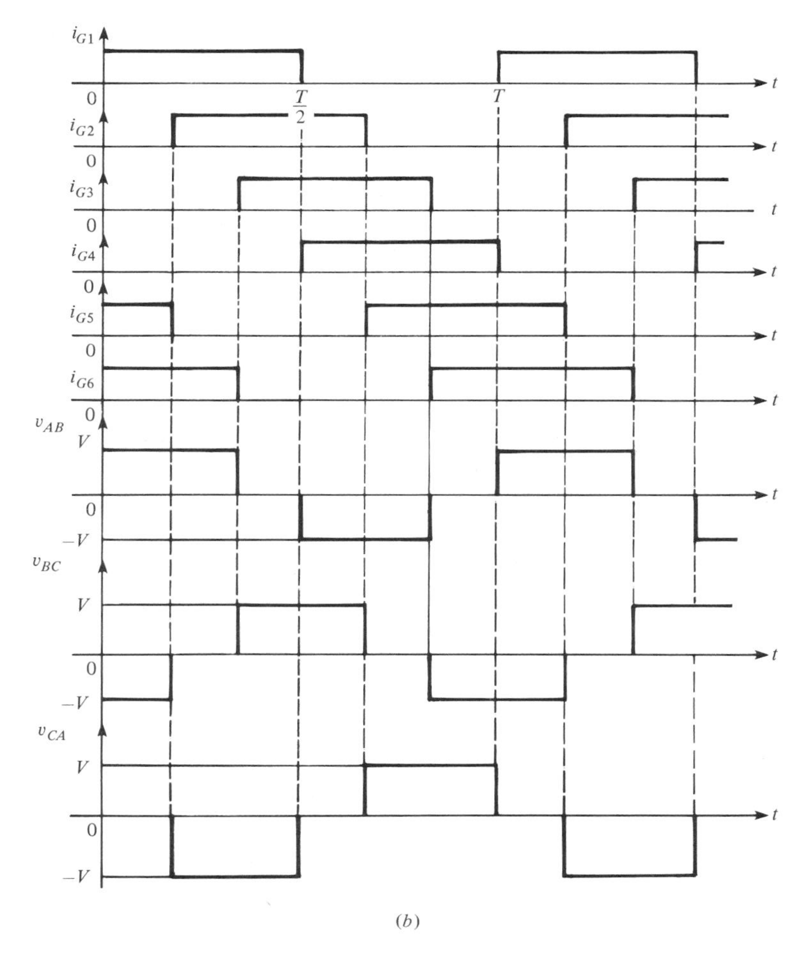

(b)

FIGURE 12-22. (a) Three-phase bridge inverter; (b) gating current and output voltage waveforms.

decreases by about 5 percent compared to a motor driven by a purely sinusoidal voltage.

Pulse-Width-Modulated and Pulse-Frequency-Modulated Inverters. The voltage control in pulse-width-modulated (PWM) and pulse-frequency-modulated (PFM) inverters is obtained in a manner similar to that for a chopper shown earlier in Fig. 12-13. In a PWM inverter, the output voltage amplitude is fixed and equal to the battery voltage. The voltage is varied by varying the width of the pulse on-time relative to the fundamental half-cycle period, as illustrated in Fig. 12-23(a) and (b). Notice that the full-voltage output wave is similar to that of the AVI output shown in Fig. 12-22(b). However, as the voltage is reduced [Fig. 12-23(b)] the harmonics vary rapidly in magnitude. Figure 12-24 shows the harmonic content of the output voltage with single-pulse modulation.

Low-frequency harmonics can be reduced by pulse frequency modulation (PFM), in which the number and width of pulses within the half-cycle period is varied. PFM waveforms for three different cases are shown in Fig. 12-25. Harmonics from the output of a PFM inverter can be reduced by increasing the number of pulses per half-cycle. But this requires a reduction in the pulse width, which is limited by the thyristor turn-off time and the switching losses of the thyristor. Furthermore, an increase in the number of pulses increases the complexity of the logic system and thereby increases the overall cost of the PFM inverter system.

In comparing an AVI system with a PWM (or PFM) inverter system, we observe that whereas the AVI system is efficient but expensive, the PWM inverter is relatively inexpensive but inefficient. Both types of inverter system are suitable for induction motors. However, for the control of a synchronous motor,

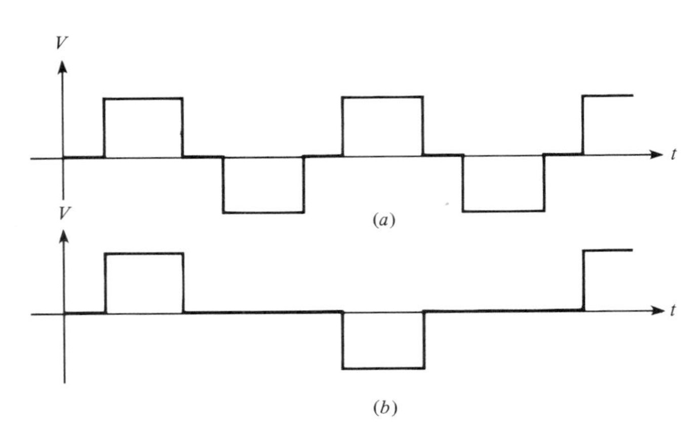

FIGURE 12-23. (a) Full voltage output and (b) half-full voltage output of a single-pulse PWM inverter.

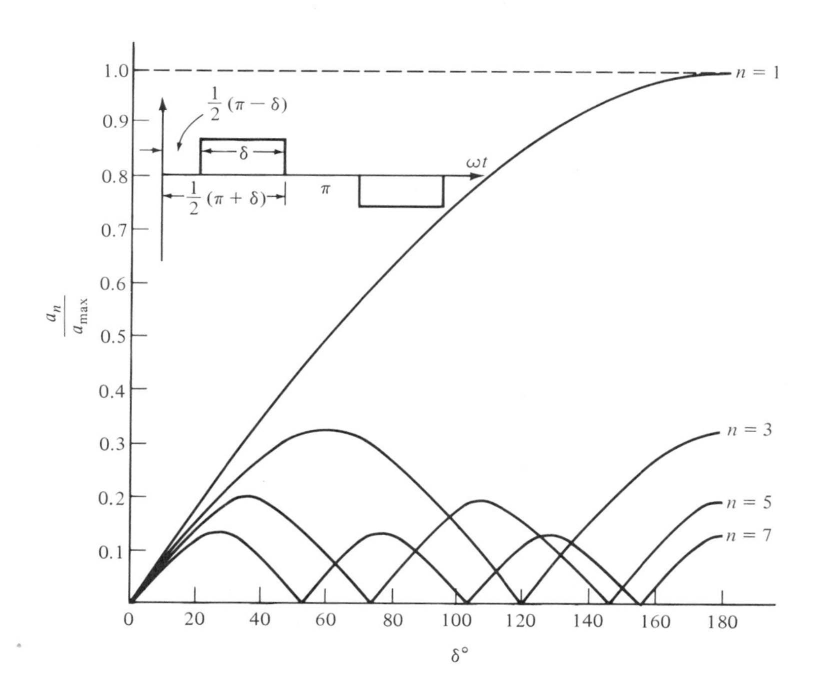

FIGURE 12-24. Harmonic content of a single pulse.

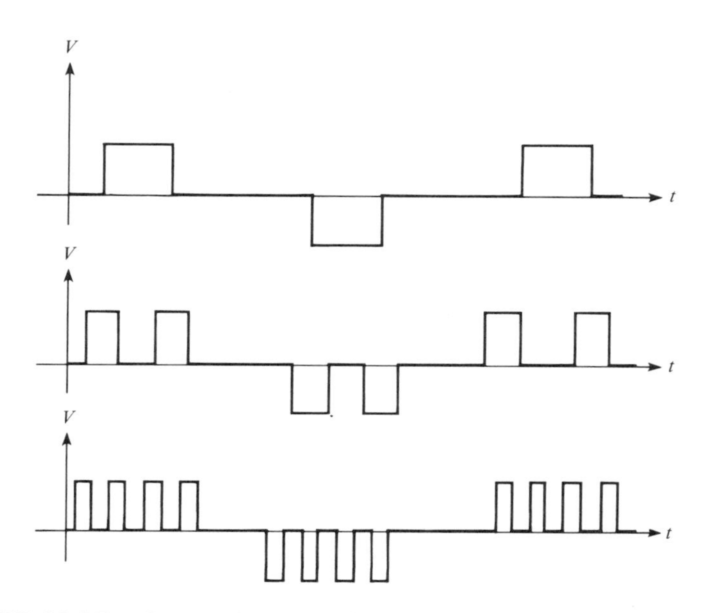

FIGURE 12-25. Output from a PFM inverter.

a thyristor inverter requires a motor voltage greater than the dc link voltage in order to turn off the thyristors. Also, no diodes are required in the ac portion of the circuit, thus avoiding uncontrolled currents. If a synchronous motor is controlled by a transistor inverter, the motor internal voltage must be less than the battery voltage; otherwise, the diodes in the inverter circuit will conduct, resulting in uncontrollable currents and high losses.

Note on Motor Performance (Fed from Inverters)

The inverter output voltage and current waveforms are rich in harmonics. These harmonics have detrimental effects on motor performance. Among the most important effects are the production of additional losses and harmonic torques. The additional losses that may occur in a cage induction motor owing to the harmonics in the input current are summarized below.

1. *Primary I^2R losses*. The harmonic currents contribute to the total rms input current. Skin effect in the primary conductors may be neglected in small wire-wound machines, but it should be taken into account in motor analysis when the primary-conductor depth is appreciable.
2. *Secondary I^2R losses*. When calculating the additional secondary I^2R losses, the skin effect must be taken into account for all sizes of motor.
3. *Core losses due to harmonic main fluxes*. These core losses occur at high frequencies, but the fluxes are highly damped by induced secondary currents.
4. *Losses due to skew-leakage fluxes*. These losses occur if there is relative skew between the primary and secondary conductors. At 60 Hz the loss is usually small, but it may be appreciable at harmonic frequencies. Since the time-harmonic mmf's rotate relative to both primary and secondary, skew-leakage losses are produced in both members.
5. *Losses due to end-leakage fluxes*. As in the case of skew-leakage losses, these losses occur in the end regions of both the primary and secondary and are a function of harmonic frequency.
6. *Space-harmonic mmf losses excited by time-harmonic currents*. These correspond to those losses which, in the case of the fundamental current component, are termed high-frequency stray-load losses.

In addition to these losses and harmonic torques, the harmonics act as sources of magnetic noise in the motor.

EXAMPLE 12-6

A single-phase half-bridge inverter is shown in Fig. 12-26(a). The thyristor triggering currents are shown in Fig. 12-26(b). Sketch the load voltage and current waveforms. Solve for the steady-state load current and determine the maximum value of the current.

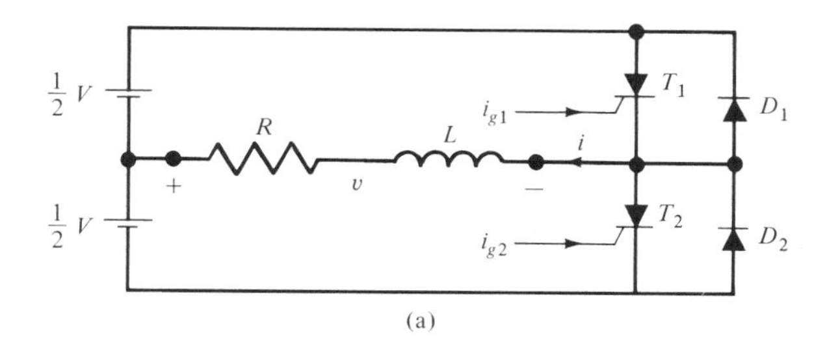

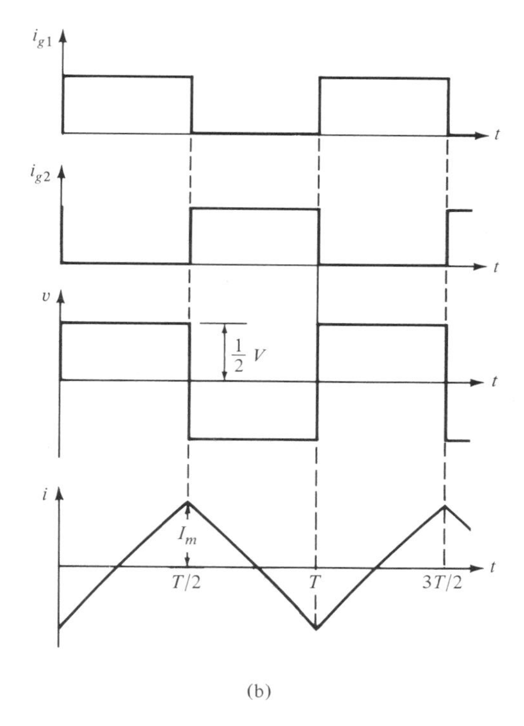

FIGURE 12-26. (a) Single-phase, half-bridge inverter; (b) voltage and current waveforms.

The circuit equation is

$$L\frac{di}{dt} + Ri = v = \begin{cases} \dfrac{V}{2} & 0 < t < T/2 \\ -\dfrac{V}{2} & T/2 < t < T \end{cases} \tag{12-52}$$

Expressing v in a Fourier series, we have

$$v = \frac{2V}{\pi} \sum_{n\text{odd}} \frac{1}{n} \sin n\omega t \qquad (12\text{-}53)$$

We solve (12-52) by superposition, to obtain the steady-state current:

$$i = \frac{2V}{\pi} \sum_{n\text{odd}} \frac{1}{nZ_n} \sin (n\omega t - \theta_n) \qquad (12\text{-}54)$$

where

$$Z_n = \sqrt{R^2 + (n\omega L)^2} \qquad \tan \theta_n = \frac{n\omega L}{R}$$

At the instants of commutation ($t = 0, T/2, T, 3T/2, 2T, \ldots$),

$$i = \pm I_m \equiv \pm \frac{2V}{\pi\omega L} \sum_{n\text{odd}} \frac{1}{n^2 + (R/\omega L)^2} = \pm \frac{V}{2R} \tanh \frac{R\pi}{2\omega L} \qquad (12\text{-}55)$$

When the plus sign holds in (12-55), forced commutation is required; it may not be necessary when the minus sign holds.

Voltage and current waveforms are shown in Fig. 12-26(b). ■

Cycloconverters

The cycloconverter is a control device used on variable-speed motors supplied by an ac power source. It is a means of converting a source at fixed (peak) voltage and fixed frequency to an output with variable voltage and variable frequency. The source frequency must be at least three to four times the maximum frequency of the output. A single-phase bridge cycloconverter is shown in Fig. 12-27(a), and the various waveforms are shown in Fig. 12-27(b), which also indicates the firing sequence of the thyristors. In Fig. 12-27(b), α_p denotes the minimum delay time for the positive group of converters, and α_n denotes the maximum delay time for the negative group. The variation of the delay controls the output voltage, in that it determines how many half-cycles of line voltage go to make up one half-cycle of the load voltage fundamental.

The cycloconverter has the advantage that it does not require a dc link, and consequently has high efficiency. Further advantages of a cycloconverter are the possibility of voltage control within the converter, and the fact that it has line commutation.

Major applications of cycloconverters are for low-speed motors rated in the range 300 to 20,000 hp (or 15,000 kW). Such a motor-drive system is capable of providing low-voltage starting, reversing, and regenerative braking.

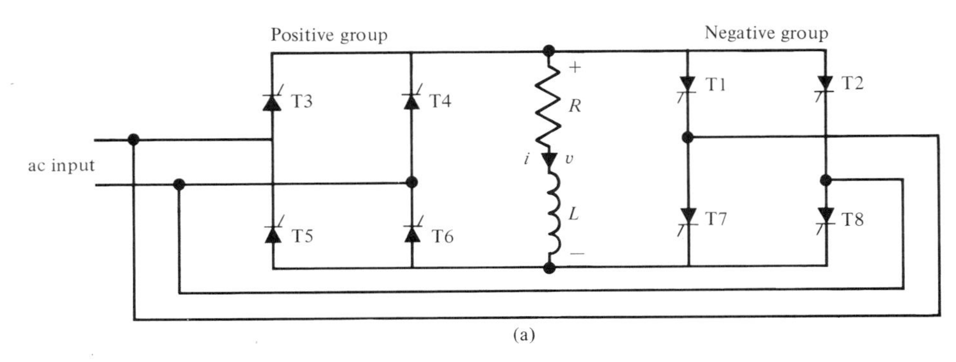

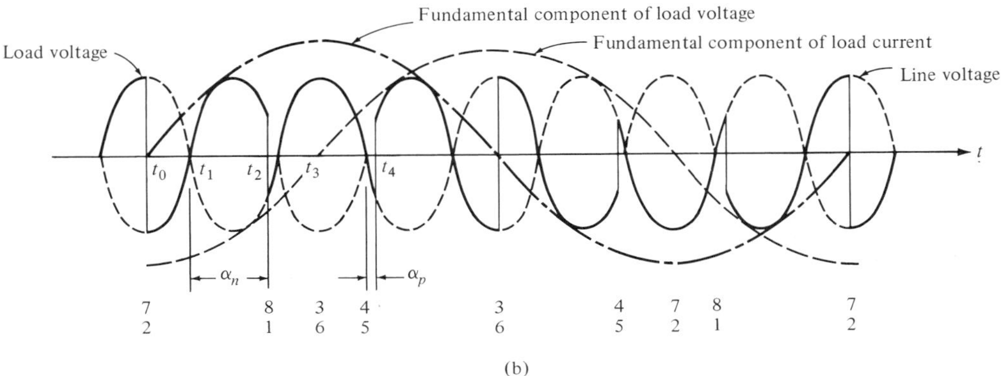

FIGURE 12-27. (a) A single-phase cycloconverter; (b) voltage and current waveforms.

12-7

CERTAIN APPLICATIONS TO POWER SYSTEMS

We know that contactors, disconnect switches, and circuit breakers are used in electric power systems to open and close circuits as desired. In power system protection, during abnormal conditions, signals are sent by relays to operate appropriate circuit breakers. Contactors and relays may be either electromechanical or solid state. The choice of the relay for a particular application is based on the overall system design. For instance, if a low contact resistance is required of the relay, the junction voltage of the semiconductor prohibits its use in the output. In such a case an electromechanical relay (EMR) is preferred. On the other hand, if a precise control of turn-on and turn-off times is required, a solid-state relay (SSR) is preferred over an electromechanical one.

Compared to an EMR, an SSR draws no arc when opened and thus presents no radio-frequency interference problems. Also, SSRs, unlike EMRs, have no contact bounce. In an SSR, surge current can be minimized by appropriate circuitry. But in an EMR, the closure of time precludes this type of a control of surges. Some of the other advantages of SSRs are quick response, resistance to mechanical shocks, low power requirement for operation, long life, and acceptance of a wide range of input voltages, including both ac and dc by any one relay. Some of the disadvantages of SSRs are that they are expensive, have high design and temperature sensitiveness, are complex in operation, have a high junction voltage drop, and have runaway possibilities due to thermal turn-ons. When a multicontact array is required, SSRs are inferior to EMRs. SSRs are less versatile than EMRs and can handle a single ac (or dc) output.

From the preceding remarks it is clear that SSRs and EMRs are not direct replacements for each other and that each has its own peculiar set of characteristics. The design advances common to most types of SSRs include dynamic performance, very low burden, flexible settings, easy testing, improved operating indicator, panel space reduction, and seismic capability.

Before we present an example of an SSR application, let us consider the principle of operation of solid-state circuit breakers and relays.

DC and AC Circuit Breakers

A flip-flop circuit, such as the one shown in Fig. 12-28, may be used as a solid-state dc circuit breaker. When the thyristor T1 is conducting, the circuit breaker is closed. The full dc voltage appears across the load and across the capacitor C. To open the circuit breaker, thyristor T2 is turned on. Thus C will discharge through T2 and T1. In doing so, the current through T1 is reversed, and T1 is turned off. This will trip the circuit breaker. At the same time, C will be charged in the reverse direction, the current through T2 will fall below its holding current value, and T2 will be turned off. The response of the circuit breaker depends on the thyristor turn-off time. Hence solid-state dc circuit breakers are suitable for high-speed applications.

The principle of an ac circuit breaker can be followed by referring to Fig. 12-29. The thyristors T1 and T2 are turned on every half-cycle such that the circuit breaker is closed for normal operation. These thyristors are turned on by firing pulses every half-cycle. To trip the breaker, the firing pulses are stopped.

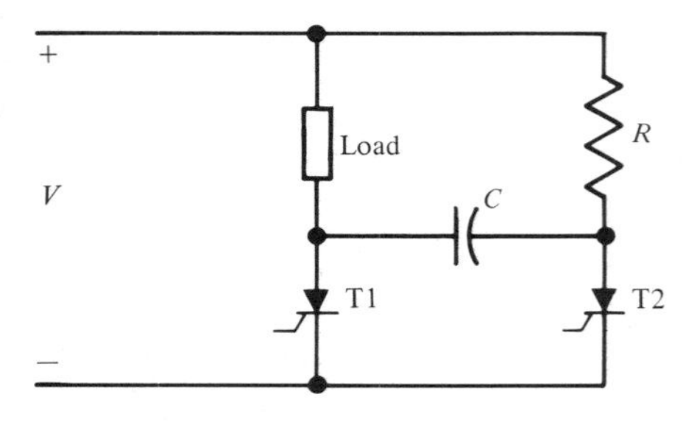

FIGURE 12-28. A solid-state dc circuit breaker.

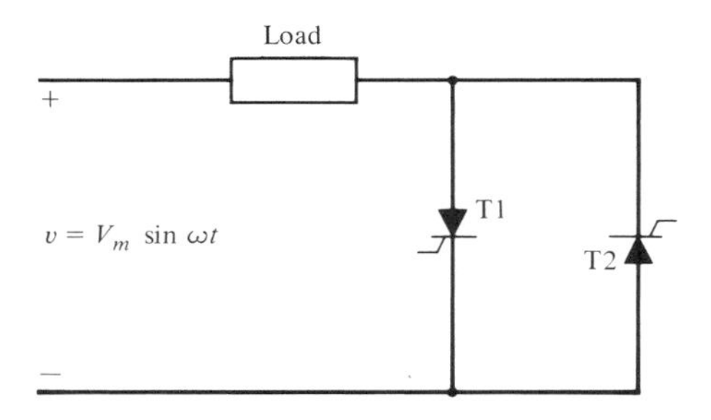

FIGURE 12-29. A solid-state ac circuit breaker.

The thyristor conducting at that instant undergoes natural commutation. Ac circuit breakers are also suitable for high-speed applications.

Relays and Contactors

As we already know, a solid-state relay (SSR) can function like an electromechanical relay (EMR); that is, as a consequence of a control signal from a control circuit, the relay can be made to control an event in the main circuit. An SSR is represented schematically in Fig. 12-30. This relay may be either normally open or normally closed. Notice from Fig. 12-30 that the signal from the control circuit turns the thyristor on (or off) and thereby controls the main circuit.

Solid-state contactors are similar to relays except for higher power ratings.

Static Relays in Power Systems

Numerous relay schemes for power system protection are given in Ref. 8, and Ref. 9 gives a good summary of the various types of solid-state relays which are commercially available. Interested readers should consult these references. Here we present an overview of static relays as applied to power systems. Note first that we have used the term "static" in place of "solid-state" because the former term is more commonly used and has a broader scope. According to the American National Standards Institute, a static relay is one "in which there is

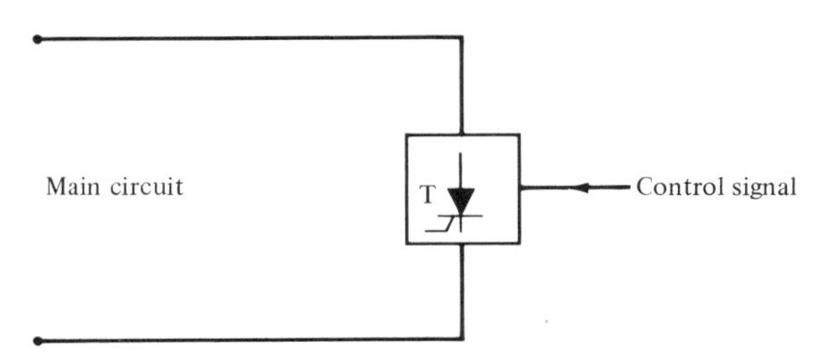

FIGURE 12-30. A solid-state relay.

TABLE 12-2 Standard Protective Functions for Induction Motors

Function	Protective Relay
Overload and locked rotor	Time overcurrent
Short-circuit protection	Instantaneous overcurrent
Ground fault	Ground sensor
	Residual overcurrent
	Negative sequence
Stator overtemperature	Temperature
Unbalanced current	Negative-sequence overcurrent
Abnormal voltage	Undervoltage
	Undervoltage and phase sequence
	Negative-sequence overvoltage
	Residual voltage
Differential	Self-balancing
	Percentage differential

Source: Ref. 9.

no armature or other moving element, the designed response being developed by electronic, solid-state, magnetic or other components without mechanical motion.'' Various types of static relays now commercially available include generator-differential, timing, phase-commutation, directional-comparison, mho-distance, overcurrent, and voltage relays. For induction motors, Table 12-2 shows the types of static relays now available to perform the various protective functions. More details are beyond the scope of this book, and Ref. 8 and 9 should be consulted for details.

High-Voltage DC Transmission

In Chapter 10 we reviewed briefly the topic of high-voltage dc transmission. Clearly, we need converters to link dc lines with ac systems. Reference 10 discusses several types of converters suitable for high-voltage dc systems. A high-voltage dc transmission line is represented schematically in Fig. 12-31. In such cases, 12-pulse thyristor converters are used to reduce harmonics. Harmonics are further reduced by the use of filters. Because of the high-voltage, high-current requirements of dc transmission lines, series–parallel combination of thyristors is used. A 12-pulse converter connection is shown in Fig. 12-32.

A major problem with the 12-pulse converter system is the simultaneous firing of all the thyristors. For short-distance, low-power dc transmission, the use of thyristors is expensive. Generation of harmonics is a further disadvantage of thyristor converters.

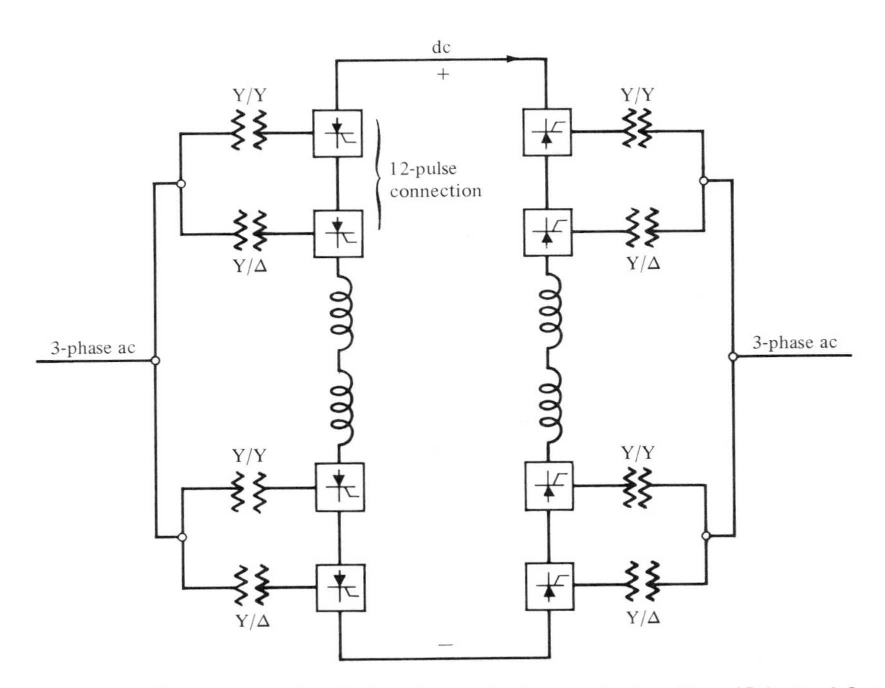

FIGURE 12-31. Schematic of a high-voltage dc transmission line. (Adapted from [3])

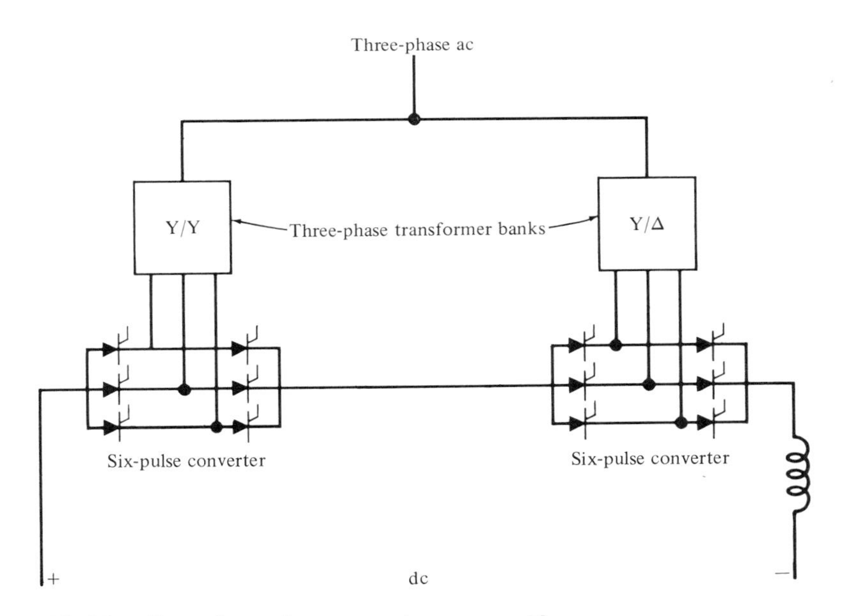

FIGURE 12-32. A twelve-pulse converter connection.

REFERENCES

1. **S. A. Nasar** and **L. E. Unnewehr,** *Electromechanics and Electric Machines,* 2nd ed., Wiley, New York, 1982, Chap. 8.
2. **S. B. Dewan** and **A. Straughen,** *Power Semiconductor Circuits,* Wiley-Interscience, New York, 1975.
3. **R. K. Sugandhi** and **K. K. Sugandhi,** *Thyristors,* Halsted Press, New York, 1981.
4. **R. S. Ramshaw,** *Power Electronics,* Chapman & Hall, London, 1973.
5. **S. A. Nasar,** *Electric Machines and Electromechanics,* Schaum's Outline Series, McGraw-Hill, New York, 1981, Chap. 8.
6. **L. E. Unnewehr** and **S. A. Nasar,** *Electric Vehicle Technology,* Wiley-Interscience, New York, 1982, Chaps. 4 and 5.
7. **P. C. Sen,** *Thyristor Drives,* Wiley-Interscience, New York, 1981.
8. **S. H. Horowitz,** ed., *Protective Relaying for Power Systems,* IEEE Press, New York, 1980.
9. **J. E. Waldron,** "Innovations in Solid-State Protective Relays," *IEEE Trans.,* vol. IA-14, no. 1, 1978, pp. 39–47.
10. **E. W. Kimbark,** *Direct Current Transmission,* Vol. 1, Wiley-Interscience, New York, 1971.

PROBLEMS

12-1 Find the average rms values of the waveforms shown in Fig. 12P-1.

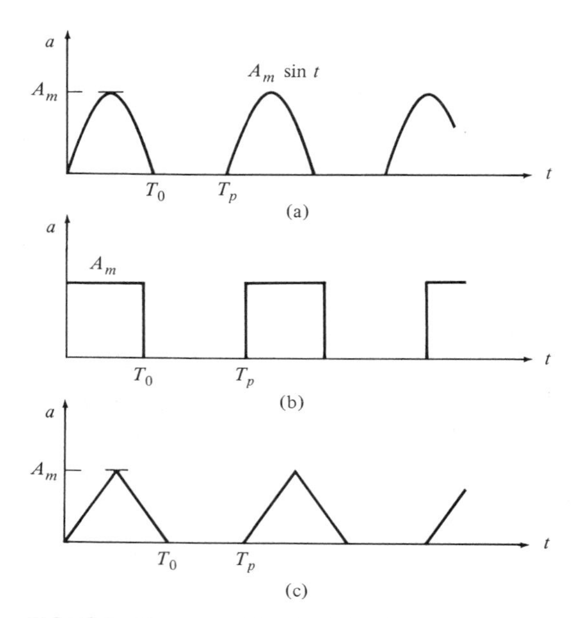

FIGURE 12-P1. PROBLEM 12-1.

12-2 Show that the solution given by (12-5) in Example 12-2 is correct.

12-3 Solve (12-13) for the three cases $\zeta > \omega_0$, $\zeta = \omega_0$, and $\zeta < \omega_0$.

12-4 Verify that (12-21) is a valid solution to (12-20) for the conditions stated in the text.

12-5 The half-wave rectifier shown in Fig. 12-5 may be used as a filtered power supply. The input is 220 V at 60 Hz. (a) Determine the value of C to limit the ripple in v_o (peak to peak) to less than 5%. (b) For this value of C, calculate the average value of the output voltage.

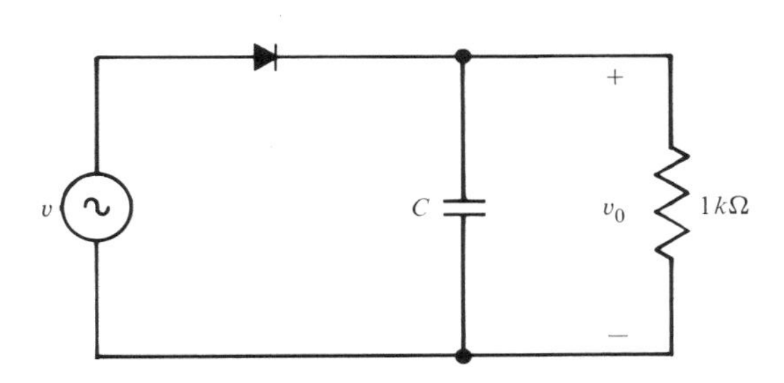

FIGURE 12-P5. PROBLEM 12-5.

12-6 A half-wave rectifier for battery charging may be modeled by the circuit shown in Fig. 12P-6. Given values are $v_s = 14 \sin 100t$, $V_0 = 10$ V, and $R = 0.2\ \Omega$. (a) Sketch v_o. (b) Find an expression for $i_o(t)$ for $0 < \omega t \leq T$. (c) Calculate the average and rms values of i_o. (d) Determine the power delivered to the battery.

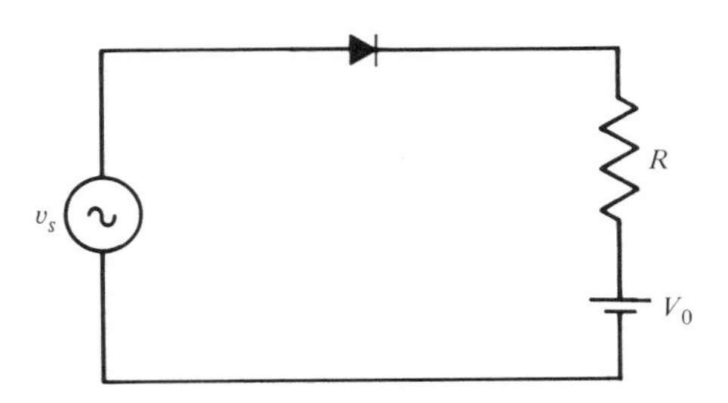

FIGURE 12-P6. PROBLEM 12-6.

12-7 For the circuit shown in Fig. 12P-7, the rms value of the input voltage is 10 V. (a) What is the minimum value of the firing angle to turn on the thyristor? (b) For a firing angle of 75°, determine the extinction angle.

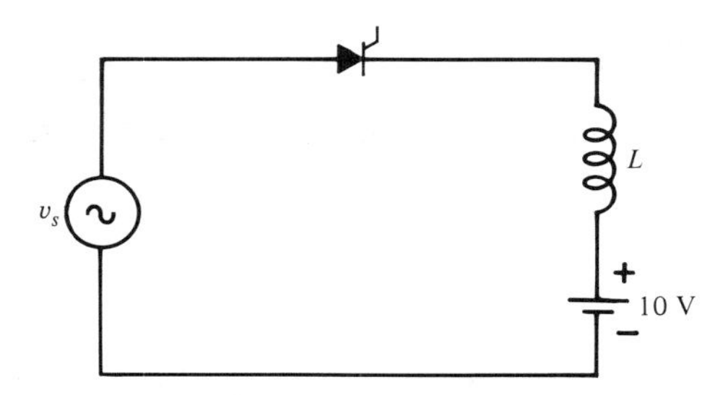

FIGURE 12-P7. PROBLEM 12-7.

12-8 Consider the circuit of Fig. 12-7, where $V = 48$ V, $L = 20$ mH, $C = 60$ μF, and $R = 50$ Ω. If the initial charge on the capacitor is zero, at what time will the thyristor turn off?

12-9 The input voltage to a chopper, supplying a 2-Ω resistive load, is 50 V. For a duty cycle of 75%, determine the average output voltage and current.

12-10 A chopper drives a separately excited dc motor having the following data:

armature resistance, $R = 0.2$ Ω

armature inductance, $L = 2$ mH

energy conversion constant, $k = 1.27$ V-s/rad

input voltage = 300 V

chopping period, $T = 0.05$ s

duty cycle, $\alpha = 0.6$

load/motor friction coefficient, $b = 0.11$ N·m-s/rad

load torque, $T_L = 893$ N·m

Determine (a) the speed at which conduction becomes marginally continuous and (b) the average motor current at the point of discontinuous conduction.

12-11 The following data pertain to a chopper-driven dc series motor: average input current to the motor = 160 A; armature-circuit resistance = 0.08 Ω; armature-circuit inductance = 15 mH; input voltage to the chopper = 300 V; duty cycle = 0.8; chopping period = 0.005 s; motor energy conversion constant =

0.017 V-s/A-rad; motor-load friction coefficient = 0.35 N·m-s/rad. Assuming continuous conduction, find (a) the average speed, (b) the rms value of the armature current, and (c) the mechanical output power.

12-12 For a dc series motor driven by a chopper, the following data are given: supply voltage = 440 V; duty cycle = 30%; armature circuit inductance = 0.04 H. Determine the chopper frequency if the maximum allowable change in the armature current is 8 A.

APPENDIX A

Unit Conversion

Symbol	Description	One: (SI Unit)	Is Equal to: (English Unit)	(CGS Unit)
B	Magnetic flux density	tesla (T)(= 1 Wb/m^2)	6.452×10^4 lines/in.^{2}	10^4 G
H	Magnetic field intensity	ampere per meter (A/m)	0.0254 A/in.	0.004π Oe
φ	Magnetic flux	weber (Wb)	10^8 lines	10^8 Mx
D	Viscous damping coefficient	newton-meter-second (N-m-s)	0.73756 lb-ft-sec	10^7 dyn-cm-sec
F	Force	newton (N)	0.2248 lb	10^5 dyn
J	Inertia	kg-square meter	23.73 lb-ft^2	10^7 g-cm^2
T	Torque	newton-meter (N-m)	0.73756 ft-lb	10^7 dyn-cm
W	Energy	joule (J)	1 W-sec	10^7 ergs

Conversion of Energy Units

To Convert From	To	Multiplied By
Quad	Btu	10^{15}
Btu	kwh	0.000293
Billion BBL of Oil	Quads	5.8
Billion Tons of Coal	Quads	22
Trillion Cubic Ft. Gas	Quads	1.031
Billion Metric Tons of Oil	Quads	42,514
Million Tons U_3O_8 (Breeder)	Quads	37,800 (approx)
Million Tons U_3O_8 (LWR)	Quads	300 (approx)
Billion Kwhe	Quads (input)	0.0114
Million BBL oil/day	Quads/yr	2.117
Million kw	Quads/yr	0.02989
Quads/yr (input)	Million kwe (capacity)	17 (approx)
1,000 MWe Power Plants	Tons Coal/wk	51,000 (approx)
Miles of Supply Train	Tons of Coal	10,000
Miles of Supply Train	BBL of Oil	50,000
BBL of Oil	Gallons	42
Kilojoules	Btu	0.9488
MTOE (Metric)	Quads	0.0425
Therm	Btu	100,000

APPENDIX B

Balanced Three-Phase Systems

Most of the commercial electrical power produced in this country is generated in *three-phase* systems. A three-phase system requires the generation of three sinusoidal voltages having the same magnitude and frequency, but separated from each other by 120° in phase. These voltages are then interconnected in one of the two types of connections shown in Fig. B-1. The result, for either type of connection, is a three-terminal source of power supplying a balanced set of three voltages to a load. The individual generator voltages are given by

$$\mathbf{V}_{aa'} = V_p \angle 0^\circ \quad \text{(B-1)}$$

$$\mathbf{V}_{bb'} = V_p \angle 120^\circ \quad \text{(B-2)}$$

$$\mathbf{V}_{cc'} = V_p \angle -120^\circ \quad \text{(B-3)}$$

Because the time axis is arbitrary, one of the phasors has been chosen to coincide with the horizontal axis of the phasor diagram, as shown in Fig. B-2, which shows all of the individual voltages. The symbol V_p stands for *phase voltage*, i.e., the rms value of the voltage of each of the phases.

In three-phase systems, it is usually the line voltages (rather than phase volt-

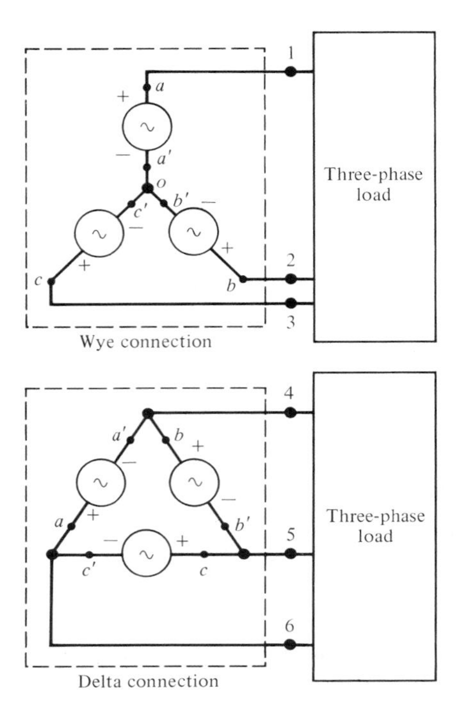

FIGURE B-1.

ages) that are of greatest interest. The line voltages $\mathbf{V}_{12}$, $\mathbf{V}_{23}$, and $\mathbf{V}_{31}$ can be determined in terms of the phase voltages, as will be demonstrated for $\mathbf{V}_{12}$.

$$\mathbf{V}_{12} = \mathbf{V}_{aa'} + \mathbf{V}_{b'b} = \mathbf{V}_{aa'} - \mathbf{V}_{bb'} = V_p \angle 0 - V_p \angle 120°$$

$$= V_p(1 + j0) - V_p\left(-0.5 + j\frac{\sqrt{3}}{2}\right) \tag{B-4}$$

$$V_p\left(\frac{3}{2} - j\frac{\sqrt{3}}{2}\right) = \sqrt{3}\, V_p \angle -30° = V_L \angle -30°$$

where V_L represents the rms value of the *line voltages*. The graphical equivalent of this computation is shown in Fig. B-2. Although we could solve for each of the line voltages in the manner above, symmetry demands that the other two voltages be given by

$$\mathbf{V}_{23} = V_L \angle 90° \tag{B-5}$$

$$\mathbf{V}_{31} = V_L \angle -150° \tag{B-6}$$

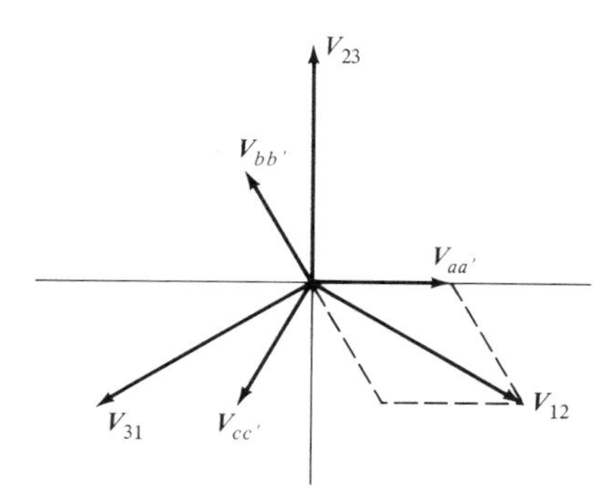

FIGURE B-2.

These are also shown on the phasor diagram of Fig. B-2. For the delta connection, the voltages between the terminals of the three-phase generator are evidently the same as the individual phase voltages, and so no special computations need be made.

When the load connected to a balanced, three-phase set of voltages is a balanced load, i.e., made up of identical impedances connected in delta or wye arrangement, then the currents which result form a balanced set of three-phase currents. Thus, suppose the currents in the wye connection are given by

$$\mathbf{I}_{a'a} = I_p \angle \theta_p \tag{B-7}$$

$$\mathbf{I}_{b'b} = I_p \angle \theta_p + 120^\circ \tag{B-8}$$

$$\mathbf{I}_{c'c} = I_p \angle \theta_p - 120^\circ \tag{B-9}$$

where I_p is the rms value of the current in each phase. The average power delivered by each phase is given by $V_p I_p \cos \theta_p$, so the total power delivered to the load is given (for this *balanced* system) by

$$P_T = 3[V_p I_p \cos \theta_p] \tag{B-10}$$

But for the wye connection, the phase current is the same as the current in the line, I_L, and we showed earlier that $V_L = \sqrt{3}V_p$, so we can express the total power in terms of line voltages and currents as follows.

$$P_T = 3\left(\frac{V_L}{\sqrt{3}}\right) I_L \cos \theta_p = \sqrt{3} V_L I_L \cos \theta_p \tag{B-11}$$

It should be noted that the angle θ_p is still the angle between the phase voltage and the phase current.

The expression for the total power in a delta-connected system is the very same. Whereas the line voltage equals the phase voltage for a delta-connected system, the line current is $\sqrt{3}$ times greater than the phase current, as we could easily show. Hence if (B-10) is converted to line quantities for the delta-connected system we obtain

$$P_T = 3V_L\left(\frac{I_L}{\sqrt{3}}\right)\cos\theta_p = \sqrt{3}V_LI_L\cos\theta_p \tag{B-12}$$

which is the same as Eq. (12-64)

In working with *balanced* three-phase systems, computations are usually made on a *per-phase* basis, and then the total results for the entire circuit are obtained on the basis of the symmetry and other factors which must apply. Thus, if a set of three identical wye-connected impedances is connected to a wye-connected, three-phase source (as shown in Fig. B-3), because of the symmetry, point 0′ turns out to have the same potential as point 0, and could therefore be connected to it. Thus each generator seems to supply only its own phase, and the computations on a per-phase basis are therefore legitimate.

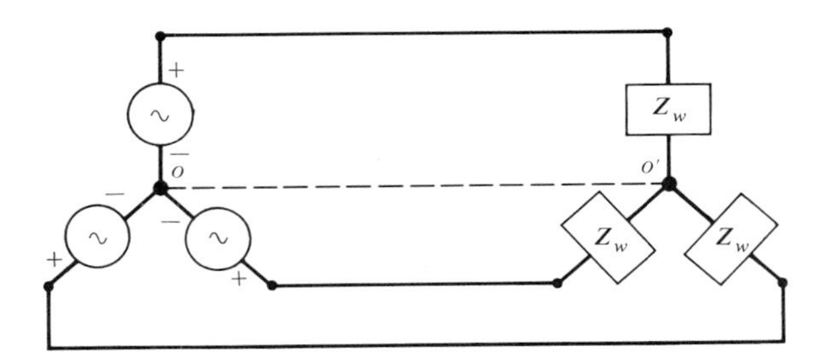

FIGURE B-3.

Sometimes it is necessary, or desirable, to mathematically convert a set of identical impedances connected in wye to an equivalent set connected in delta, or vice versa (Fig. B-4). In order for this transformation to be valid, the impedances seen between any two of the three terminals must be the same. Thus, getting the impedance $\mathbf{Z}_{ab}$ for both the wye and the delta, and equating them, we get.

$$2\mathbf{Z}_w = \frac{\mathbf{Z}_d(\mathbf{Z}_d + \mathbf{Z}_d)}{\mathbf{Z}_d + \mathbf{Z}_d + \mathbf{Z}_d} = \frac{2}{3}\mathbf{Z}_d \tag{B-13}$$

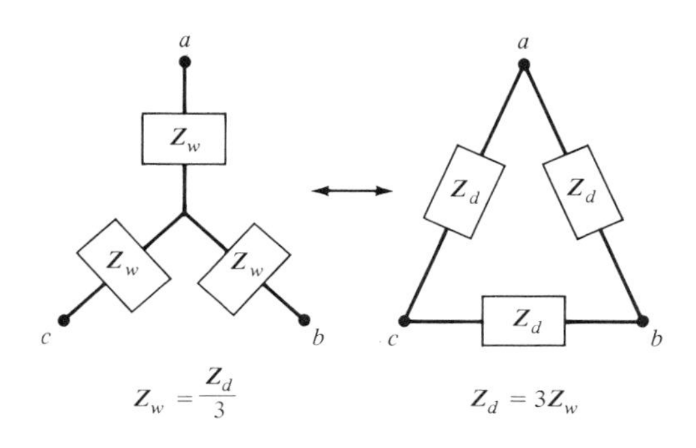

FIGURE B-4.

From this we can solve for the wye impedances $\mathbf{Z}_w$ in terms of the delta impedances $\mathbf{Z}_d$, or vice versa, such that

$$\mathbf{Z}_w = \frac{\mathbf{Z}_d}{3}$$

and (B-14)

$$\mathbf{Z}_d = 3\mathbf{Z}_w$$

Hence we can switch back and forth between balanced wye and delta as desirable.

Heat Content of Certain Fuels

Heat Content of Various Solid Fuels

Coal	
Lignite	6,000- 7,500 Btu/lb
Sub-bituminous	9,000-11,000 Btu/lb
Bituminous	12,000-15,000 Btu/lb
Anthracite	13,500-15,000 Btu/lb
Uranium (1,000ppm U in Ore)	
In LWR's	175,000 Btu/lb Ore
In Breeder reactors	17,800,000 Btu/lb Ore
Urban Refuse (avg.)	4,000 Btu/lb
Automobile Tires	15,000 Btu/lb
Wood	20-35 million Btu/cord

Heat Content of Various Liquid and Gaseous Fuels

Petroleum	
Crude Oil	5,800 million Btu/BBL
Refined Products	
Imports (avg.)	6,000 million Btu/BBL
Consumption (avg.)	5,517 million Btu/BBL
Gasoline	5,248 million Btu/BBL
Jet Fuel (avg.)	5,592 million Btu/BBL
Naphtha type	5,355 million Btu/BBL
Kerosene type	5,670 million Btu/BBL
Distillate Fuel Oil	5,825 million Btu/BBL
Residual Fuel Oil	6,287 million Btu/BBL
Natural Gas Liquids	4,031 million Btu/BBL
Natural Gas	
Wet	1,093 Btu/cubic foot
Dry	1,021 Btu/cubic foot

APPENDIX D

Data for Pure Copper Wire

AWG Number	Diameter (in.)	Area, d^2 (circular mils; 1 mil = 0.001 in.)	Pounds per 1000 ft Bare Wire	Length (ft/lb)	Resistance at 77°F (Ω/1000 ft)
	1.151	1000000.	3090.	0.3235	0.0108
	1.029	800000.	2470.	0.4024	0.0135
	0.963	700000.	2160.	0.4628	0.0154
	0.891	600000.	1850.	0.5400	0.0180
	0.814	500000.	1540.	0.6488	0.0216
	0.726	400000.	1240.	0.8060	0.0270
Stranded	0.574	250000.	772.	1.30	0.0341
0000	0.4600	211600.	640.5	1.56	0.490
000	0.4096	167800.	507.9	1.97	0.0618
00	0.3648	133100.	402.8	2.48	0.0871
0	0.3249	105500.	319.5	3.13	0.0983
1	0.2893	83690.	253.3	3.95	0.1239
2	0.2576	66370.	200.9	4.96	0.1563
3	0.2294	52630.	159.3	6.28	0.1970
4	0.2043	41740.	126.4	7.91	0.2485
5	0.1819	31100.	100.2	9.98	0.3133

AWG Number	Diameter (in.)	Area, d^2 (circular mils; 1 mil = 0.001 in.)	Pounds per 1000 ft Bare Wire	Length (ft/lb)	Resistance at 77°F (Ω/1000 ft)
6	0.1620	26250.	79.46	12.59	0.3951
7	0.1443	20820.	63.02	15.87	0.4982
8	0.1285	16510.	49.98	20.01	0.6282
9	0.1144	13090.	39.63	25.23	0.7921
10	0.1019	10380.	31.43	31.82	0.9989
11	0.09074	8234.	24.92	40.12	1.260
12	0.08081	6530.	19.77	50.59	1.588
13	0.07196	5178.	15.68	63.80	2.003
14	0.06408	4107.	12.43	80.44	2.525
15	0.05707	3257.	9.86	101.4	3.184
16	0.05082	2583.	7.82	127.9	4.016
17	0.04526	2048.	6.20	161.3	5.064
18	0.04030	1624.	4.92	203.4	6.385
19	0.03589	1288.	3.90	256.5	8.051
20	0.03196	1022.	3.09	323.4	10.15
21	0.02846	810.1	2.45	407.8	12.80
22	0.02535	642.4	1.95	514.2	16.14
23	0.02257	509.5	1.54	648.4	20.36
24	0.02010	404.0	1.22	817.7	25.67
25	0.01790	320.4	0.970	1031.0	32.37
26	0.01594	254.1	0.769	1300.0	40.81
27	0.01420	201.5	0.610	1639.0	51.47
28	0.01264	159.8	0.484	2067.0	64.90
29	0.01126	126.7	0.384	2607.0	81.83
30	0.01003	100.5	0.304	3287.0	103.2
31	0.00893	79.70	0.241	4145.0	130.1
32	0.00795	63.21	0.191	5227.0	164.1
33	0.00708	50.13	0.152	6591.0	206.9
34	0.00631	39.75	0.120	8310.0	260.9
35	0.00562	31.52	0.095	10480.0	329.0
36	0.00500	25.00	0.076	13210.0	414.8

Answers to Selected Problems

CHAPTER 1

1.1 135.3 MWH. **1.2** 2.175 billion GWh. **1.3** Gas: 19.238×10^{15} cu. ft.; Coal: 575 million tons. **1.4** 27.46 yrs. **1.5** 6.93%. **1.6** 9.78×10^{10} tons. **1.7** 1.317 MW. **1.8** 36.8 years. **1.9** 42.53 kW. **1.10** 126.8 sq m.

CHAPTER 2

2.1 404.41 kWh. **2.2** $= 405.8 \times 10^6$ cal. $= 205.15 \times 10^6$ cal. **2.3** 58%. **2.4** $= -58.8$ kilocal $= 1.275$ v. **2.5** -36.96 cal **2.6** 2.37×10^{-4} V/°K. **2.7** a) 400 V; b) 9.6 MW; c) 13.4 MW. **2.8** V = 5000 V; I = 4000 A; P_{loss} = 10 MW. **2.9** For Prob. 2-7 $= 0.48 \times 10^6$ N; For Prob. 2-8 $= 0.4 \times 10^6$ N. **2.11** 347.6 N·m. **2.12** 1310.4 V. **2.13** 6 poles. **2.14** 19.73 MWH. **2.15** 480×10^3 cal, 505.26×10^3 cal. **2.16** 3924 kW.

CHAPTER 3

3.1 27.27A, 403.33 Ω. **3.2** (a) 0.4 pu, 0.6 pu; (b) 4 percent; 6 percent; 863 percent; and 0.42 percent. **3.3** 12.5 kV, 3125 kVA. **3.4** 31.5 Ω, 24 A, 2240 V, 645.12 kVA. **3.9** (a) base VA = 100,000 VA; base voltage = 20,000 V; base current $= \frac{100}{20} = 5$ A; base impedance = 4000 Ω; ohmic impedance = 400 Ω; (b) base VA = 100,000 VA; base voltage = 5,000 V; base current = 20A; base impedance = 250 Ω; ohmic impedance = 25 Ω.

CHAPTER 4

4.1 (a) 800 At.; (b) 2546.5 At/m. **4.2** (a) 0.32 T; (b) 0.512 mWb; (c) 1.562 $\times 10^6$ H^{-1}; (d) 0.64 $\times 10^{-6}$ H. **4.4** (a) 100; (b) 94; (c) 108 $\times 10^{-4}$ H/M. **4.5** ϕ = 80 MoWb; B_{g2} = 0.2 T; $\phi_{g1} = \phi_{g3}$; B_{g1} = 0.1 T; B_{g3} = 0.05 T, **4.6** 0.1 T. **4.8** L_{11} = 1.34 mH; L_{22} = 0.75 mH; L_{33} = 3 mH; L_{12} = 0.5 mH; L_{13} = 1 mH. **4.9** 115.2 W. **4.10** (a) 2; (b) (1) I_2 = 45.45 A; (2) I_2 = 34.09 A; **4.11** 5.5 A, 40 Ω, 1210 W. **4.12** 16.5 mWb. **4.13** 59.87 W, 187.6 Ω, 202.1 Ω. **4.15** 98.4%. **4.19** 97.4%, 96.8%. **4.20** 96.4%. **4.21** (a) 98.39%; (b) 98.12%. **4.22** R_c'' = 322.67 Ω; R_e' = 1290.67 Ω; I_c = .6818 A; I_m = .987 A; X_m' = 891.6 Ω; 0.0794 Ω; X_e' = 0.4815 Ω. **4.23** $\overline{V}_0$ = 448.06 V; $I_1^2 R_e^1$ = 41 W; V_0^2/R_e = 155.5 W; P_0 = 8000 W; η = 97.58%. **4.24** η75% = 94.31%. **4.25** 65.45 A. **4.26** I^2R loss = 37.1 W; η = 98.09%; I_1 = 23.17 A. **4.27** 57.47 kW, 38.53 kW. **4.28** I_B' = 1.14 A; I_B'' = 20A. **4.29** I_{pp} = 1.82 A; I_{pi} = 3.15 A. **4.30** $P_0/_p$ = 96.36%. **4.31** 86.6 kVA. **4.34** V_1 = 346.4 V, V_2 = 220 V; V_2/V_1 = 0.635. **4.35** 48 kW.

CHAPTER 5

5.1 36 V, 67.5 V. **5.2** Z = 960; ϕ = 0.03 Wb. **5.3** 114.6 N·m. **5.4** 250.24 V. **5.5** (a) P_L = 96,875 W; (b) P_L = 98,009 W. **5.6** 396.6 rpm. **5.7** R_a = 0.1215 Ω. **5.9** P_L = 11428 W. **5.10** P_h = 1000 W; P_e = 2000 W. **5.13** 254 V, 226 V. **5.14** (a) 46.89 kW; (b) 53309 kW. **5.15** η = 90.6%. **5.16** (a) 188.6 N·m; (b) 93.17%. **5.17** 802.8 rpm. **5.18** 1035.8 rpm. **5.19** 692.8 rpm. **5.20** (a) 1129 rpm; (b) 108.9 N·m; (c) 88.3%. **5.21** η_2 = 1297.6 rpm. **5.22** n = 1020 rpm; n = 84.3%. **5.23** I_{F2} = 0.65375, R_{ext} = 226.52 Ω. **5.24** 514 rpm. **5.25** (a) Rx = 0.704 Ω; (b) Ts = 479.5 N·m. **5.26** N_2 = 753.5 rpm. **5.27** (a) 510.5 V; (b) 777.93 kW; (c) 16,508 N·m; (d) 95.56%.

CHAPTER 6

6.1 1000 rpm. **6.2** K_p = 0.966; K_d = 0.966. **6.3** 1841.4 V. **6.4** ϕm = 15 mWb. **6.6** 4583.3 V. **6.7** 382.86 V, 50.72%. **6.8** 7.44°; % regulation = −8.16%. **6.9** 164.7%. **6.10** V_t = 6664.3 V or 11.54 kV line to line. **6.11** (a) P_m = 22 MW; (b) I_m = 1566.4 A; (c) Cos ϕ = 0.737. **6.12** η = 82.17%. **6.13** 260.5 V per phase. **6.14** (a) V_0 = 6035 or 10.48 kV; (b) V_0 = 6372 or 11.04 kV; (c) V_0 = 6676.8 or 11.56 kV. **6.17** (a) 786.4 kW; (b) 6,260 N·m; (c) δ = 40.5°. **6.18** Total: kW = 117; KVAR = 0; pf = 1. **6.19** Motor KW = 82.9; KVAR = 107; KVA = 135.36 KVA. **6.20** (a) $\tan\delta = \dfrac{64.95 \times 0.8 \times 0.9}{\dfrac{400}{\sqrt{3}} - 64.95 \times 0.8 \times 0.436}$;(b) 16.3 kW.

6.21 I_2 = 25.48 A.

CHAPTER 7

7.1 1200 rpm, 4%. **7.2** 25%. **7.3** 144.35 Nm. **7.4** 0.4 Ω. **7.5** 960 rpm. **7.6** 3341 rpm. **7.7** 500 W. **7.8** 76.95%. **7.9** 3.27%. **7.10** 24084 W; 187.5 Nm; 191.7 Nm. **7.11** r = 3.2 Ω. **7.12** 20888 W. **7.13** (a) 14.61 A; (b) 0.84. **7.15** $R_2 = 0.1433\ \Omega$. **7.17** 298.14 V; $I = 111.8$ A. **7.18** 41.67 percent of T_F. **7.19** $S_F = 4.69\%$; $V_{st} = 0.5\ V_f$. **7.20** 17.9.

CHAPTER 8

8.1 16.67%; 183.33%. **8.2** 78.32 W; torque = 0.219 Nm. **8.4** 48.1%. **8.5** .92. **8.6** $X_1 = 9.57\ \Omega$; $X_m = 13.68\ \Omega$. **8.7** 0.021 N·m. **8.8** 0.993 N·m. **8.9** 12,000 rpm. **8.10** 0.794 hp; 1.57 N·m. **8.11** 0.38 N·m. **8.12** $\eta = 88.37\%$.

CHAPTER 9

9.1 326 turns. **9.2** 326 turns; no. **9.3** 83.33 N; 333.3 V. **9.4** 166.67 $\sin^2 377t$; 83.33 N. **9.5** (a) decrease X; (b) 250 N. **9.7** -14.9×10^3 N. **9.9** (a) $\mathrm{Te} = \dfrac{2\mathrm{Wm}}{2\theta} = 0$; (b) $\mathrm{Te} = -CI_0^2 \sin\theta$; (c) $\mathrm{Te} = -CI_0I_m \sin\omega t \sin\theta$; (d) $\mathrm{Te} = -CI_m^2 \sin^2\omega t \sin\theta$; (e) $\mathrm{Te} = \dfrac{2\omega\mathrm{M}}{2\theta} = \dfrac{I_0^2}{A} C^2 \cos\theta \sin\theta$. **9.10** Mechanical loss = 15.67 W; Total loss = 24 W. **9.12** 4 A; 56 W; 0.127. **9.13** 1.4 J; 0.7 J. **9.14** 106 N; 109.2 V.

CHAPTER 10

10.1 $40.17 \times 10^{-3}\ \Omega$. **10.2** $L = 35.5$ mH; $C = 0.203\ \mu$F. **10.3** Power loss = 907.4 W; $V_R = 66.4$ kV; $V_S = 66988$ V; regulation = 0.88%. **10.4** (a) $\eta = 98.78\%$; (b) $V_S = 79.67 + 226.4\angle{-36.87}\ (2 + J4) \times 10^{-3} \simeq 80.57/0.32$. **10.5** (a) Hence, $V_R = 87.29$ kV/phase or 151.18 kV line-to-line; (b) 99.66%. **10.6** 839.2 MW. **10.7** 51 KM. **10.9** 202.56 kV/phase or 350.8 kV, line-to-line. **10.10** 207.46 kV/phase or 359.3 kV line-to-line. **10.11** (a) 97.3%; (b) 97%. **10.12** sending end power = 240.8 MW. **10.13** $\overline{A} = 0.7147\angle{0^\circ}$; $\overline{B} = 387.3 \times 0.6994\angle{90^\circ}$; $\overline{C} = 1.8 \times 10^{-3}\angle{90^\circ}$; $\overline{D} = 0.714\angle{0^\circ}$ and $\overline{V}_R = 199.2\angle{0^\circ}$.

CHAPTER 11

11.1 $X = 30\%$. **11.2** (a) 44%; 68.18 MVA; (b) 30%; 100 MVA. **11.3** The short-circuit MVA at B ≃ short circuit MVA at A = 0.218 MVA. **11.4** $I_{a1} = 276\angle{25.6^\circ}$ A; $I_{a2} = 307\angle{86.7^\circ}$ A; $I_{a0} = 137.4\angle{76^\circ}$ A. **11.5** $I_{a0} = 0$; $I_{a1} = 5.38 - j1.44 = 5.57\angle{-15^\circ}$; $I_{a2} = 5.73 - j5.5 = 7.94\angle{224^\circ}$; $I_{ab1} = 3.1 + j0.83$; $I_{ab2} = -4.45 - j1.11$; $I_{ab} = 1.38$ A. **11.6** $I_a = 100\angle{0^\circ}$; $I_b = 100\angle{180^\circ}$; $I_c = 0$; $I_{a0} = 0$; $I_{a1} = 50 - j28.86$; $I_{a2} = 50 + j28.86$. **11.7** $V_a = 0$; $V_b = 4.922\angle{254.74}$ KV; $V_c = 4.922\angle{-74}$ KV. **11.8** $V_a = 5.35$ KV; $V_b = V_c = 2.67$ KV.

11.9 Load on station 1 = 16 + j15.36 pu; Power Factor at station 1 = 0.72 lagging; Load on station 2 = 24 + j13.36 pu; Power Factor at station 2 = 0.87 lagging. **11.10** $S_1 = 1.043 + j0.569$; $P_1 = 1.043$ pu, Q = 0.569 pu. **11.11** 0.259 pu. **11.12** $\frac{X}{R} = \sqrt{3}$. **11.13** (a) $(P_R)_{MAX}$ = 275.5 mW/phase or 826.5 mW (total); (b) Increase in load within stable limits = 433.2 mW/phase (or 1299.6 mW total). **11.15** KE stored = 3553 MJ; H = 35.53 MJ/MVA; M = 18.85 MJ − rad/sec. **11.16** 471.25°/sec^2. **11.17** 3678.5 rpms. **11.18** 0.882 Hz.

CHAPTER 12

12.5 C = 0.1162 F; $V_{0AVE} = 0.974\ Vm$. **12.6** (c) $I_{0AVE} = 3.25$ A; I^2_{0RMS} = 7.2 A; (d) $P_{AVE} = 42.94$ W. **12.7** (a) 45°; (b) For $\alpha = 75°$, by trial and error $\beta \simeq 176°$. **12.8** 9.7 ms. **12.9** 37.5 V; 18.75 A. **12.10** (a) Ωm = 30.58 r/s; (b) $I_{AVE} = 750.8$ A. **12.11** (a) 83.5 r/s; (b) $I_{RMS} = 160$ A; (c) $T_L = 45.5$ HP. **12.12** 289 pulse/sec.

INDEX

E

F